SystemVerilog Assertions and Functional Coverage

Ashok B. Mehta

SystemVerilog Assertions and Functional Coverage

Guide to Language, Methodology and Applications

Third Edition

Ashok B. Mehta
DefineView Consulting
Los Gatos, CA, USA

Additional material to this book can be downloaded from https://www.springer.com

ISBN 978-3-030-24736-2 ISBN 978-3-030-24737-9 (eBook)
https://doi.org/10.1007/978-3-030-24737-9

This Springer imprint is published by the registered company Springer Nature Switzerland AG
The registered company address is: Gewerbestrasse 11, 6330 Cham, Switzerland

To
My dear wife, Ashraf Zahedi,
And
My dear parents, Rukshamani Behn
and Babubhai Mehta

Foreword

Louis H. Sullivan, an American architect, considered the father of the modern skyscraper, and mentor to Frank Lloyd Wright, coined the phrase "form follows function." The actual quote is "form ever follows function" which is a bit more poetic and assertive than the version that has found its way into the common vernacular. He wrote those words in an article written for Lippincott's Magazine #57 published in March 1896. Here is the passage in that article that contains the famous quote:

"Whether it be the sweeping eagle in his light or the open apple-blossom, then toiling work horse, the blithe swan, the branching oak, the winding stream at its base, the drifting clouds—over all the coursing sun, form ever follows function, and this is the law. Where function does not change, form does not change. The granite rocks, the ever-brooding hills, remain for ages; the lightning lives, comes into shape, and dies, in a twinkling.

It is the pervading law of all things organic and inorganic, of all things physical and metaphysical, of all things human and all things superhuman—of all true manifestations of the head, of the heart, of the soul—that the life is recognizable in its expression, that form ever follows function. This is the law."

Earlier in the article, Sullivan foreshadows his thought with this passage:

"All things in nature have a shape, that is to say, a form, an outward semblance, that tells us what they are, that distinguishes them from ourselves and from each other."

The precise meaning of this pithy phrase has been debated in art and architecture circles since Sullivan's article was first published. However, it is widely accepted to mean that the form of something—its shape, color, size, etc.—is related to what it does. Water flows, rocks sit, birds fly.

In his book "The Design of Everyday Things," (Basic Books, 1988) Don Norman discusses a similar concept, the notion of affordances. Norman defines the term as "… the perceived and actual properties of the thing, primarily those fundamental properties that determine just how the thing could possibly be used." He cites some examples: "A chair affords ("is for") support and, therefore, affords sitting. A chair can also be carried. Glass is for seeing through, and for breaking. Wood is normally used for solidity, opacity, support or carving."

Norman's idea turns Sullivan's upside down. He is saying function follows form. The shape, color, size, etc. of an object affects what it does. Nonetheless, both men would likely agree that form and function, whichever drives the other, are inextricably linked.

Software designers have the luxury of choosing the form to fit the function. They are not as constrained by the laws of physics as say, a cabinetmaker. The cabinetmaker must choose materials that will not only look nice but will withstand the weight of books or dishes or whatever is to be placed on the shelves. Software designers have some constraints with regard to memory space and processing time, but beyond that they have a lot of freedom to build whatever comes to mind.

Sullivan referred to "all things physical and metaphysical." Without much of a stretch we can interpret that to include software, a most abstract human creation. The form of a piece of software is linked to its function. The complex software that verification engineers build, called a testbench, must be designed before it can be built. The verification engineer, like an architect, must determine the form of his creation.

The architecture space is wide open. Computer code, while much more abstract than say, a staircase or a door handle on a car, has a form and a function. The form of computer code is the set of syntactic elements strung together in a program. The function is what the program does when executed, often referred to as its semantics.

A verification engineer is typically presented a set of requirements, often as a design specification, and asked to build a testbench that meets these requirements. Because of the tremendous flexibility afforded by the software medium he must choose the form carefully to ensure that not only meets the requirements, but is easy to use, reusable, and robust. He must choose a form that fits the function.

Often an assertion is just the right thing to capture the essence of some part of a design. The *form* of an assertion is short sequence of text that can be inserted easily without disrupting the design. With their compact syntax and concise semantics assertions can be used to check low level invariants, protocols, or end-to-end behavior.

The *function* of an assertion, in a simulation context, is to assert that something is always (or never) the case. It ensures that invariants are indeed invariant. Assertions can operate as checkers or as coverpoints. The fact that they can be included in-line in RTL code or in separate checkers, they can be short or long for simple or complex checking makes them invaluable in any testbench.

The wise verification engineer uses all the tools as his disposal to create an effective and easy to use testbench. He will consider the function of the testbench and devise a form that suits the required function. Assertions are an important part of any testbench.

Ashok Mehta has written a book that makes assertions accessible. His approach is very pragmatic, choosing to show you how to build and use assertions rather than engage in a lot of theoretical discussion. Not that theoretical discussion is irrelevant—it's useful to understand the theoretical underpinnings of any technology. However, there are many other books on that topic. This book fills a gap for practicing engineers where before no text provided the how-to of building and using assertions in a real-world context.

Ashok opens up the world of assertions to verification engineers who may have thought them too opaque to consider using in a real testbench. He does an especially nice job of deconstructing assertions to show how they work and how to write them. Through detailed examples he shows all the pieces that go into creating assertions of different kinds, and how they fit together. Ashok completes the picture by demonstrating how assertions and coverage fit together.

Part of the book is devoted to functional coverage. He deconstructs the sometimes awkward SystemVerilog syntax of covergroups and coverpoints. Like he has with assertions, he takes the mystery out of building a high-quality coverage model.

With the mysteries of assertions unmasked, you can now include them in your personal vocabulary of testbench forms. This will enable you to create testbenches with more sophisticated function.

nVidia corporation
February 2013

Mark Glasser

Preface

Having been an end user of EDA tools for over 20 years, I have seen that many new technologies stay on way side because either the engineers don't have time to learn of these new technologies/languages or the available material is too complex to digest. A few years back I decided to tackle this problem by creating a very practical, application oriented down-to-earth SystemVerilog Assertions (SVA) and Functional Coverage (FC) class for professional engineers. The class was well received, and I received a lot of feedback on making the class even more useful. That culminated in over 600 slides of class material just on SVA and FC. Many suggested that I had collected enough material for a book. That is how I ended up on this project with the same goal that the reader should understand the concept clearly in an easy and intuitive manner and be able to apply the concepts to real life applications right away.

The style of the book is such that the concepts are clarified directly in a slide style diagram with talking points. This will hopefully make it easy to use the book as a quick reference as well. Applications immediately following a topic will further clarify the subject matter and my hope is that once you understand the semantics and applications of a given topic, you are ready to apply that to your daily design work. These applications are modeled such that you should be able to use them in your design with minimal modifications.

This book is meant for both design and verification engineers. As a matter of fact, I have devoted a complete section on the reasons and practicality behind having micro level assertions written by the design engineers and macro level assertions written by verification engineers. Gone are the days when designers would write RTL and throw it over the wall for the verification engineer to quality check.

The book covers both IEEE 1800-2005 and IEEE 1800-2009/2012 standard SVA language.

Chapter 1 is Introduction to SVA and FC giving a brief history of SVA evolution. It also explains how SVA and FC fall under SystemVerilog umbrella to provide a complete assertions and functional coverage driven methodology.

PART I: System Verilog Assertions (SVA)

Chapter 2 goes in-depth on SVA based methodology providing detail that you can right away use in your project execution. Questions like "How do I know I have added enough assertions?", "What type of assertions should I add", etc. are explained with clarity.

Chapter 3 gives a high-level overview of Sequential Domain coverage namely 'cover' property and 'cover' sequence.

Chapter 4 simply describes the conventions used throughout the book.

Chapter 5 describes Immediate Assertions. These are non-temporal assertions allowed in procedural code.

Chapter 6 goes into the fundamentals of Concurrent Assertions to set the stage for the rest of the book. How the concurrent multi-threaded semantics work, when and how assertions get evaluated in a simulation time tick, formal arguments, disabling, etc. are described here.

Chapter 7 describes the so-called sampled value functions such as $rose, $fell, $stable, $past etc.

Chapter 8 is the big one! This chapter describes all the operators offered by the language including Clock Delay with and without range, Consecutive repetition with and without range, non-consecutive repetition with and without range, 'throughout', 'within', 'and', 'or', 'intersect', 'first_match', 'if…else', etc. Each of the operator description is immediately followed by examples and applications to solidify the concept.

Chapter 9 describes the System Functions and Tasks such as $isunknown, $onehot, etc.

Chapter 10 discusses a very important aspect of the language that being properties with multiple clocks. There is not a single design now a day that uses only a single clock. A simple asynchronous FIFO will have a Read Clock and a Write Clock which are asynchronous. Properties need to be written such that check in one clock domain triggers a check in another clock domain. The chapter goes in plenty detail to demystify semantics to write assertions that cross clock domains. The so-called CDC (Clock Domain Crossing) assertions are explained in this chapter.

Chapter 11 is probably the most useful one describing Local Variables. Without this multi-threaded feature many of the assertions would be impossible to write. There are plenty of examples to help you weed through the semantics.

Chapter 12 is on recursive properties. These are rarely used but are very handy when you want to know that a property holds until another becomes true or false.

Chapters 13 describe detecting end point of a sequence. The .triggered and .matched end-points of sequences are indeed very practical features. Note that .ended (of LRM 2005) is now deprecated and replaced with .triggered.

Chapter 14 describes the ‘expect’ statement and how it applies to formal and simulation verification.

Chapter 15 describes the ‘assume’ and ‘restrict’ statements as applied to formal verification and simulation.

Chapter 16 explains the use of multi-clocked properties in the verification of clock domain crossing (CDC) logic.

Chapter 17 is entirely devoted to very powerful and practical features that do not quite fit elsewhere. Features such as concurrent assertions in procedural code, sequence in Verilog ‘always’ block sensitivity list, cyclic dependency, empty sequences and the phenomenon of a ‘vacuous pass’ !

Chapter 18 shows the example/test bench for asynchronous FIFO checks

Chapter 19 is solely devoted to Asynchronous assertions. The example in this chapter shows why you need to be extremely careful in using such assertions.

Chapter 20 is entirely devoted to IEEE 1800 2009-2012 features, such as strong and weak sequences, global clocking, always, eventually, until, nexttime, inferred clock and disable, abort properties, etc. There are many useful features added by the language designers.

Chapter 21 is devoted to ‘let’ declarations and its usage.

Chapter 22 is devoted to ‘Checkers’.

Chapter 23 describes 6 LABs for you to try out. The LABs start with simple example moving gradually onto complex ones.

Note: The LABs are available on Springer download site. All required Verilog files, test benches and run scripts are included for both PC and Linux OS.

Chapter 24 provides answers to the LABs of Chapter 17

PART II: System Verilog Functional Coverage (FC)

Chapter 25 provides introduction to Functional Coverage and explains differences with Code Coverage

Chapter 26 is fully devoted to Functional Coverage language including in-depth detail on Covergroups, Coverpoints, bins, bins filtering, systemverilog class based covergroup usage, wildcard bins, ignore bins, binsof, binsof intersect, etc., including transition and cross coverage

Chapter 27 provides practical hints to performance implications of coverage methodology. Don’t try to cover everything all the time.

Chapter 28 describes Coverage Options, which you may keep in your back pocket as reference material for a rainy day!

Los Gatos, CA, USA Ashok B. Mehta

Preface to the Third Edition

The first and second editions of this book were well received, and the readers provided many a good suggestion for further elaboration of language semantics. The readers also pointed out some errata on the language syntax. I am greatly indebted to the readers and colleagues for their input and support. This edition elaborates on features such as SystemVerilog class-based functional coverage, CDC (clock domain crossing) verification, global clocking, and "let" and "checker" nuances and adds plenty more examples throughout the book. Over 100 pages of new material have been added (over the second edition).

Pleasant reading.

Preface to the Second Edition

The first edition of this book was well received, and the readers provided many a good suggestion on further elaboration of language semantics. Readers also pointed out some Errata on the language syntax. I am greatly indebted to the readers and colleagues for their input and support. In addition, the IEEE 1800-2012 LRM came along. Many features of the 2012 LRM were missing in the first edition, since the LRM was not ready yet. This edition incorporates the Errata/suggestions from readers as well as the IEEE 1800-2012 feature set. Among many, features such as 'checkers', 'let declarations', past and future global clock sampled value functions, strong and weak properties, abort properties, '.triggered' end point detection method, etc. are included. Furthermore, this edition adds many more examples and adds further clarification of the semantic nuances of the language.

Pleasant reading.

Acknowledgments

I am very grateful to many who helped with reviewing and editing of this book. In particular, Mark Glasser (nVidia) for his excellent foreword and in-depth review of the book, Vijay Akkati (Qualcomm) for his detailed review of the chapters, Dr. Sandeep Goel (TSMC) for the motivation and for editing the book, and Bob Slee for his sustained support throughout the endeavor and for facilitating close cooperation with EDA vendors. I would also like to thank Norbert Eng for all things verification and many other reviewers for their valuable input on various chapters.

Last but certainly not the least, I would like to thank my wife, Ashraf Zahedi, for her enthusiasm and encouragement throughout the writing of this book and putting up with long nights and weekends required to finish the book. She is the cornerstone of my life always with a positive attitude to carry the day through up and down of life.

Contents

About the Author

Ashok B. Mehta has been working in the ASIC/SoC design and verification field for over 30 years. He started his career at Digital Equipment Corporation (DEC) as a CPU design engineer. He then worked at Data General, at Intel (first Pentium design team), and after a route of a couple of startups at Applied Micro Circuits Corporation and then TSMC.

He was a very early adopter of Verilog and participated in Verilog, VHDL, iHDL (Intel HDL), and SDF (standard delay format) technical subcommittees. He has also been a proponent of ESL (electronic system level) designs, and at TSMC, he released two industry standard reference flows that allow complete reuse of the verification environment and code from ESL design verification to gate verification. Lately, he has been involved with 3DIC designs where SystemVerilog Assertions play an instrumental role in 3D stacked die SoC design verification.

He earned his MSEE from the University of Missouri. He holds 18 US patents in the field of SoC, 3DIC, and ESL design verification. In his spare time, he is an amateur photographer and likes to play drums on the 1970s rock music driving his neighbors up the wall ☺.

List of Figures

List of Tables

Chapter 1
Introduction

Introduction: This chapter introduces the reader to the SystemVerilog Assertions language and its role under SystemVerilog IEEE-1800 umbrella and the roadblocks to design verification productivity and solutions thereof and explains SVA evolution and sets the stage for the rest of the book.

A. B. Mehta, *SystemVerilog Assertions and Functional Coverage*,
https://doi.org/10.1007/978-3-030-24737-9_1

As is well known in the industry, the design complexity at 7 nm node and below is exploding. Small form factor requirements and conflicting demands of high performance and low power and small area result in ever so complex design architecture. Multi-core, multi-threading and Power, Performance and Area (PPA) demands exacerbate the design complexity and functional verification thereof.

The burden lies on functional and temporal domain verification to make sure that the design adheres to the specification. Not only is RTL (and Virtual Platform) functional verification important but so is silicon validation. Days when engineering teams would take months to validate the silicon in the lab are over. What can you do during pre-silicon verification to guarantee post-silicon validation a first pass success?

Note that the verification complexity applies both to ASIC designs and to FPGA designs. Specifically, FPGA designs are essentially SoC designs with multiple well-placed and routed cores in the design. The days of burn and learn strategy employed by FPGA design and verification engineers are over. In burn, the FPGA design and debug in the lab requires that the FPGA design is ready (to some extent) *before* you burn the FPGA. If the FPGA design was not well verified, then the debug time in lab increases exponentially. This is the reason a robust verification methodology is essential for FPGA designs as well.

The biggest challenge that the companies face is short time-to-market to deliver first pass working silicon of increasing complexity. Functional design verification is the long poll to design tape-out. Here are two key problem statements.

1. *Design Verification Productivity*: 40–50% of project resources go to functional design verification. The chart in Fig. 1.1 shows design cost for different parts of a design cycle. As is evident, the design verification cost component is about 40+% of the total design cost. In other words, this problem states that we must increase the productivity of functional design verification and shorten the design

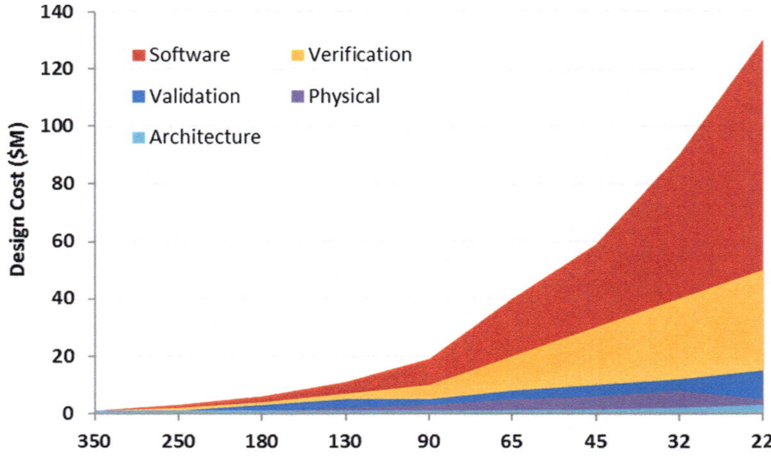

Fig. 1.1 Verification cost increases as the technology node shrinks

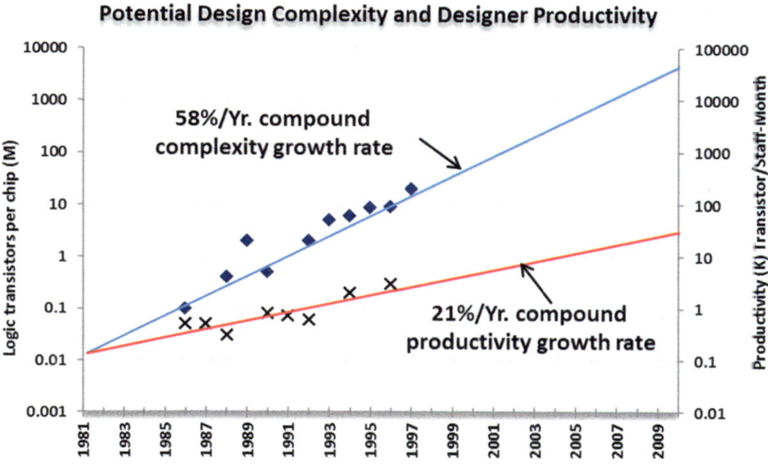

Fig. 1.2 Design productivity and design complexity

⇔ simulate ⇔ debug ⇔ cover loop. This is a productivity issue, which needs to be addressed.

Continuing with the productivity issue, the chart in Fig. 1.2 shows that the compounded complexity growth rate per year is 58% while the compounded productivity growth rate is only 21%. There is a huge gap between what *needs* to get done and what *is* getting done. This is another example of why the productivity of design cycle components such as functional design verification must be improved.

2. *Design Coverage*: The second problem statement states that more than 50% of designs require re-spin due to functional bugs. One of the factors that contribute to this is the fact that we did not objectively determine *before* tape-out that we had really *covered* the entire design space with our test-bench. The motto "If it's not verified, it will not work" seems to have taken hold in design cycle. Not knowing if you have indeed covered the entire design space is the real culprit towards escaped bugs and functional silicon failures.

So, what's the solution to each problem statement?

1. Increase Design Verification Productivity.
   (a) *Reduce Time to Develop.*
      - Raise abstraction level of tests. Use TLM (Transaction Level Modeling) methodologies such as UVM and SystemVerilog/C++/DPI. The higher the abstraction level, easier it is to model and maintain verification logic. Modification and debug of transaction level logic is much easier, further reducing time to develop test-bench, reference models (scoreboard), peripheral models, and other such verification logic.
      - Use constrained random verification (CRV) methodologies to reach exhaustive coverage with fewer tests. Fewer tests mean less time to develop and debug.

- Develop Verification Components (UVM agents, for example that are reusable). Make them parameterized for adoptability in future projects.
- Use System Verilog Assertions to reduce time to develop complex temporal domain and combinatorial checks. As we will see, assertions are intuitive and much simpler to model, especially for complex sequential (temporal domain) checks. SystemVerilog code for a given assertion will be much lengthier, hard to model, and hard to debug. SVA indeed reduces time to develop and debug.

(b) *Reduce Time to Simulate.*

- Again, higher level of abstraction simulates much faster than pure RTL test bench which is modeled at signal level. Use transaction level test bench.
- Use System Verilog Assertions to directly point to the root cause of a bug. This reduces the simulate ⇔ debug ⇔ verify loop time. Debugging the design is time consuming as is, but not knowing where the bug is, and trial and error simulations further exacerbate the already lengthy simulation time.

(c) *Reduce Time to Debug.*

- Use System Verilog Assertion Based Verification (ABV) methodology to quickly reach to the source of the bug. As we will see, assertions are placed at various places in design to catch bugs where they occur. Traditional way of debug is at IO level. You see the effect of a bug at primary output. You then trace back from primary output until you find the cause of the bug resulting in lengthy debug time. In contrast, an SVA assertion points directly at the source of the failure (for example, a FIFO assertion will point directly to the FIFO condition that failed and right away help with debug of the failure) drastically reducing the debug effort.
- Use Transaction level methodologies to reduce debugging effort (and not get bogged down into signal level granularity).
- Constraint Random Verification allows for fewer tests. They also narrow down the cone of logic to debug. CRV indeed reduces time to debug.

2. *Reduce Time to Cover* and build confidence in taping out a fully verified design.

(a) Use "*cover*" feature of SystemVerilog Assertions to cover complex *temporal* domain specification of your design. As we will see further in the book, "cover" helps with making sure that you have exercised low-level temporal domain conditions with your test-bench. *If an assertion does not fire, that does not necessarily mean that there is no bug.* One of the reasons for an assertion to not fire is that you probably never really stimulated the required condition (antecedent) in the first place. If you do not stimulate a condition, how would you know if there is indeed a bug in the design logic under simulation? "cover" helps you determine if you have indeed exercised the required temporal domain condition. More on this can be seen in later chapters.
(b) Use SystemVerilog *Functional Coverage* language to measure the "*intent*" of the design. How well have your test bench verified the "intent" of the design.

For example, have you verified all transition of Write/Read/Snoop on the bus? Have you verified that a CPU1-snoop occurs to the same line at the same time that a CPU2-write invalid occurs to the same line? Code Coverage will not help with this. We will cover Functional Coverage in plenty detail in the book.

(c) Use Code Coverage to cover *structural* coverage (yes, code coverage is still important as the first line of defense even though it simply provides structural coverage). As we will see in detail in the section on SV Functional Coverage, structural coverage does not verify the intent of the design, it simply sees that the code that you have written has been exercised (e.g., if you have verified all "case" items of a "case" statement, or toggled all possible assigns, expressions, states, etc.). Nonetheless, code coverage is still important as a starting point to measure coverage of the design.

As you notice from above analysis, SystemVerilog Assertions and Functional Coverage play a key role in about every aspect of Functional Verification. Note that in this book, I use Functional Verification to include both the "function" domain functional coverage and the "temporal" domain functional coverage.

1.1 How Will This Book Help You?

This book will go systematically through each of SystemVerilog Assertions (SVA) and Functional Coverage (FC) language features and methodology components with practical applications at each step. These applications are modeled such that you should be able to use them in your design with minimal modifications. The book is organized using power point style slides and talking points to make it very easy to grasp the key fundamentals. Advanced applications are given for those users who are familiar with the basics. For most part, the book concentrates on the in-depth discussion of the features of the languages and shows examples that make the feature easily understandable and applicable. Simulation logs are frequently used to make it easier to understand the underlying concepts of a feature or method.

The book is written by a design engineer for (mainly) design and verification engineers with the intent to make the languages easy to grasp avoiding decipher of lengthy verbose descriptions. The author has been in System and Chip design field for over 20 years and knows the importance of learning new languages and methodologies in shortest possible time to be productive.

The book concentrates on SVA features of the IEEE 1800–2005 standard. Author believes that the features of this standard are plenty to designing practical assertions for the reader's project(s). However, the author has indeed covered the entire IEEE 1800–2009/2012 feature set in a standalone chapter (Chap. 20) to give an in-depth look at the new standard. Note that some of the 2009/2012 features were not supported by popular simulators as of this writing and the examples provided were not simulated. Please do send your suggestions/corrections to the author (ashok_mehta@yahoo.com).

1.2 SystemVerilog Assertions and Functional Coverage Under IEEE 1800 SystemVerilog Umbrella

SystemVerilog assertions (SVA) and Functional Coverage (FC) are part of IEEE 1800 SystemVerilog standard. In other words, SVA and FC are two of the four distinct language subsets that fall under the SystemVerilog umbrella.

1. SystemVerilog Object Oriented language for functional verification (using UVM style libraries).
2. SystemVerilog language for Design.
3. SystemVerilog Assertions (SVA) language.
4. SystemVerilog Functional Coverage (FC) language to see that the verification environment/test-bench have fully verified your design.

As shown in Fig. 1.3, SVA and FC are two of the important language subsets of SystemVerilog. Note that SystemVerilog assertions is orthogonal to OOP, UVM and Functional Coverage languages. In other words, SVA has its own syntax and semantics. In yet more words, knowledge of OOP/UVM does *not* mean you know SVA. It's a distinct language that needs to be learnt on its own. You can deploy assertions without the knowledge of OOP or UVM. The same applies to Functional Coverage language. It is orthogonal to SVA and OOP/UVM. The Functional Coverage language stands alone and needs to be learnt on its own. Albeit, once you learn Functional Coverage language, you can then use it in Class based OOP subset of SystemVerilog.

In any design, there are three main components of verification: (1) Stimulus Generators to stimulate the design, (2) Response Checkers to see that the device

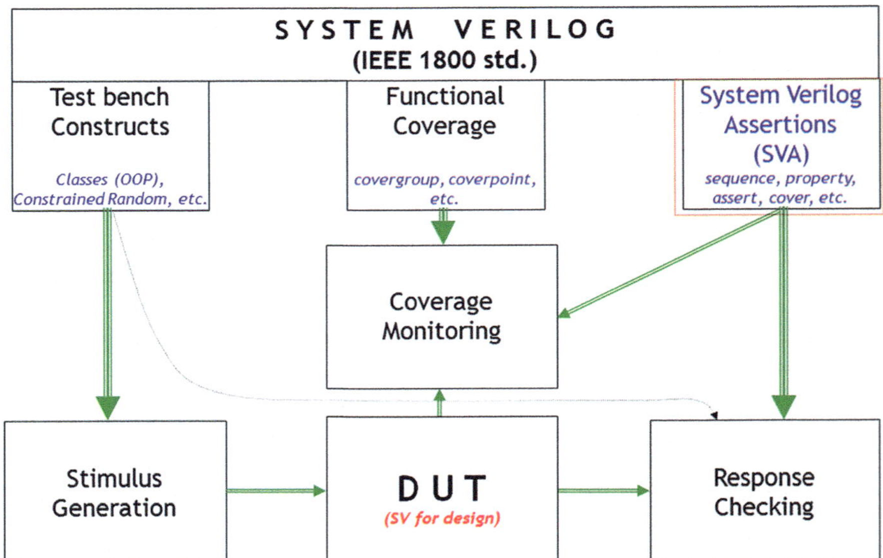

Fig. 1.3 SystemVerilog Assertions and functional coverage components under SystemVerilog umbrella

adheres to the device specifications, and (3) Coverage components to see that we have indeed structurally and functionally covered everything in the DUT according to the device specifications.

1. *Stimulus Generation*: This entails creating different ways in which a DUT needs to be exercised. For example, a peripheral (e.g., USB) maybe modeled as a Bus Functional Mode (or a UVM (Universal Verification Methodology) agent) to drive traffic through SystemVerilog transactions to the DUT. Different techniques are deployed to achieve exhaustive coverage of the design. For example, constrained random, transaction based, UVM based, memory based, etc. These topics are beyond the scope of this book.
2. *Response checking*: Now that you have stimulated the DUT, you need to make sure that the device has responded to that stimulus according to the device specs. Here is where SVA comes into picture along with UVM monitors, scoreboards, and other such techniques. SVA will check to see that the design not only meets high-level specifications but also low-level combinatorial and temporal design rules.
3. *Functional Coverage*: How do we know that we have exercised everything that the device specification dictates? Code Coverage is one measure. But code coverage is only structural. For example, it will point out if a conditional has been exercised. But code coverage has no idea if the conditional itself is correct, which is where Functional Coverage comes into picture (more on this later when we discuss Functional Coverage—See chapter (Chap. 25 **Functional Coverage**). Functional coverage gives an objective measure of the design coverage (e.g., have we verified all different cache access transitions (for example, write followed by read from the same address) to L2 from CPU? Code Coverage will not give such measure). We will discuss entire coverage methodology in detail in Chap. 25.

1.3 SystemVerilog Assertions Evolution

To set the stage, here is a brief history of Verilog to SystemVerilog evolution (Figs. 1.4 and 1.5). Starting with Verilog 95, we reached Verilog 2001 with Multi-dimensional arrays and auto variables, among other useful features. Meanwhile, functional verification was eating up ever more resources of a given project. Everyone had disparate functional verification environments and methodologies around Verilog. This was no longer feasible.

Industry recognized the need for a standard language that allowed the design *and* verification of a device and a methodology around which reusable components can be built avoiding multi-language cumbersome environments. Enter Superlog, which was a language with high-level constructs required for functional verification. Superlog was donated (along with other language subset donations) to create SystemVerilog 3.0 from which evolved SystemVerilog 3.1, which added new features for design but over 80% of the new language subset was dedicated to functional verification. We can only thank the Superlog inventor (the same inventor as that for Verilog—namely, Phil Moorby) and the Accelera technical subcommittees

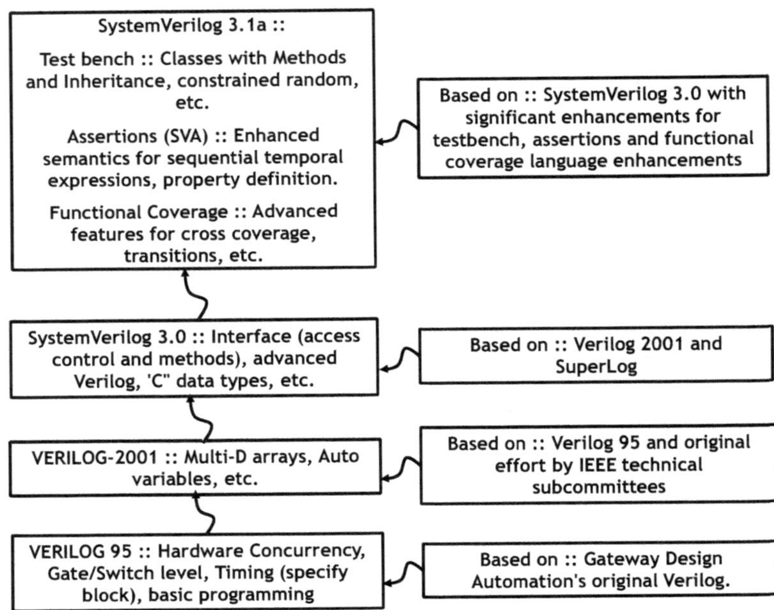

Fig. 1.4 SystemVerilog evolution

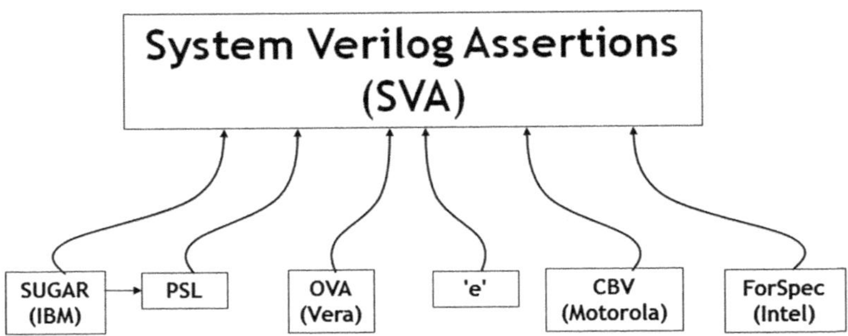

Fig. 1.5 SystemVerilog Assertion evolution

for having a long-term vision to design such a robust all-encompassing language. No multi-language solutions were required any more. No more reinventing of the wheel with each project was required anymore.

As shown in Fig. 1.5, SystemVerilog Assertion language is derived from many different languages. Features from these languages either influenced the language or were directly used as part of the language syntax/semantic.

Sugar from IBM led to PSL. Both contributed to SVA. The other languages that contributed are Vera, “e,” CBV from Motorola, and ForSpec from Intel.

In short, when we use SystemVerilog Assertions language, we have the benefit of using the latest evolution of an assertions language that benefited from many other robust assertions languages.

Part I
System Verilog Assertions (SVA)

Chapter 2
System Verilog Assertions

Introduction: This chapter will start with definition of an assertion with simple examples, moving on to its advantages as applied to real-life projects, what types of assertions need to be added for a given SoC project, and the methodology components to successfully adopt assertions in your project.

A. B. Mehta, *SystemVerilog Assertions and Functional Coverage*,
https://doi.org/10.1007/978-3-030-24737-9_2

2.1 What Is an Assertion?

An assertion is simply a check against the specification of your design that you want to make sure never violates. If the specs are violated, you want to see a failure.

A simple example is given below. Whenever FRAME_ is de-asserted (i.e., goes High), the Last Data Phase (LDP_) must be asserted (i.e., goes Low) within the next two clocks. Such a check is imperative to correct functioning of the given interface. The SVA code is very self-explanatory. There is the property "ldpcheck" that says "at posedge clock, if FRAME_ rises, it implies that within the next 2 clocks LDP_ falls." SVA language is precisely designed to tackle such sequential temporal domain scenarios. As we will see in Sect. 2.3, modeling such a check is far easier in SVA than in Verilog. Note also that assertions work in temporal domain (and we will cover a lot more on this later), and are concurrent as well as multi-threaded. These attributes are what makes SVA language so suitable for writing temporal domain checks.

Figure 2.1 shows the assertion for this simple bus protocol. We will discuss how to read this code and how this code compares with Verilog in the immediately following Sect. 2.3.

2.2 Why Assertions? What Are the Advantages?

As we discussed in the introductory section, we need to increase productivity of the design/debug/simulate/cover loop. Assertions help exactly in these areas. As we will see, they are easier to write than standard Verilog or SystemVerilog (thereby

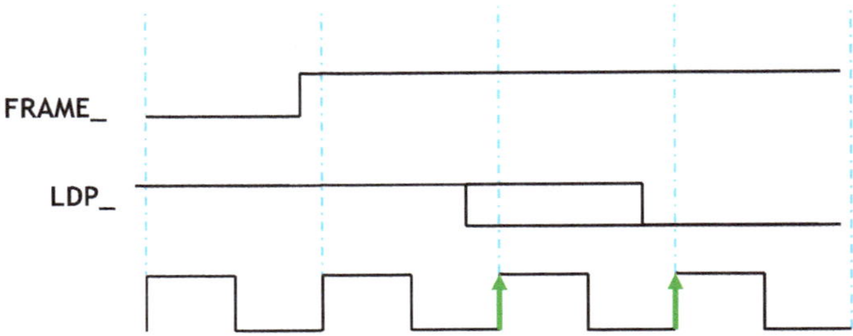

When FRAME_ is de-asserted (high), LDP_ (last data phase) (low) must be asserted within the next 2 clocks

```
property ldpcheck;
  @(posedge clk) $rose (FRAME_) |-> ##[1:2] $fell (LDP_);
endproperty
aP: assert property (ldpcheck) else $display("ldpcheck FAIL");
cP: cover property (ldpcheck) $display("ldpcheck PASS");
```

Fig. 2.1 A simple bus protocol design and its SVA property

increasing design productivity), easier to debug (thereby increasing debug productivity), provide functional coverage, and simulate faster compared to the same assertion written in Verilog or SystemVerilog. Let us see these advantages one by one

2.3 Assertions Shorten Time to Develop

Referring to the timing diagram in Fig. 2.1, let us see how SVA shortens time to develop. The SVA code is very self-explanatory. There is the property "ldpcheck" that says "at posedge clock, if FRAME_ rises, it implies that within the next 2 clocks LDP_ falls." This is almost like writing the checker in English. We then "assert" this property, which will check for the required condition to meet at every posedge clk. We also "cover" this property to see that we have indeed exercised the required condition. But we are getting ahead of ourselves. All this will be explained in detail in coming chapters. For now, simply understand that the SV assertion is easy to write, easy to read, and easy to debug.

Now examine the Verilog code for the same check (Fig. 2.2). There are many ways to write this code. One of the ways at behavioral level is shown. Here you "fork" out two procedural blocks, one that monitors LDP_ and another that waits for 2 clocks. You then disable the entire block ("ldpcheck") when either of the two procedural blocks complete. As you can see that not only is the checker very hard to read/interpret but also very prone to errors. You may end up spending more time debugging your checker than the logic under verification.

2.4 Assertions Improve Observability (Fig. 2.3)

One of the most important advantages of assertions is that they fire at the source of the problem. As we will see in the coming chapters, assertions are located local to logic in your design. In other words, you don't have to back trace a bug all the way

Verilog Code

```
always @(posedge FRAME_)
begin:ldpcheck
  @(posedge clk);
  fork
    begin
      @(negedge LDP_) disable ldpcheck;
    end
    begin
      repeat (2) @(posedge clk); $display("ldpcheck FAIL");
      disable ldpcheck;
    end
  join
end
```

Fig. 2.2 Verilog code for the simple bus protocol

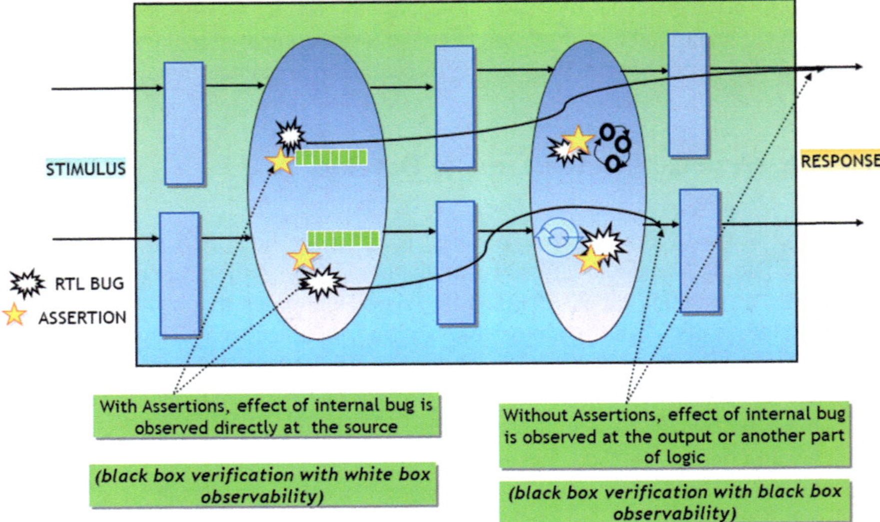

Fig. 2.3 Assertions improve observability

from primary output to somewhere internal to the design where the bug originates. Assertions are written such that they are close to logic (e.g., @ (posedge clk) state0 |-> Read); such an assertion is sitting close to the state machine and if the assertion fails, we know that when the state machine was in state0 that Read did not take place. Some of the most useful places to place assertions are FIFOs, Counters, block-to-block interface, block-to-IO interface, State Machines, etc. These constructs are where many of the bugs originate. Placing an assertion that checks for local condition will fire when that local condition fails, thereby directly pointing to the source of the bug. This can be called black box verification with white box observability.

Traditional verification can be called Black Box verification with Black Box observability, meaning, you apply vectors/transactions at the primary input of the "block" without caring for what's in the block (blackbox verification) and you observe the behavior of the block only at the primary outputs (blackbox observability). Since you don't have observability in the design under test, you basically start debugging from primary output to internal logic and with lengthy waveform based debug you find the bug. Assertions on the other hand allow you to do black box verification with white box (internal to the block) observability.

2.5 Assertions in Static Formal

The same assertions that you write for design verification can be used with static functional verification or the so-called hybrid static functional plus simulation algorithms. Figure 2.4 shows (on LHS) SVA *Assumptions* and (on RHS/Center) SVA

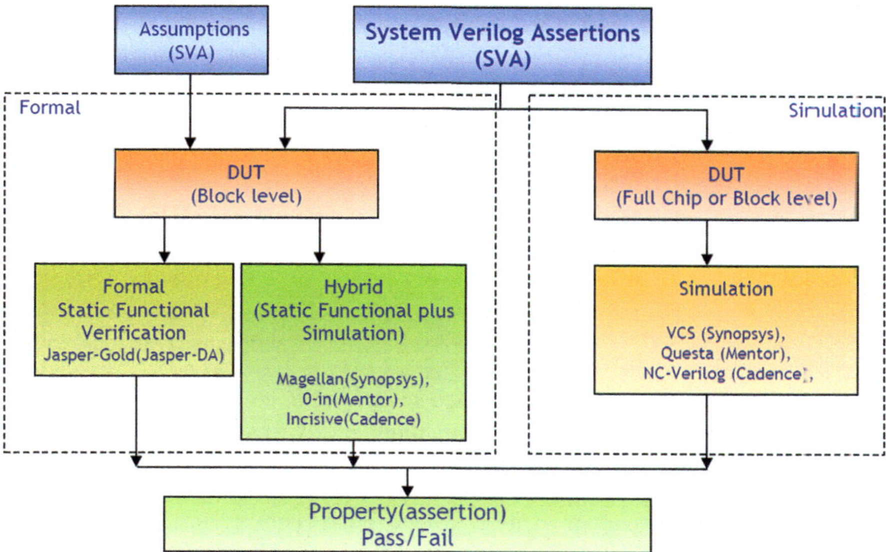

Fig. 2.4 Assertions and Assumptions in Formal (static functional) and Simulation

Assertions. As you see, the assumptions are most useful to *Static Functional Verification (aka Formal*) (even though assumptions can indeed be used in Simulation as well, as we will see in later sections) while SVA assertions are useful in both Formal and Simulation.

So, what is Static Functional Verification (also called Static Formal Functional or simply Formal)? In plain English, static formal is a method whereby the static formal algorithm applies all possible combinational and temporal domain stimulus possibilities to exercise all possible "logic cones" of a given logic block and see that the assertion(s) are not violated. This eliminates the need for a test-bench and also makes sure that the logic never fails under *any* circumstance. This provides 100% comprehensiveness to the logic under verification. So as a side note, why do we ever need to write a test-bench? The static formal (as of this writing) is limited by the size of the logic block (i.e., gate equivalent RTL) especially if the temporal domain of inputs to exercise is large. The reason for this limitation is that the algorithm has to create different logic cones to try and prove that the property holds. With larger logic blocks, the number of these so-called logic cones explode. This is also known as "state space explosion" problem. To counter this problem, simulation experts came up with the *Hybrid Simulation* technique. In this technique, simulation is deployed to reach "closer" to the assertion logic and then employ the static functional verification algorithms to the logic under test. This reduces the scope of the number of logic cones and their size and you may be successful in seeing that the property holds. Since static functional or hybrid is beyond the scope of this book, we'll leave it at that.

2.6 Assertion Synthesis

Yes, you can synthesize assertions, well, at least the simpler ones. This effort is picking up steam as more engineers turn towards hardware acceleration and emulation (FPGA or EDA Tool based) for verifying their design. Long latency and massive random tests need acceleration/emulation tools. These tools are beginning to support synthesizable assertions.

This section is to point out that assertions are not only useful in software-based simulation but also hardware-based emulation. The reason you can use assertions to fire directly in hardware is because assertions are synthesizable. Even though assertion synthesis has ways to go, there is enough of a subset covered by synthesis and that is enough to deploy assertions in hardware.

Especially for an FPGA, a synthesized assertion can be very useful during field deployment of FPGA. In the field, if something goes wrong, the synthesized assertion will fire, quickly pointing to the source of the bug.

Even though emulation speeds measure in megahertz, the debug cycle on emulated platform is horrendous. This is due to poor visibility inside the emulated design. You emulate a design in minutes and then spend hours debugging the failures. This is where synthesizable assertions come into picture. As we know, assertions fire at the root of the failure. So, if an assertion can be synthesized, it can become part of the logic that ends up on emulated platform. Once the synthesized logic is part of the hardware design in the emulated platform, the synthesized assertion will fire at the root of the failure and you'll know right away the cause of failure, thereby drastically reducing the debug of an emulated design.

Let us first look at a simple assertion module that is taken from the OVL library (OVL) (http://www.accellera.org/downloads/standards/ovl). Open Verification Library is a publicly available source from accelera.org website (Accellera).

We will see how the OVL assertion gets synthesized and how the resulting logic looks like.

We take a module "ovl_always" from the OVL library. It simply checks to see that an input "test_expr" holds true at every posedge clock. That's all. Here's a block diagram of "ovl_always."

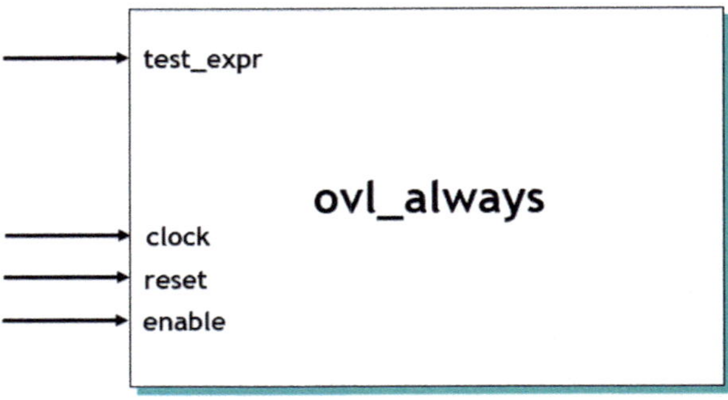

Here's the module definition of ovl_always (from the OVL library). I've removed code that is not of significance to our exercise.

```
// Accellera Standard V2.8.1 Open Verification Library (OVL).
// Accellera Copyright (c) 2005-2014. All rights reserved.

`include "std_ovl_defines.h"
module ovl_always (clock, reset, enable, test_expr);
  input           clock, reset, enable;
  input           test_expr;

  parameter assert_name = "OVL_ALWAYS";


  property ASSERT_ALWAYS_P;
  @(posedge clk)
  disable iff (`OVL_RESET_SIGNAL != 1'b1)

    !($isunknown(test_expr)) |->  test_expr;
  endproperty
endmodule
```

Now, let's instantiate this module in a test-bench and pass it a specific "test_expr" to see that it holds. The 'test_expr' we are passing is "reg_a < reg_b."

The test-bench and the synthesized circuit are provided by Mike McGregor, a colleague of mine.

```
`include "std_ovl/ovl_always.v"
module regTest (
input clk, reset, enable;
output logic [2:0] fire;
output logic [31:0] error = 32'h0;
)

logic [7:0] reg_a = 8'h0, reg_b=8'h10;
always @(posedge clk) begin
    reg_a <= reg_a = 8'h1;
end
ovl_always reg_a_lt_reg_b (
     clk,
     reset,
     enable,
     reg_a < reg_b);  //This is the test expression
endmodule
```

And here's the synthesized circuit (McGregor) for "reg_a_lt_reg_b." This proves that assertions are indeed synthesizable.

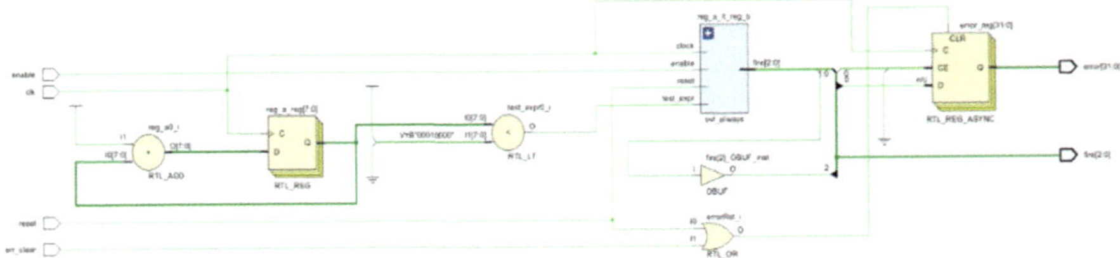

The synthesized circuit is provided by Mike McGregor. (McGregor).

Ok, so we now have an assertion that will be part of hardware design being emulated. But what happens when it fires? How do we know that it fired in the first place? In simulation, you get PASS or FAIL display. Well, there is no such display in hardware.

There are multiple ways you can access the fact that an assertion fired and failed or passed.

Have a register in your design that gets updated every time an assertion fires. Encode various assertions to distinguish them from one another. Then scan out this register and decode it to see which assertion fired and if it failed or not.

Another way is to directly encode the assertion on your SoC's GPIO. Enable GPIO for assertion encoded output and decode the output to see how assertion behaved.

Or create a simple FIFO to store assertion results and flush it out to see assertion behavior.

There are many ways one can "access" the behavior of an assertion in hardware.

A generic emulation system is shown in Fig. 2.5. Synthesizable assertions are part of the design that get synthesized and get partitioned to the emulation hardware. During emulation, if the design logic has a bug, the synthesized assertion will fire and trigger a stop/trace register to stop emulation and directly point to the cause of failure.

Anyone who has used emulation as part of their verification strategy very well know that even though emulator may take seconds to "simulate" the design, it takes hours thereafter to debug failures. Assertions will be a great boon to the debug effort. Many commercial vendors are beginning to support synthesizable assertions.

Figure 2.5 shows a generic emulation system representative of emulation systems available from EDA vendors. In the figure, you see a synthesizable SVA assertion embedded in the emulated design. On the firing of the assertion a trigger output is asserted. That can help you start and stop trace for the logic under test.

On the same line of thought, assertions can be synthesized in silicon as well. During post-silicon validation, a functional bug can fire, and a hardware register can record the failure. This register can be reflected on GPIO of the chip or the register can be scanned out using JTAG boundary scan. Without such facility, it takes hours of debug time to pinpoint the cause of silicon failure. This technique is now being used widely. The "area" overhead of synthesized assertion logic is negligible compared to the overall area of the chip, but the debug facilitation is of immense value. Note that such assertions can make it easier to debug silicon failures in field as well.

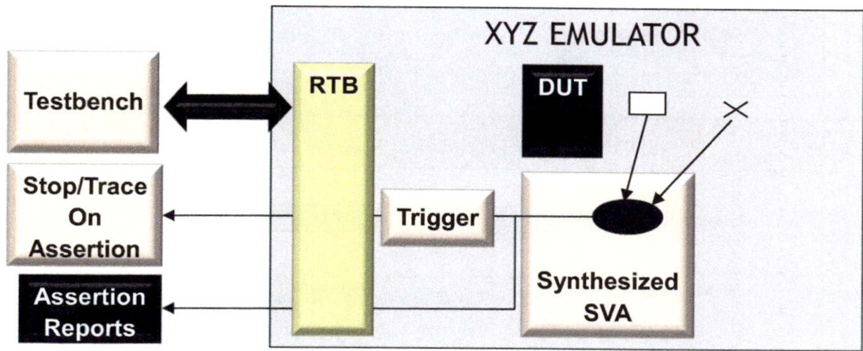

- Synthesizable assertions supported by XYZ Emulator
- Embedded and bound assertions are automatically synthesized and mapped into the emulator
- Assertion reporting and triggering at run-time, similar to simulation
- Compile time and Run time control over Assertion compilation and reporting/triggering

Fig. 2.5 Assertions for hardware emulation

2.7 One-Time Effort, Many Benefits

Figure 2.6 shows the advantage of assertions. Write them once and use them with many tools.

We have discussed at high level the use of assertions in Simulation, Formal, Coverage, and Emulation. But how do you use them for Test-bench Generation/Checker and what is OVL assertions library? *Test-bench Generation/Checker*: With ever-increasing complexity of logic design, the test benches are getting ever so complex as well. How can assertions help in designing test-bench logic? Let us assume that you need to drive certain traffic to a DUT input under certain condition. You can design an assertion to check for that condition and upon its detection the FAIL or PASS action block triggers, which can be used to drive traffic to the DUT. Checking for a condition is far easier with assertions language than with SystemVerilog alone. Second benefit is to place assertions on verification logic itself. Since verification logic (in some cases) is even more complex than the design logic, it makes sense to use assertions to check test-bench logic also.

OVL Library: Open Verification Library. This library of predefined checkers was written in Verilog before PSL and SVA became mainstream. Currently the library includes SVA (and PSL) based assertions as well. The OVL library of assertion checkers is intended for use by design, integration, and verification engineers to check for good/bad behavior in simulation, emulation, and formal verification. OVL contains popular assertions such as FIFO assertions, among others. OVL is still in use and you can download the entire standard library from Accellera website. http://www.accellera.org/downloads/standards/ovl

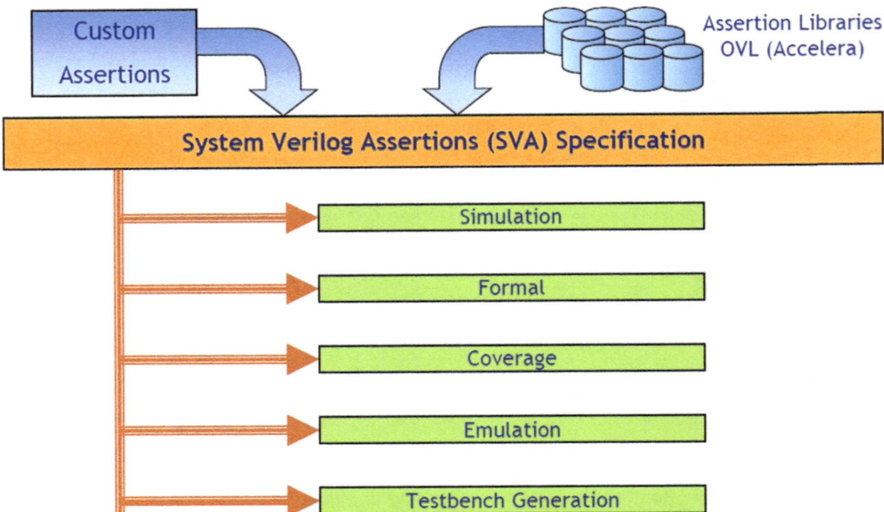

Fig. 2.6 Assertions and OVL for different uses

We will not go into the detail of OVL since there is plenty of information available on OVL on net. OVL code itself is quite clear to understand. It is also a good place to see how assertions can be written for "popular" checks (e.g., FIFO) once you have better understood assertion semantics.

2.8 Assertions Whining

Maybe the paradigm has now shifted but as of this writing there is still a lot of hesitation on adopting SVA in the overall verification methodology. Here are some popular objections.

- *I don't have time to add assertions. I don't even have time to complete my design. Where am I going to find time to add assertions?*
  - That depends on your definition of "completing my design." If the definition is to simply add all the RTL code without—any—verification/debug features in the design and then throw the design over the wall for verification, it will take significantly longer to debug your design for it to work as specified.
  - During design you are already contemplating and assuming many conditions (state transition assumptions, inter-block protocol assumptions, etc.). *Simply convert your assumptions into assertions as you design*. They will go a long way in finding those corner case bugs even with your simple sanity test benches.

- *I don't have time to add assertions. I am in the middle of debugging the bugs already filed against my design.*
  - Well, actually you will be able to debug your design in shorter time, if you *did* add assertions as you were designing (or at least add them now) so that if a failing test fires an assertion, your debug time will be drastically short.
  - Assertions point to the source of the bug and significantly reduce time to debug as you verify your block level, chip level design.
  - In other words, this is a bit of chicken & egg problem. You don't have time to write assertions but without these assertions you will spend a lot more time debugging your design!
- *Isn't writing assertions the job of a verification engineer?*
  - Not quite. Design Verification (DV) engineers do not have insight into the micro-architectural level RTL detail. But the real answer is that BOTH Design and DV engineers need to add assertions. We will discuss that in detail in upcoming section.
- *DV engineer says I am new to assertions and will spend more time debugging my assertions than debugging the design.*
  - Well, don't you spend time in debugging your test bench logic? Your reference models? What's the difference in debugging assertions? If anything, assertions have proven to be very effective in finding bugs and cutting down on debug time.
  - In my personal experience (over the last many SoC and Processor projects), approximately 25% of the total bugs reported for a project were DV/Testbench bugs. There are significant benefits to adding assertions to your testbench that outweigh the time to debug them.
- *The designer cannot be the verifier also. Doesn't asking a designer to add assertions violate this rule?*
  - As we will discuss in the following sections, assertions are added to check the "intent" of the design and validate your own assumptions. *You are not writing assertions to duplicate your RTL.* Following example makes it clear that the designer does need to add assertions.
  - For example,

    For every "req" issued to the next block, I will indeed get an "ack" and that I will get only 1 "ack" for every "req." This is a cross module assumption which has nothing to do with how you have designed your RTL. You are not duplicating your RTL in assertions.

    My state machine should never get stuck in any "state" (except "idle") for more than 10 clocks.

2.9 Who Will Add Assertions? War Within!

Both Design and Verification engineers need to add assertions...

- Design Engineers:
  - Micro architectural level decisions/assumptions are not visible to DV engineers. So, designers are best suited to guarantee uArch level logic correctness.
  - Every assumption is an assertion. If you assume that the “request” you send to the other block will always get an “ack” in 2 clocks, that’s an assumption. So, design an assertion for it.
  - Add assertions as you design your logic, not as an afterthought.
- DV Engineers:
  - Add assertions to check macro functions and Chip/SoC level functionality.

    Once the packet has been processed for L4 layer, it will indeed show up in the DMA queue.
    A machine check exception indeed sets PC to the exception handler address.
  - Add assertions to check Interface IO logic.

    After Reset is de-asserted none of the signals ever go “X..”
    If the processor is in Wait Mode and no instructions are pending that it must assert a SleepReq to memory subsystem within 10 clocks.
    On Critical Interrupt, the external clock/control logic block must assert CPU_wakeup within 10 clocks.

2.10 A Simple PCI Read Example—Creating an Assertions Test Plan

Let us consider a simple example of PCI Read. Given the specification in Fig. 2.7, what type of assertions would the design team add and what type would the verification team add? The tables below describe the difference. I have only given few of the assertions that could be written. There are many more assertions that need to be written by verification and design engineers. However, this example will act as a basis for differentiation.

Designers add assertions at micro-architecture level (Table 2.2) while verification engineers concentrate at system level (Table 2.1), specifically the interface level in this example.

We will model the assertions for this PCI protocol later in the book under LAB6 exercise. It is too early to jump into writing assertions without knowing the basics at this stage.

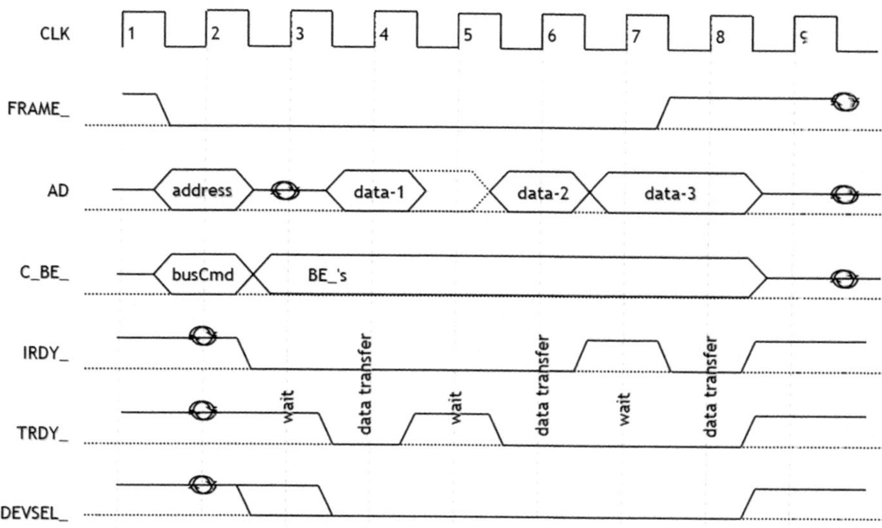

Fig. 2.7 A simple PCI Read Protocol

Table 2.1 PCI Read Protocol Test Plan by Functional Verification Team

Property Name	Description	Property FAIL?	Property Covered?
Protocol Interface Assertions			
checkPCI_AD_CBE (check1)	On falling edge of FRAME_, AD and C_BE_ bus cannot be unknown		
checkPCI_ DataPhase (check2)	When both IRDY_ and TRDY_ are asserted, AD or C_BE_ bus cannot be unknown		
checkPCI_Frame_ Irdy (check3)	FRAME can be de-asserted only if IRDY_ is asserted		
checkPCI_ trdyDevsel (check4)	TRDY_ can be asserted only if DEVSEL_ is asserted		
checkPCI_CBE_ during_trx (check5)	Once the cycle starts (i.e. at FRAME_ assertion) C_BE_ cannot float until FRAME_ is de-asserted.		

PCI: Basic Read Protocol Test Plan: Verification team

The PCI protocol is for a simple READ. With FRAME_ assertion, AD address and C_BE_ have valid values. Then IRDY_ is asserted to indicate that the master is ready to receive data. Target transfers data with intermittent wait states. Last data transfer takes place a clock after FRAME_ is de-asserted.

Let us see what type of assertions need to be written by design and verification engineers.

Table 2.2 PCI Read Protocol Test Plan by Design Team

Property Name	Description	Property FAIL?	Property Covered?
Microarchitectural Assertions			
check_pci_ adrcbe_St	PCI state machine is in "adr_cbe" state the first clock edge when FRAME_ is found asserted		
check_pci_ data_St	PCI state machine is in "data_transfer" state when both IRDY_ and TRDY_ are asserted		
check_pci_ idle_St	PCI state machine is in "idle" state when both FRAME_ and IRDY_ are de-asserted		
check_pci_ wait_St	PCI state machine is in "wait" state if either IRDY_ or TRDY_ is de-asserted		

PCI: Basic Read Protocol Test Plan: Design Team

Note that in Table 2.2 there are two columns: (1) Did the property FAIL? and (2) Did the property get covered? There is no column for the property PASS. That is because, "cover" in an assertion triggers only when a property is exercised but does not fail; in other words, it passes. Hence, there is no need for a PASS column. This "cover" column tells you that you indeed covered (exercised) the assertion and that it did not fail. When the assertion FAILs, it tells you that the assertion was exercised (covered) and that it Failed during the exercise.

2.11 What Type of Assertions Should I Add?

It is important to understand and plan for the types of assertions one needs to add. Make this part of your verification plan. It will also help you partition work among your team members.

Note the "performance implication" assertions. Many miss on this point. Coming from processor background, I have seen that these assertions turn out to be some of the most useful assertions. These assertions would let us know of the (e.g.) cache read latency upfront and would allow us enough time to make architectural changes.

- RTL Assertions (design intent)
  - Intra Module

    Illegal state transitions; deadlocks; livelocks;
    FIFOs, onehot, etc.

- Module interface Assertions (design interface intent)
  - Inter-module protocol verification; illegal combinations (ack cannot be "1" if req is "0"); steady state requirements (when slave asserts write_queue_full, master cannot assert write_req);
  - Every design assumption on inter-module protocol is an assertion.

- Chip functionality Assertions (chip/SoC functional intent)
  - A PCI transaction that results in Target Retry will indeed end up in the Retry Queue.
- Chip interface Assertions (chip interface intent)
  - Commercially available standard bus assertion VIPs can be useful in comprehensive check of your design's adherence to std. protocol such as PCIe, AXI, etc.
  - Every design assumption on IO functionality is an assertion.
- Performance Implication assertions (performance intent)
  - Cache latency for read; packet processing latency; etc. to catch performance issues before it's too late. This assertion works like any other. For example, if the "Read Cache Latency" is greater than 2 clocks, fire the assertion. This is an easy-to-write assertion with very useful return.

2.12 Protocol for Adding Assertions

- Do not duplicate RTL
  - White box observability does not mean adding an assertion for each line of RTL code. This is a very important point, in that if RTL says "req" means "grant," don't write an assertion that says the same thing!! Read on.
  - Capture the intent

  For example, a Write that follows a Read to the same address in the request pipe will always be allowed to finish before the Read. This is the intent of the design. How the designer implements reordering logic is not of much interest. So, from verification point of view, you need to write assertions that verify the chip design intent.

  A note here that the above does not mean you do not add low-level assertions. Classic example here is FIFO assertions. Write FIFO assertions for all FIFOs in your design. FIFO is low-level logic, but many of the critical bugs hang around FIFO logic and adding these assertions will provide maximum bang for your buck.
- Add assertions throughout the RTL design process
  - They are hard to add as an afterthought.
  - Will help you catch bugs even with your simple block level test bench.
- If an assertion did not catch a failure...
  - If the test failed and none of the assertions fired, see if there are assertions that need to be added which would fire for the failing case.
  - The newly added assertion is now active for any other test that may trigger it.

Note: This point is very important towards making a decision if you have added enough assertions. In other words, if the test failed and none of the assertions fired, there is a good chance you still have more assertions to add.

- Reuse
  - Create libraries of common "generic" properties with formal arguments that can be instantiated (reused) with "actual" arguments. We will cover this further in the book.
  - Reuse for the next project.

2.13 How Do I Know I Have Enough Assertions?

- It's the "Test plan, test plan, test plan…".
  - Review and re-review your test plan against the design specs.
  - Make sure you have added assertions for every "critical" function that you must guarantee works.
- If tests keep failing but assertions do not fire, you do not have enough assertions.
  - In other words, if you had to trace a bug from primary outputs (of a block or SoC) without any assertions firing that means that you did not put enough assertions to cover that path.
- "formal" (aka static formal aka static functional verification) tool's ability to handle assertions
  - What this means is that if you don't have enough "assertion density" (meaning if a register value does not propagate to an assertion within 3–5 clocks—resulting in assertions sparsely populated within design), the formal analysis tool may give up due to the state/space explosion problem. In other words, a static functional formal tool may not be able to handle a large temporal domain. If the assertion density is high, the tool has to deal with smaller cones of logic. If the assertion density is sparse, the tool has to deal with larger cones of logic in both temporal and combinatorial space and it may run into trouble.

2.14 Assertion-Based Methodology Allows for Full Random Verification

Huh! What does that mean? This example is what I learnt from real-life experience. In our projects, we always do full random concurrent verification (i.e., all initiators of the design fire at the same time to all targets of the design) after we are done with directed and constrained random verification. The idea behind this is to find any deadlocks (or livelock for that matter) in the design. Most of the initiator tests are well crafted (i.e., they won't clobber each other's address space) but with such massive randomness, your target models (scoreboards) may not be able to predict response to randomly fired transactions. In all such cases, it is best to disable scoreboards in your target models (unless the scoreboards are full proof in that they can survive total randomness of transactions). BUT keep assertions alive. Now, fire concurrent random transactions, the target models will respond the best they can, but assertions will pinpoint to a problem if it exists (such as simulation hang (deadlock) or simply keep bouncing between two state machines without advancing functionality (livelock)).

In other words, assertions are always alive and regardless of transaction stream (random or directed), they will fire as soon as there is the detection of an incorrect condition.

- Example Problem Definition:
  - Your DUT has Ethernet Receive and Video as Inputs and is also a PCI target.
  - It also has internal initiators outputting transactions to PCI targets, SDRAM, Ethernet Transmit, and Video outputs.
  - After you have exhausted constrained random verification, you now want to simulate a final massive random verification, blasting transactions from all input interfaces and firing transactions from internal masters (DMA, Video Engine, Embedded processors) to all the output interfaces of the design.
  - BUT there's a good chance your reference models, self-checking tests, scoreboards may not be able to predict the correct behavior of the design under such massive randomness.
- Solution:
  - Turn off all your checking (reference models, scoreboards, etc.) unless they are full proof to massive random transaction streams.
  - BUT keep Assertions alive.
  - Blast the design with massive randomness (keep address space clean for each initiator).
  - If any of the assertions fire, you have found that corner case bug.

2.15 Assertions Help Detect Bugs Not Easily Observed at Primary Outputs

This is a classic case that we encountered in a design and luckily found before tape-out. Without the help of an assertion, we would not have found the bug and there would have to be a complex software workaround. I will let the following example explain the situation.

- The Specification:
  - On a store address Error, the address in Next Address Register (NAR) should be frozen the same cycle that the Error is detected.
- The Bug:
  - On a store address error, the state machine that controls the NAR register actually froze the next address the *next* clock (instead of freezing the current address the same clock when store address error occurred). In other words, an incorrect address was being stored in NAR at an incorrect clock.
- So, why were the tests passing with this bug?
  - The tests that were triggering this bug used the same address back to back. In other words, even though the incorrect "next" address was being captured in NAR, since the "next" and the "previous" addresses were the same, the logic would *seem* to behave correct.
  - The Assertion: An assertion was added to check that when a store address error was asserted the state machine should not move to point to the next address in pipeline. Because of the bug, the state machine actually did move to the next stage in pipeline. The assertion fired and the bug was caught.

2.16 Other Major Benefits

- SVA language supports Multi-Clock Domain Crossing (CDC) logic.
  - SVA properties can be written that cross from one clock domain to another. Great for data integrity checks while crossing clock domains. There is not a single ASIC/FPGA design that I know of that does not have multiple clock domains. So, this feature is of utmost importance.
- Assertions are Readable: Great for documenting and communicating design intent.
  - Great for creating executable specs.
  - Process of writing assertions to specify design requirements and conducting cross design reviews identify.

    Errors, Inconsistencies, omissions, vagueness.
  - Use it for design verification (test plan) review.
- Reusability for future designs.
  - Parameterized assertions (e.g., for a 16-bit bus interface) are easier to deploy with the future designs (with a 32-bit bus interface).

  - Assertions can be modeled outside of RTL and easily bound (using "bind") to RTL keeping design and DV logic separate, easy to maintain and reusable for the next design project.

- Assertions are always ON.
  - Assertions never go to sleep (until you specifically turn them off).
  - In other words, active assertions take full advantage of every new test/stimulus configuration added by monitoring designs behavior against the new stimulus.

- Acceleration/Emulation with Assertions.
  - Long latency and massive random tests need acceleration/emulation tools. These tools are beginning to support synthesizable assertions. Assertions are of great help in quick debug of long/random tests. We will discuss this further in coming sections.

- Global Severity Levels ($Error, $Fatal, etc.)
  - Helps maintain a uniform Error reporting structure in simulation.

- Global turning on/off of assertions (as in $dumpon/$dumpoff).
  - Easier code management (no need to wrap each assertion with an on/off condition).

- Formal Verification depends on Assertions.
  - The same "assert" ions used for design simulation are also used directly by formal verification tools. Static formal applies its algorithms to make sure that the "assert"ion never fails.
  - "assume" and "restrict" allow for correct design constraint important to formal.

- One language, multiple usage.
  - "assert" for design check and for formal verification
  - "cover" for temporal domain coverage check
  - "assume" and "restrict" for design constraint for formal verification.

2.17 Use Assertions for Specification and Review

- Use assertions (properties/sequences) for specification.
  - DV (Design Verification) Team:

    Document as much of the "response checking" part of your test plan as practical directly into executable properties.
    Use it for verification plan review and update.

  - Design Team:

    Document micro-arch. Level assertions directly into executable properties. Use it for design reviews.

- Assertions Cross-Review.
  - Review:

    DV team reviews macro, chip, interface level assertions with the design team.

  - Cross-Review.

    Block A designer reviews "module B" interface assertions.
    Block B designer reviews "module A" interface assertions.

  - Misassumptions, incorrect communication are detected early on.

2.18 Assertion Types

There are three types of assertions supported by SVA. In brief, here's their description. We will discuss them in plenty detail throughout this book.

- Immediate Assertion.
- Concurrent Assertion.
- Deferred Immediate Assertion (introduced in IEEE 1800–2009).

Immediate Assertions

- Simple *non-temporal domain assertions* that are executed like statements in a procedural block,
- Interpreted the same way as an expression in the conditional of a procedural "if" statement.
- Can be specified only where a procedural statement is specified.

Concurrent Assertions

- These are *temporal domain assertions* that allow creation of complex sequences using clock (sampling edge) based semantics.
- They are edge sensitive and not level sensitive. In other words, they must have a "sampling edge" on which it can sample the values of variables used in a sequence or a property. The sampling edge can be synchronous or asynchronous.
- A concurrent assertion is "assert," "cover," "assume," or "restrict." We will discuss each type in coming chapters.

Deferred Assertions (introduced in IEEE 1800–2009)

- Deferred assertions are a type of Immediate assertions. Note that immediate assertions evaluate *immediately* without waiting for variables in its combinatorial expression to *settle* down. This also means that the immediate assertions are very prone to glitches as the combinatorial expression settles down and may fire multiple times. On the other hand, deferred assertions do not evaluate their sequence expression until the end of time stamp when all values have settled down (or in the reactive region of the time stamp). Detailed explanation is in Sect. 1.1.

If some of this does not quite make sense, that's OK. That is what the rest of the book will explain. Let us start with Immediate assertions and understand its semantics. We then move on to Concurrent assertions and lastly Deferred assertions. The book focuses on concurrent assertions because that is really the main gist of SystemVerilog Assertions Language.

Chapter 3
Sequential Domain Coverage ("Cover" Property and "Cover" Sequence)

Introduction: This chapter explores the role of SVA in coverage of sequential domain coverage of your design with "cover" semantics. "cover" is a component of the overall coverage methodology, which comprises functional coverage, code coverage, and SVA "cover".

A. B. Mehta, *SystemVerilog Assertions and Functional Coverage*,
https://doi.org/10.1007/978-3-030-24737-9_3

Assertions not only help you find bugs but also help you determine if you have covered (i.e., exercised) design logic, mainly temporal domain conditions aka sequential logic. They are very useful in finding temporal domain coverage of your test-bench. Here is the reason why this is so important.

Let us say, you have been running regressions 24∗7 and have stopped finding bugs in your design. Does that mean you are done with verification? No. Not finding a bug could mean one of two things. (1) There is indeed no bug left in the design or (2) you have not exercised (or covered) the conditions that exercise the bugs. You could be continually hitting the same piece of logic in which no further bugs remain. In other words, you could be reaching a wrong conclusion that all the bugs have been found.

In brief, *functional coverage includes three components* (we will discuss this in detail in the chapter on Functional Coverage).

1. Code Coverage (which is structural) which needs to be 100% (excluding certain unreachable code; unused state transitions, etc.)
2. Functional Coverage that need to be designed to cover *functionality* (i.e., intent) of the entire design and must completely cover the design specification. This includes "covergroup" and "coverpoint" (Chap. 25).
3. Sequential (temporal) domain coverage (using SVA "cover" feature) which need to be carefully designed to fully cover all required temporal domain conditions of the design (Fig. 3.1).

You can "cover" a property as well as a sequence. Let's see each one in detail.

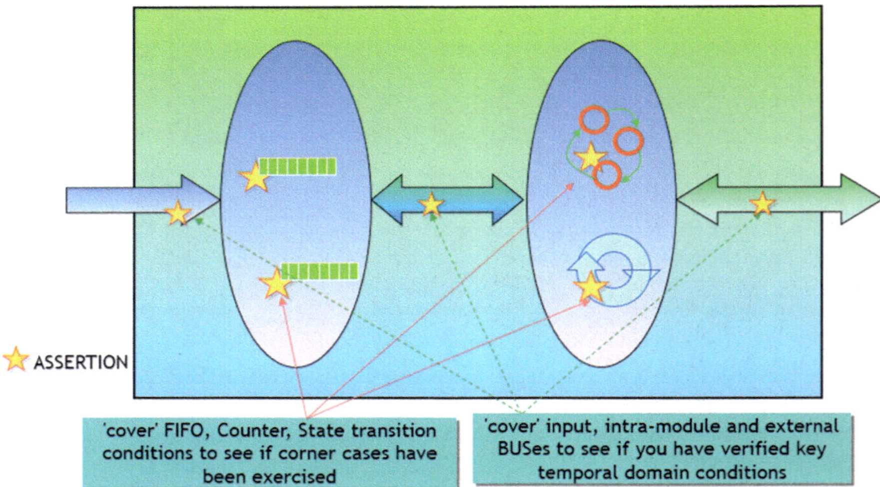

Fig. 3.1 SystemVerilog Assertions provide temporal domain functional coverage

3.1 "Cover" Property

Ok, let us go back to the simple bus protocol assertion that we saw in the previous section. Let us see how the "cover" statement in that SVA assertion works. The code is repeated here for easy reference.

```
property ldpcheck;
@(posedge clk) $rose (FRAME_) |-> ##[1:2] $fell (LDP_);
endproperty
aP: assert property (ldpcheck) else $display("ldpcheck FAIL");
cP: cover property (ldpcheck) $display("ldpcheck PASS");
```

The formal syntax of the "cover" property is as follows (LRM, 2012):

```
cover property ( property_spec ) statement_or_null
```

In the code presented above, you see that there is a "cover" statement. It covers a property. What it tells you is "did you exercise this condition" or "did you cover this property"? In other words, and as discussed above, if the assertion never fails, that could be because of two reasons. (1) You don't have a bug or (2) You never exercised the condition to start with! With the "cover" statement, if the condition gets exercised but does *not* fail you get that indication through the "pass" action block associated with the "cover" statement. If the property does not fail but passes with "cover" you know that you do not have a bug associated with that property and that you have exercised the condition. Note that you can have a PASS action block with "assert" itself but then you need to worry about vacuous pass phenomena. Since we haven't yet discussed the assertions in any detail, you may not completely understand this concept, but *determination of temporal domain coverage of your design is an extremely important aspect of verification and must be made part of your verification coverage plan.*

Note that a "cover" property cannot be used in a Class. It can only be used in structural code (module, program, or interface).

"cover property" provide you the following:

1. Number of times attempted.
2. Number of times succeeded (maximum of one per attempt).
3. Number of times succeeded because of vacuity (we will discuss vacuous pass in Chap. 1).
4. If an evaluation attempt is successful, the pass action is executed only once for each evaluation attempt.

3.2 "Cover" Sequence

The formal syntax of "**cover sequence**" is as follows (SystemVerilog 3.1a LRM):

```
cover sequence (
clocking_event ] [ disable iff ( expression_or_dist ) ]
sequence_expr )
statement_or_null
```

The body of cover sequence must be a sequence expression; that is, it may only contain an expression constructed using clock specifications, Boolean expressions, sequence instances, sequence operators, and possibly sequence match items. It may optionally contain a "disable iff" specification. A clocking event is optional if it can be inferred from the context. The optional pass action is executed when the sequence matches. In this case, the pass action may be any procedure and is not limited to a single subroutine call.

Results of coverage for a sequence will include the following:

1. Number of times matched (each attempt can generate multiple matches).
2. The coverage database for coverage on a sequence contains the number of evaluation attempts started and the total number of sequence matches.
3. The pass statement specified in *statement_or_null* will be executed, with multiplicity, for each match that is counted towards the total for the attempt. This is in contrast to 'cover property' where for each successful attempt, the pass action is executed *only once* for each evaluation attempt.

In practice, property coverage is much more useful than sequence coverage. The main use of sequence coverage is to react on *each* sequence match in the pass action block to trigger some test-bench actions.

To reiterate, SVA supports the "cover" construct that tells you if the assertion has been exercised (covered). Without this indication and in the absence of a failure, you have no idea if you indeed exercised the required condition. You can indeed use the pass action block to indicate a pass but then you need to worry about vacuous pass (or turn off vacuous pass with $vaccuouspassoff system function). See Sect. 17.18 to understand vacuous pass.

In our example, if FRAME_ never rises, the assertion won't fire and obviously there won't be any bug reported. So, at the end of simulation if you do not see a bug or you do not even see the "ldpcheck PASS" display, you know that the assertion never fired (i.e., never got covered). In other words, you must see the "cover property" statement executed in order to know that the condition did get exercised. We will discuss this further in coming chapters. Use "cover" to full extent as part of your verification methodology.

Chapter 4
Conventions Used in the Book

Introduction: This chapter describes the signal level sensitive and edge sensitive figure annotation conventions that are applied throughout the book (Table 4.1).

A. B. Mehta, *SystemVerilog Assertions and Functional Coverage*,
https://doi.org/10.1007/978-3-030-24737-9_4

Note that the level sensitive attribute of a signal is shown as a "fat" High and Low symbol. I could have drawn regular timing diagrams but saw that they look very cumbersome and does not easily convey the point. Hence, I chose the fat arrow to convey that ***when the fat arrow is high, the signal was high before the clock, at the clock, and after the clock. The same applies for the fat low arrow.***

For edge sensitive assertions, I chose the regular timing diagram to distinguish them from the level sensitive symbol.

A high green arrow is for PASS and a low red arrow is for FAIL.

Table 4.1 Conventions used in this book

(fat up arrow)	**LEVEL SENSITIVE HIGH:** This symbol means that the signal is detected HIGH (level sensitive) at the clock edge noted in a timing diagram. It could have been high or low the previous clock and may remain high or low after the clock edge. *It does NOT however mean that a 'posedge' is expected on this signal at the noted clock edge.*
(fat down arrow)	**LEVEL SENSITIVE LOW:** This symbol means that the signal is detected LOW (level sensitive) at the clock edge noted in a timing diagram. It could have been high or low the previous clock and may remain high or low after the clock edge. *It does NOT however mean that a 'negedge' is expected on this signal at the noted clock edge.*
(rising edge)	**EDGE SENSITIVE HIGH:** This symbol means that a posedge is expected on this signal.
(falling edge)	**EDGE SENSITIVE LOW:** This symbol means that a negedge is expected on this signal.
(green up arrow)	**PROPERTY PASSes:** This symbol means that a sequence/property match is detected here (i.e. the sequence/property PASSes).
(red down arrow)	**PROPERTY FAILs:** This symbol means that a sequence/property did not match here (i.e. the sequence/property FAILs).

Chapter 5
Immediate Assertions

Introduction: This chapter will introduce the "Immediate" assertions (immediate "assert," "cover," "assume") starting with a simple definition and leading to detailed nuances of its semantics and syntax. There are two types of immediate assertions, namely Immediate Assertion and Deferred Immediate Assertion. We will cover both in this chapter.

A. B. Mehta, *SystemVerilog Assertions and Functional Coverage*,
https://doi.org/10.1007/978-3-030-24737-9_5

Immediate assertions are simple non-temporal domain assertions that are executed like statements in a procedural block. Interpret them as an expression in the condition of a procedural "if" statement. Immediate assertions can be specified only where a procedural statement is specified. The evaluation is performed immediately with the values taken at that moment for the assertion condition variables. The assertion condition is non-temporal, which means its execution computes and reports the assertion results at the *same* time.

Figure 5.1 describes the basics of an immediate assertion. It is so called because it executes immediately at the time it is encountered in the procedural code. It does not wait for any temporal time (e.g., "next clock edge") to fire itself. The assertion can be preceded by a level sensitive or an edge sensitive statement. As we will see, concurrent assertions can only work on a "sampling/clock" edge sensitive logic and not on level sensitive logic.

We see in Fig. 5.1 that there is an immediate assertion embedded in the procedural block that is triggered by @ (posedge clk). The immediate assertion is triggered after @ (posedge d) and checks to see that (b || c) is true.

We need to note a couple of points here. First, the very preceding statement in this example is @ (posedge d), an edge sensitive statement. However, it does not have to be. It can be a level sensitive statement also or any other procedural statement.

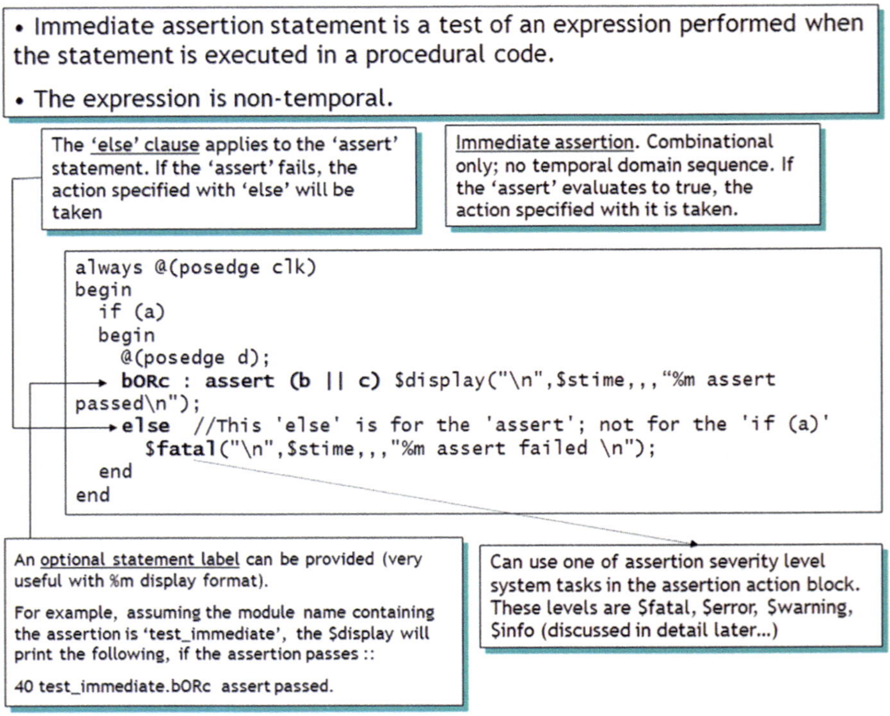

- Immediate assertion statement is a test of an expression performed when the statement is executed in a procedural code.
- The expression is non-temporal.

The 'else' clause applies to the 'assert' statement. If the 'assert' fails, the action specified with 'else' will be taken

Immediate assertion. Combinational only; no temporal domain sequence. If the 'assert' evaluates to true, the action specified with it is taken.

```
always @(posedge clk)
begin
  if (a)
  begin
    @(posedge d);
    bORc : assert (b || c) $display("\n",$stime,,,"%m assert
passed\n");
    else  //This 'else' is for the 'assert'; not for the 'if (a)'
      $fatal("\n",$stime,,,"%m assert failed \n");
  end
end
```

An optional statement label can be provided (very useful with %m display format).

For example, assuming the module name containing the assertion is 'test_immediate', the $display will print the following, if the assertion passes ::

40 test_immediate.bORc assert passed.

Can use one of assertion severity level system tasks in the assertion action block. These levels are $fatal, $error, $warning, $info (discussed in detail later...)

Fig. 5.1 Immediate assertion—basics

The reason I am pointing this out is that concurrent assertions can work only off of a sampling "edge" and not off of a level sensitive control. Keep this in your back pocket because it will be very useful to distinguish immediate assertions from concurrent assertions when we cover the latter. Second, the assertion itself cannot have temporal domain sequences. In other words, an immediate assertion cannot consume "time." It can only be combinatorial which can be executed in zero time. In other words, the assertion will be computed, and results will be available at the *same* time that the assertion was fired. If the "assert" statement evaluates to 0, X, Z then the assertion will be considered to FAIL else it will be considered to PASS.

We also see in the figure that there is (what is known as) an Action Block associated with FAIL or PASS of the assertion. This is no different than the PASS/FAIL logic we design for an "if...else" statement.

From syntax point of view, an immediate assertion uses only "assert" as the keyword in contrast to a concurrent assertion that requires "assert property."

One key difference between immediate and concurrent assertions is that concurrent assertions always work off of the sampled value in preponed region (see Sect. 6.3) of a simulation tick while immediate assertions work immediately when they are executed (as any combinatorial expression in a procedural block) and do not evaluate its expression in the preponed region. Keep this thought in your back pocket for now since we haven't yet discussed concurrent assertions and how assertions get evaluated in a simulation time tick. But this key difference will become important to note as you learn more about concurrent assertions.

Finally, as we discussed above, the immediate assertion works on a combinatorial expression whose variables are evaluated "immediately" at the time the expression is evaluated. These variables may transition from one logic value to another (e.g., 1 to 0 to 1) within a given simulation time tick and the immediate assertion may get evaluated multiple times before the expression variable values "settle" down. This is why immediate assertions are also known to be "glitch" prone. This is where the "deferred immediate" assertions come into picture. We will discuss those in Sect. 1.1.

To complete the story, there are three types of immediate assertions.

immediate **assert.**
immediate **assume.**
immediate **cover.**

Note also that the book contains a lot more information on other types of assertions that can be called from a procedural block (just as you call immediate assertions). For example, you can call a "property" or a "sequence" (or "restrict" for formal verification) from a procedural block.

"assume" and "cover" are too early to discuss. And also, I haven't discussed "property" and "sequence" yet.

Moving on,

Figure 5.2 points out a couple of finer points. First, do not put anything in the so-called action block (PASS or FAIL) of the immediate assertion. Most synthesis tools simply ignore the entire immediate assertion with its action blocks (which

```
always @(posedge clk)
begin
  if (busAck)
  begin
    checkbusReq: assert (busReq && !reset) $display("\n",$stime,,,"%m passed\n");
    else
    begin
          $fatal("\n",$stime,,,"%m failed \n");
          machineCheck = 1'b1; //DON'T PUT EXECUTABLE RTL HERE..
    end
  end
end
```

ENTIRE 'assert' block is ignored by synthesis. So, DO NOT PLACE ANY EXECUTABLE RTL CODE IN THE ASSERT 'pass' OR 'fail' ACTION BLOCK

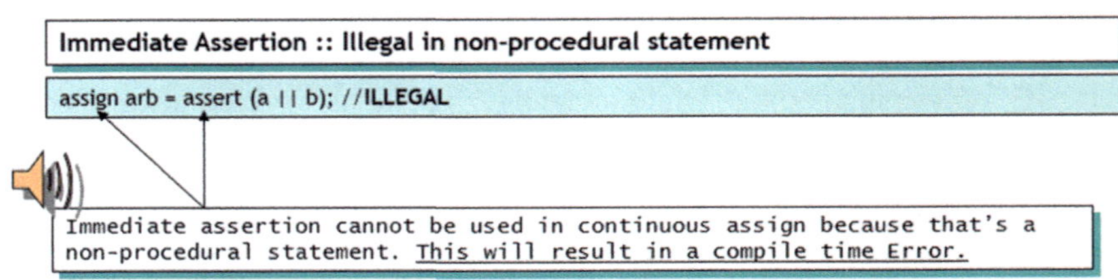

Fig. 5.2 Immediate assertions: finer points

makes sense) and with it will go your logic that (if) you were planning on putting in your design. This is rather obvious but easy to miss.

Note that an immediate assertion cannot be used in a continuous assignment statement because continuous assign is not a procedural block.

5.1 Deferred Immediate Assertions

Deferred immediate assertions are a type of "immediate" assertions. Recall that "immediate" assertions evaluate immediately without waiting for variables in its combinatorial expression to settle down. This also means that the immediate assertions are very prone to simulation glitches as the combinatorial expression settles down (for example, an expression evaluates to "0" then to "1" then back to "0" to settle down on "0"), the immediate assertion may fire multiple times. On the other hand, deferred assertions do not evaluate their sequence expression until the end of time tick when all values have settled down (or in the reactive region of the time tick).

The syntax for deferred immediate assertion is "assert #0" or "assert final." It's the #0 (or "final") that distinguishes deferred immediate assertion from the immediate assertion.

Note that there are two types of deferred immediate assertions. The difference between the two is identified by the keywords "assert #0" for observed deferred assertions and "assert final" for final deferred assertions.

For all practical purpose, I use "assert final" for deferred immediate assertion. That's because the "observed immediate," in certain circumstances may still be glitch prone. The difference between observed and final deferred assertions is in the extent of glitch filtering. Observed assertions filter glitches that occur in a single scheduling region set, Active or Reactive. In contrast, final assertions filter glitches that are created by the interaction between Active and Reactive regions. In that case the glitch spans both regions and would not be filtered by the observed deferred assertion.

I'll leave the discussion of the differences between the two at this and focus on "assert final." I'll be using deferred immediate assertion terminology to mean "assert final," unless I specifically talk about the Observed immediate assertion.

Let us examine the following example.

```
assign not_a = !a;
always_comb begin:b1
 a1: assert (not_a != a)  //immediate
 a2: assert #0 (not_a != a);    //Observed Deferred immediate
 a3: assert final (not_a !=a)  //Final Deferred immediate
end
```

Let us examine the difference between immediate and deferred immediate assertions in this example. As soon as "a" changes, always_comb wakes up and both the immediate and deferred assertions fire right away. When the immediate assertion fires, the continuous assignment "not_a =!a" may not have completed its assignment. In other words, "a" has not been inverted yet. But the immediate assertion expects an inverted "a" on "not_a" immediately. The assertion will fail. This is why immediate assertions are known to be glitch prone.

On the other hand, the deferred assertion will wait until all expressions in the given time stamp have settled down (in other words, it was put in the deferred assertion report queue). In our case, the continuous assign would have completed its evaluation by the end of time stamp (meaning when the deferred assertion queue will be flushed), and "not_a" would indeed be =!a. This is the value the deferred assertion will take into account when evaluating its expression. The deferred assertion will pass.

To reiterate, in a simple immediate assertion, pass and fail actions take place immediately upon assertion evaluation. In a deferred immediate assertion, the actions are delayed until later in the time step, providing some level of protection against unintended multiple executions on transient or "glitch" values.

Note that there is a limitation on the action block that a deferred immediate assertion has. The action block can only be a single subroutine (a task or a function). The requirement of a single subroutine call also implies that no begin-end block can surround the pass or fail statements, as begin is itself a statement that is not a subroutine call. A subroutine argument may be passed by value as an input or passed by reference as a ref. or const ref. Actual argument expressions that are passed by value, including function calls, will be fully evaluated at the instant the deferred assertion expression is evaluated. It is also an error to pass automatic or dynamic variables as actuals to a ref. or const ref. formal.

For example, following action blocks are illegal with deferred assertions because either they contain more than one statement or are not subroutine calls.

```
frameirdy: assert final (!frame_ == irdy) else begin interrupt=1;
$error ("FAILure"); end //ILLEGAL
frameirdy: assertfinal (!frame==irdy) elsebegin $error("FAILure");
end //ILLEGAL
```

Following are legal.

```
frameirdy: assert final (!frame == irdy) else $error("FAILure");
//LEGAL (no begin-end)
frameirdy: assert final (!frame == irdy) $info("PASS"); else
$error("FAILure"); //LEGAL
frameirdy: assert final (!frame == irdy); //LEGAL - no action
block
```

Another feature of deferred immediate assertion to note is that it can be declared both in the procedural block as well as *outside* of it (recall that immediate assertion can only be declared in procedural block). For example, following is legal for deferred immediate assertion but not for immediate assertion.

```
module (x, y, z);
        .....
        z1: assert final (x == y || z);
endmodule

This is equivalent to

module (x, y, z);
        always_comb begin
        z1: assert final (x == y || z);
        end
endmodule
```

Analogous to "assert," we also have deferred "cover" and "assume." Again, the idea is the same as that for immediate "cover" and "assume" that you want to use the final values of the combinatorial logic before evaluating "cover" or "assume."

A deferred "assume" will often be useful in cases where a combinational condition is checked in a function but needs to be used as an assumption rather than a proof target by formal tools. A deferred cover is useful to avoid crediting tests for covering a condition that is only met in passing by glitch values.

```
assign a = c || d;
assign b = e || f;

always_comb begin:b1
        a1: cover (b != a) //immediate cover
        a2: cover #0 (b != a); //Observed deferred cover
        a3: cover final (b != a) //Final deferred cover
end

and for 'assume'
assign a = c || d;
assign b = e || f;

always_comb begin:b1
        a1: assume (b !=a) //immediate assume
        a2: assume #0 (b != a); //deferred assume
        a3: assume final (b !=a) //deferred assume
end
```

Some more nuances of deferred assertions are further explained below.

5.2 Disabling a Deferred Assertion

The following example illustrates how user code can explicitly flush a pending assertion report. In this case, failures or successes of "a1" are only reported in time steps where "NoGo" does not settle at a value of 1.

```
always @(NoGo or Go) begin : b1
           a1: assert final (Go) else $fatal(1, "Sorry");
           if (NoGo)
                  begin
                     disable a1;
                  end
end
```

On the similar line of thought, the following example illustrates how user code can explicitly flush *all* pending assertion reports on the deferred assertion queue of process “b2”:

```
always @(a or b or c)
begin : b2
        if (c == 8'hff)
        begin
                a2: assert final (a && b);
        end
        else begin
                a3: assert final (a || b);
        end
end

always @(posedge NoGo)
begin : b3
        disable b2;
end
```

Finally, as I mentioned before and unlike immediate assertions, the deferred immediate assertion can be placed outside of procedural code. When declared outside a procedural block, the deferred immediate assertion is treated semantically as if the assertion were enclosed with an always_comb block.

For example, here’s the code with always_comb:

```
assign Frame_ = !Frame_ && IRDY;
assign ACK = Req && Gnt;
always_comb a1: assert final (Frame_ || ACK);
```

The same deferred immediate assertion can be written as follows:

```
assign Frame_ = !Frame_ && IRDY;
assign ACK = Req && Gnt;
a1: assert final (Frame_ || ACK);
```

5.3 Deferred Assertion in a *Function*

Finally, a deferred assertion can also be defined in a function. And a simple function call from a single process is rather intuitive to understand. But what happens when you call the same function from two *different* procedural blocks (or different processes)?

Let us examine this with an example:

```
module ReqAck;

bit ReqPending, currentReq, AckPending, currentAck;
logic ReqCheck, AckCheck;

function bit checkReqAck (bit a, bit b);
     A1: assert final (a == b);
endfunction
......
always_comb begin : ReqC
     ReqCheck = checkReqAck (ReqPending, currentReq);
end

always_comb begin : AckC
     AckCheck = checkReqAck (AckPending, currentAck);
end
endmodule
```

Let us examine this code carefully, since it does get a bit complicated.

There are two different procedural blocks, namely “ReqC” and “AckC.” Both call the same function “checkReqAck” with different actuals. The function “checkReqAck” has a single deferred assertion “A1.”

Now, let’s say following events take place in different time steps.

In the first time step: block “ReqC” executes with “ReqPending” *not* equal to “currentReq.” Similarly, block “AckC” executes with “AckPending” *not* equal to “currentAck.”

This is what will happen in the first time step. Since “A1” fails independently for processes “ReqC” and “AckC,” it’s failure will be reported twice.

In the second (different) time step: block “ReqC” executes first with “ReqPending” not equal to “currentReq.” But since this block has a combinational trigger, “ReqPending” and “currentReq” may settle down to being equal.

This is what will happen in the second time step. First, since “ReqPending” is *not* equal to “currentReq,” a failure will be pending to be reported. But in *the same time step*, both these variables then settle to being equal. Note that the failure will not be reported until the end of the time step. So, the *pending* failure report will be “flushed.” Now when both these variables settle down to being equal at the end of the same time step, the failure will *not* be reported.

Chapter 6
Concurrent Assertions: Basics

Introduction: This chapter introduces basics of concurrent assertions, namely "sequence," "property," "assert," "cover," and "assume." It discusses fine grained nuances of Clocking of concurrent assertions and also implication operators, multi-threaded semantics, formal arguments, "bind"ing of assertions, formal arguments, severity levels, and disable iff, among other topics.

A. B. Mehta, *SystemVerilog Assertions and Functional Coverage*,
https://doi.org/10.1007/978-3-030-24737-9_6

Concurrent assertions are temporal domain assertions that allow creation of complex sequences which are based on *clock* (*sampling*) *edge* semantics. This is in contrast to the immediate assertions that are purely combinatorial and do not allow temporal domain sequences.

Concurrent assertions are the gist of SVA language. They are called concurrent because they execute in *parallel* with the rest of the design logic and are multi-threaded. Let us start with basics and move onto the complex concepts of concurrent assertions.

Let us first learn the basic syntax of a concurrent assertion and then study its semantics.

In Fig. 6.1 we have declared a property "pr1" and asserted it with a label "reqGnt" (label is optional but highly recommended). The figure explains various parts of a concurrent assertion including a property, a sequence, and assertion of the property.

The "assert property (pr1)" statement triggers property "pr1." "pr1" in turn waits for the antecedent "cStart" to be true at a (posedge clk) and on it being true implies (fires) a sequence called "sr1." "sr1" checks to see that "req" is high when it is fired and that 2 "clocks" later "gnt" is true. If this temporal domain condition is satisfied, then the sequence "sr1" will PASS and so will property "pr1" and the "assert property" will be a PASS as well. Let us continue with this example and study other key semantics.

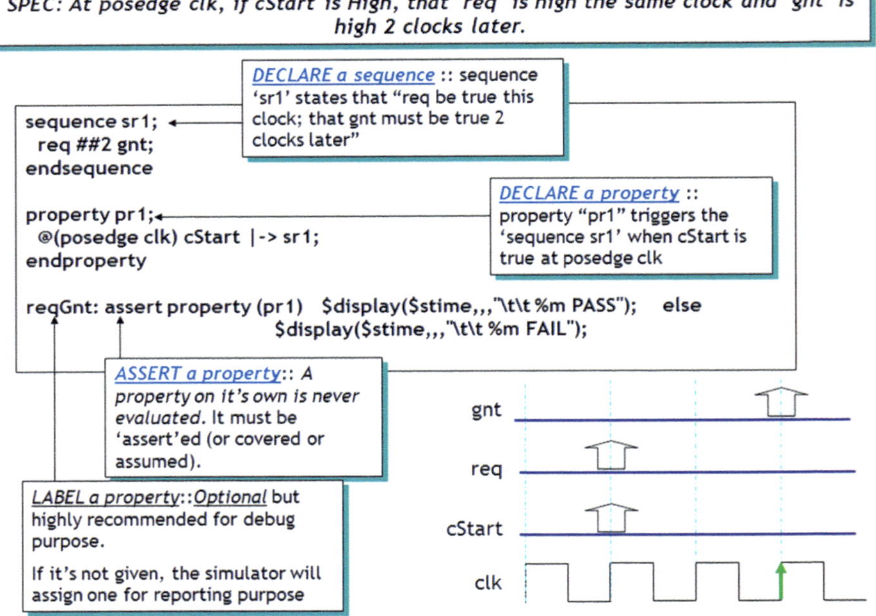

Fig. 6.1 Concurrent assertion—basics

As explained in Fig. 6.2, following are the basic and mandatory parts of an assertion. Each of these features will be further explored as we move along.

1. "assert"—you have to assert a property, i.e., invoke or trigger it.
2. There is an action block associated with either the pass or fail of the assertion.
3. "property pr1" is edge triggered on posedge of clk (more on the fact that you *must* have a sampling edge for trigger is explained further on).
4. "property pr1" has an *antecedent* which is a signal called cStart, which if sampled high (in the preponed region) on the posedge clk, will imply that the *consequent* (sequence sr1) be executed. More on preponed region coming up shortly.
5. Sequence sr1 samples "req" to see if it is sampled high the same posedge of clk when the sequence was triggered because of the *overlapping implication* operator and then waits for 2 clocks and sees if "gnt" is high.
6. Note that each of "cStart," "req," "gnt" are *sampled* at the edge specified in the property which is the posedge of "clk." In other words, even though there is no edge specified in the sequence, the edge is inherited from property pr1.

Note also that we are using the notion of sampling the values at posedge clk which means that the "posedge clk" is the "*sampling edge*." In other words, the sampling

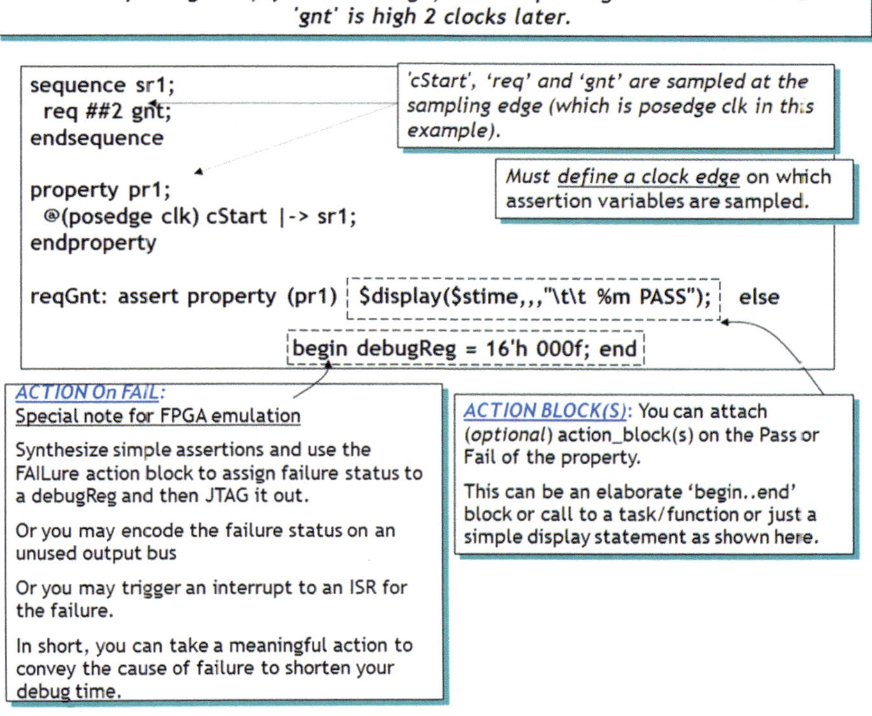

Fig. 6.2 Concurrent assertion—sampling edge and action blocks

edge can be anything (as long as it's an edge and is not a level), meaning it does not necessarily have to be a synchronous edge such as a clock. It can be an asynchronous edge as well. However, *be very careful about using an asynchronous edge* unless you are sure what you want to achieve. I have devoted a complete example (see Chap. 19) on the pitfalls of using an asynchronous edge as the sampling edge. It's too soon to get into that. This is a very important concept in concurrent assertions and should be well understood. However, do not worry, you will get much more insight as we move further.

Now, let us slightly modify the sequence "sr1" to highlight Boolean expression in a sequence or a property and study some more key elements of a concurrent assertion.

As shown in Fig. 6.3 there are three main parts of the expression that determines when an assertion will fire, what it will do once fired, and time duration between the firing event and execution event.

The condition under which an assertion will be fired is called an "*antecedent*." This is the LHS of the implication operator.

RHS of the assertion that executes once the antecedent matches is called the "*consequent*."

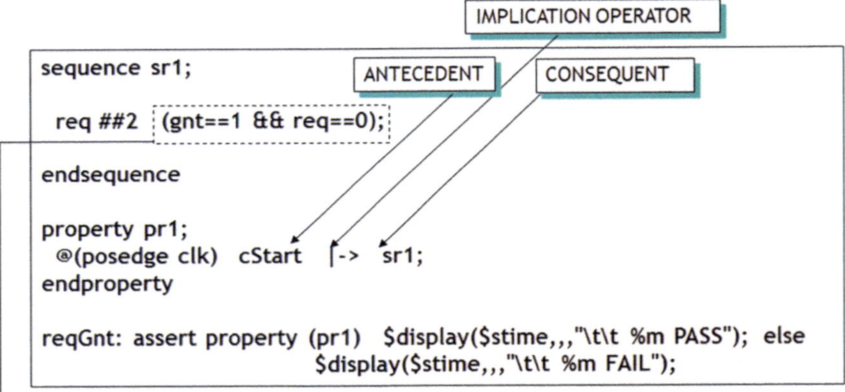

Boolean Expressions

- *Variables in the expression are all sampled values at the sampling edge (in Preponed region)*
- *The outcome of the eval of the expression is a boolean and is interpreted the same as the conditional expression of the procedural 'if' statement.*
- *If the expression evaluates to 'x' or 'z' or '0', the expression is evaluated to False; else it is True*
- *Expressions can include function calls*

Fig. 6.3 Concurrent assertion—implication, antecedent, and consequent

The way to "read" the implication operator is "if there is a match on the antecedent that the consequent will be executed". If there is no match, consequent will not fire and the assertion will continue to wait for a match on the antecedent. The "*implication*" operator also determines the time duration that will lapse between the antecedent match and the consequent execution. In other words, the implication operator ties the antecedent and the consequent in one of two ways. It ties them with an "*overlapping*" implication operator or a "*nonoverlapping*" implication operator. More on this coming up…

One additional note on Boolean Expressions before we move on. Following types are not allowed for the variables used in a Boolean Expression.

- Dynamic Arrays
- class
- string
- event
- real, shortreal, realtime
- Associative Arrays
- chandle

Here are explicit rules that govern Boolean Expressions

- An expression must result in a type that is cast compatible with an integral type. Subexpressions need not meet this requirement as long as the overall expression is cast compatible with an integral type.
- Elements of dynamic arrays, queues, and associative arrays that are sampled for assertion expression evaluation may get removed from the array or the array may get resized before the assertion expression is evaluated. These specific array elements sampled for assertion expression evaluation must continue to exist within the scope of the assertion until the assertion expression evaluation completes.
- Expressions that appear in *procedural* concurrent assertions may reference automatic variables. Otherwise, expressions in concurrent assertions shall not reference automatic variables. Procedural concurrent assertions are discussed in Sect. 6.1.
- Expressions must not reference non-static class properties or methods.
- Expressions must not reference variables of the **chandle** data type.
- Functions that appear in expressions must not contain output or ref arguments (const ref is allowed).

Figure 6.4 further explains the antecedent and consequent. As shown, you don't have to have a sequence in order to model a property. If the logic to execute in consequent is simple enough, then it can be declared directly in consequent as shown. But please note that *it is always best to break down a property into smaller sequences to model complex properties/sequences.* Hence, consider this example only as describing the semantics of the language. Practice should be to divide and conquer. You will see many examples, which seem very complex to start with, but once you break them down into smaller chunks of logic and model them with smaller sequences, tying all those together will be much easier then writing one long complex assertion sequence.

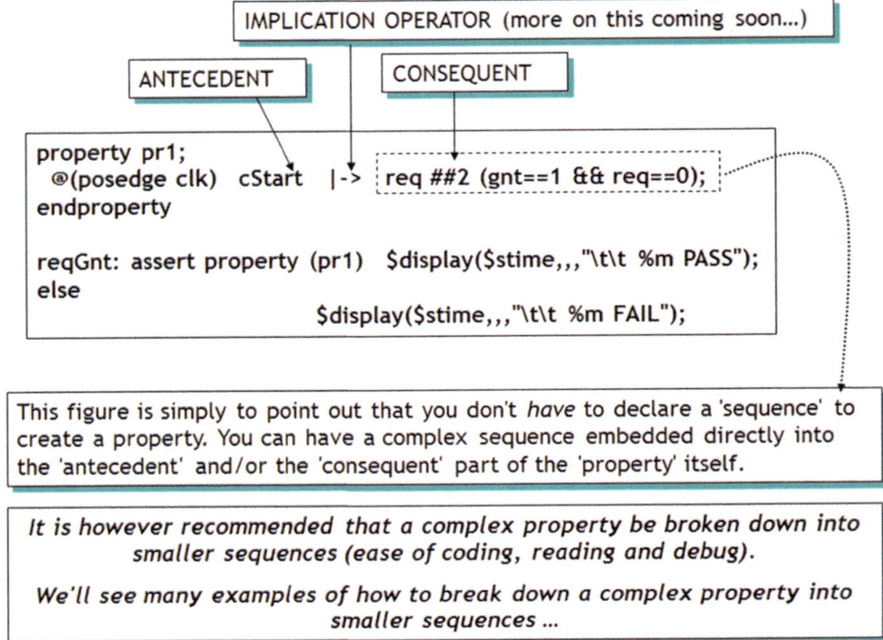

Fig. 6.4 Property with an embedded sequence

6.1 Implication Operator, Antecedent and Consequent

Implication operator ties the antecedent and consequent. If antecedent holds true it implies that the consequent should hold true.

There are two types of implication operators as shown in Fig. 6.5

1. *Overlapping Implication Operator*: Referring to Fig. 6.5, the top most property shows the use of an overlapping operator. Note its symbol (|->), which differs from that of the nonoverlapping operator (|=>). Overlapping means that when the antecedent is found to be true, that the consequent will start its execution (evaluation) the "*same*" clk. As shown in the figure, when cStart is sampled High at posedge of clk that the req is required to be High at the "*same*" posedge clk. This is shown in the timing diagram associated with the property.

   (a) So, what happens if "req" is sampled true at the next posedge clk after the antecedent (and false before that)? Will the overlapping property pass?

2. *Nonoverlapping Implication Operator*: In contrast, nonoverlapping means that when the antecedent is found to be true that the consequent should start its execution, *one clk later*. This is shown in the timing diagram associated with the property.

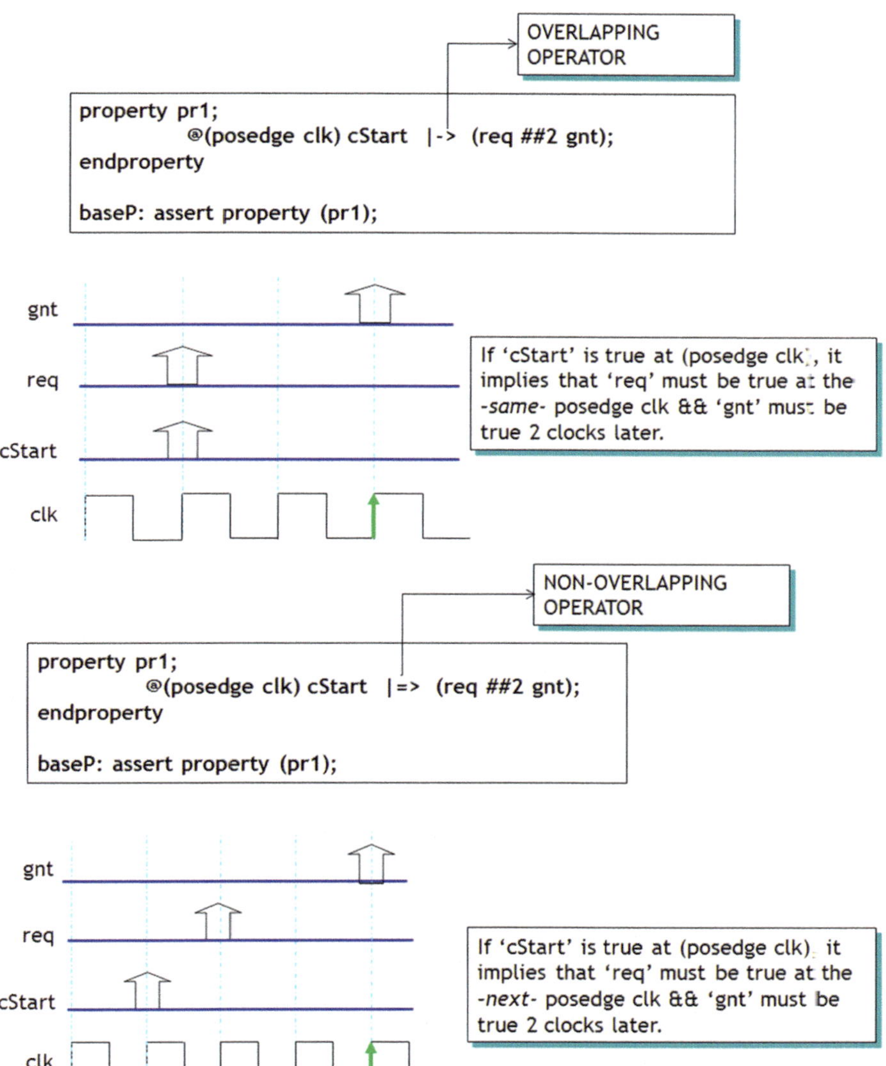

Fig. 6.5 Implication operator—overlapping and nonoverlapping

(a) So, what happens if "req" is sampled true at the same posedge clk as the antecedent (and False after that)? Will the nonoverlapping property pass?

Answer to both 1.(a) and 2.(a) is NO.

In 1.(a), the property strictly looks for "req" to be true the *same* clock when "cStart" is true. It does not care if "req" is true the next clock. So, if "req" is not true when "cStart" is true, the property will fail.

In 2.(a), the property strictly looks for "req" to be true the *next* clock after "cStart" is true. It does not care if "req" is true the same clock. So, if "req" is not true 1 clock after "cStart" is true, the property will fail.

Figure 6.6 further shows the equivalence between overlapping and nonoverlapping operators. "|=>" is equivalent to "|-> ##1." Note that ##1 is not the same as Verilog's #1 delay. ##1 means one clock edge (sampling edge). Hence "|-> ##1" means the same as "|=>."

Suggestion: To make debugging easier and have project wide uniformity, use the overlapping operator in your assertions. Reason? Overlapping is the common denominator of the two types of operator. You can always model nonoverlapping from overlapping, but you cannot do vice versa. What this means is that during debug everyone would know that all the properties are modeled using overlapping and that the # of clocks are exactly the same as specified in the property. You do not have to add or subtract from the # of clocks specified in the chip specification. More important, if everyone uses his or her favorite operator, debugging would be very messy not knowing which property uses which operator.

Finally, do note that concurrent assertions can be placed:

1. in "always" procedural block
2. in "initial" procedural block
3. standalone (static)—outside of the procedural block—which is what we have seen so far

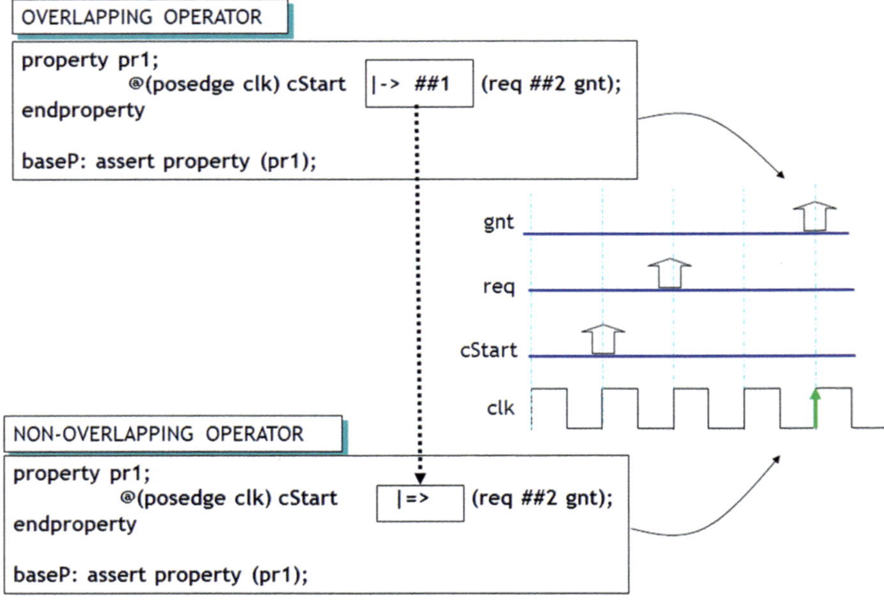

Fig. 6.6 Equivalence between overlapping and nonoverlapping implication operators

6.2 Clocking Basics

As mentioned before, a concurrent assertion is evaluated only on the occurrence of an "edge," known as the "sampling edge." Most often it is a synchronous "edge" that you may be using. But you can indeed have an asynchronous edge as well. BUT be very careful. I have devoted a complete example precisely to explain how an assertion with an asynchronous edge works. In Fig. 6.7, we are using a nonoverlapping implication operator, which means that at a posedge of clk if cStart is high, then one clock later sr1 should be executed.

Let us revisit "sampling" of variables. The expression variables cStart, req, and gnt are all sampled in the *preponed region* (see Sect. 6.3) of posedge clk. In other words, *if (e.g.) cStart = 1 and posedge clk change at the same time, the sampled value of cStart in the "preponed region" will be equal to "zero" and not "one."* We will soon discuss what "preponed region" really means in a simulation time tick and how it affects the evaluation of an assertion, especially when the sampling edge and the sampled variable change at the same time.

Note again that "sequence sr1" does not have a clock in its expression. The clock for "sequence sr1" is inherited from the "property pr1." This is explained next using Fig. 6.8.

As explained in Fig. 6.8, the "clk" as an edge can be specified either directly in the assert statement or in the property or in the sequence. Regardless of where it is declared, it will be inherited by the entire assertion (i.e., the assert, property, and sequence blocks).

> *Suggestion*: As noted in the Fig. 6.8, *my recommendation is to specify the "clk" in a property*. Reason being you can keep sequences void of sampling edge (i.e., "clk") and thus make them reusable. The sampling edge can change in the property but sequence (or cascaded sequences) remain untouched and can change their logic without worrying about the sampling edge. Note that it is also more readable when the sampling edge "clk" is declared in a property.

Note that a clock can be contextually inferred from a procedural block for a concurrent assertion. For example,

```
always @(posedge clk) assert property (not (FRAME_ ##2 IRDY));
```

Here the concurrent assertion is fired from a procedural block. The clock "@ (posedge clk)" is inferred from the "**always** @(posedge clk)" statement.

There is also an entire Sect. 10.1 devoted to multi-clock properties. It is too early to delve into its detail.

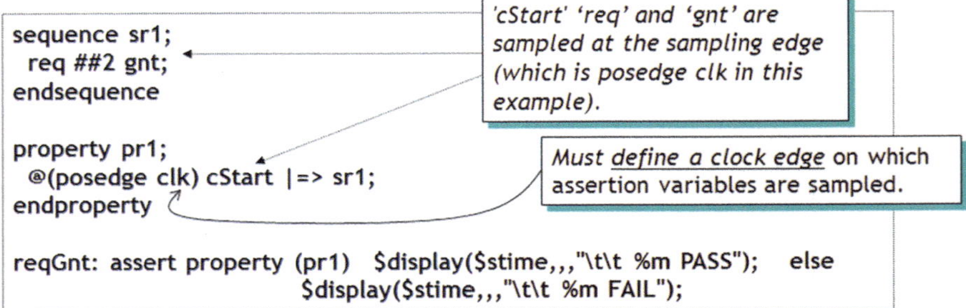

Fig. 6.7 Clocking basics

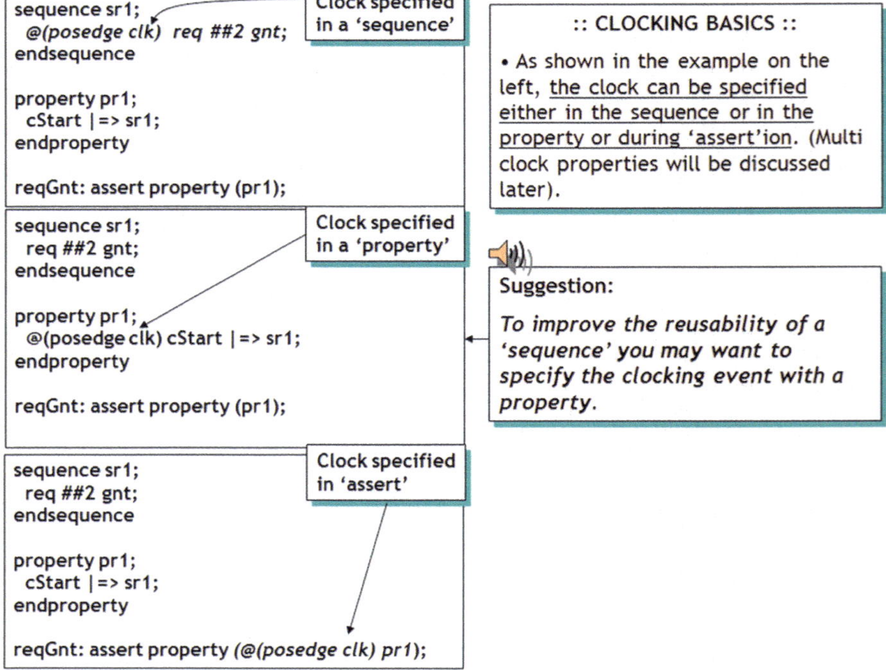

Fig. 6.8 Clocking basics—lock in "assert," "property," and "sequence"

Finally, note that a named event can also act as a clock. For example:

```
module eventtrig;
event e;
      always @(posedge clk) -> e;
      a1: assert property (@e a |=> b);
endmodule
```

6.3 Sampling Edge (Clock Edge) Value: How Are Assertions Evaluated in a Simulation Time Tick?

Let me first give a brief description of the Active, Observed, and Reactive region. As far as the sampling edge is concerned, the preponed region comes into picture and you need to understand it very carefully. Coming up...

6.3.1 *Active Region*

In this region, the assertion "clock ticks" are detected, and assertions are scheduled (*not* executed) in the Observed region. Events originating in Observed region get scheduled into the Active region (or Reactive Region) for execution.

6.3.2 *Observed Region*

The Observed region is meant for the evaluation of sequences, properties, and concurrent assertions. Signal values remain constant during the Observed region. Events originated in this region get scheduled into the Active and Reactive regions. In general, here's what the Observed Region does.

- Determine match of sequences.
- Start new attempts for sequences.
- Start new attempts for assertions.
- Resume evaluation of previous attempts.
- Schedule action blocks (*not* execute them – that's done in Reactive region).

6.3.3 *Reactive Region*

The Reactive region executes statements from programs and checkers and action blocks. Programs are intended for writing test-benches, as external environments for designs, feeding stimuli, observing design evaluation results, and building tests to exercise the design.

6.3.4 Preponed Region

This region is of importance in terms of understanding how the so-called sampling semantics of assertions work.

How does the so-called *sampling edge* sample the variables in a property or a sequence is one of the most important concept you need to understand when designing assertions? As shown in Fig. 6.9 the important thing to note is that the variables used in assertions (property/sequence/expression) are sampled in the *preponed* region. What does that mean? It means (for example) if a sampled variable changes the same time as the sampling edge (e.g., clk) that the value of the variable will be the value it held—*before*—the clock edge.

```
@ (posedge clk) a |=> !b;
```

In the above sequence, let us say that variable "a" changes to "1" the same time that the sampling edge clock goes posedge clk (and assume "a" was "0" before it went to a "1"). Will there be a match of the antecedent "a"? No! Since "a" went from "0" to "1" the same time that clock went posedge clk, the sampled value of "a" at posedge clk will be "0" (from the preponed region) and not "1." This will not cause the property to trigger because the antecedent is not evaluated to be a "1." This will confuse you during debug. You would expect "1" to be sampled and the property triggered thereof. However, you will get just the opposite result.

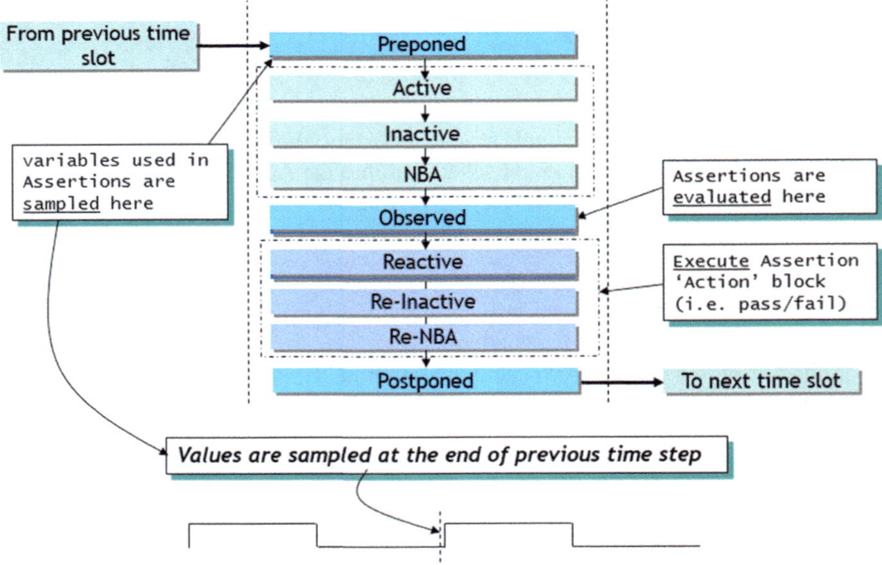

Fig. 6.9 Assertions variable sampling and evaluation/execution in a simulation time tick

This is a very important point to understand because in a simulation waveform (or for that matter with Verilog $monitor or $strobe) you will see a "1" on "a" with posedge clk and would not understand why the property did not fire or why it failed (or passed for that matter). *Always remember that at the sampling edge, the "previous" value (i.e., a delta before the sampling edge in the preponed region) of the sampled variable is used.* To reiterate, preponed region is a precursor to the time slot, where only sampling of the data values take place. No value changes or events occur in this region. Effectively, sampled values of signals do not change through the time slot.

Following is to establish what is *not* sampled. As you read the book further, you will better understand what this means. For now, keep it in your back pocket.

Following are NOT sampled in the preponed region of the time tick.

- Assertion *local* variables.
- Assertion action blocks (pass and fail).
- Clocking event expression (e.g., @(posedge clk); here clk is not sampled—but the *current* value of clock is used).
- Disable condition (variables of conditional expression) of "disable iff" (more on this in Sect. 6.8—this concept is important to understand).
- Actual arguments passed to "ref" or "const ref" arguments of subroutines attached to sequences.
- The sampled value of a const cast expression is defined as the current value of its argument. For example, if "*a*" is a variable, then the sampled value of const'(a) is the current value of a. When a past or a future value of a const cast expression is referenced by a sampled value function, the current value of this expression is taken instead.
- The default sampled value of a static variable is the value assigned in its declaration, or, in the absence of such an assignment, it is the default (or uninitialized) value of the corresponding type.
- The default sampled value of any other variable or net is the default value of the corresponding type. For example, the default sampled value of variable "y" of type logic is 1'bx.
- The default sampled value comes into picture at time "0." After that, it'll always be the value from the preponed region.
- Sampled value of the .triggered event property and the sequence methods .triggered and .matched (see Chap. 13) is defined as the current value returned by the event property or sequence method. For example, if "a" is a static module variable, "s" is a sequence, and "f" is a function, the sampled value of f (a, s.triggered) is the result of the application of "f" to the sampled values of "a" and s. triggered, i.e., to the value of "a" taken from the Preponed region and to the *current* value of s.triggered.

Here is a complete example including the test-bench and comments that explain how sampling of variables in the preponed region affect assertion results.

```
module assert1;
reg A, B, C, D, clk;

property ab;
  @ (posedge clk) !A |-> B;
endproperty

aba: assert property (ab) else $display ($stime,,, "ab FAIL");
abc: cover property (ab) $display($stime,,, "ab PASS");

initial begin
     clk=0; A=0; B=0; //Note: A and B are equal to '0' at
     time 0.
     forever #10 clk=! clk;
 end

initial begin
`ifdef PASS

/* Following sequence of events will cause property 'ab' to
PASS because even though A=0 and B=1 change simultaneously
they had settled down because of #1 before posedge clk. Hence
when @ (posedge clk) samples A, B; A=0 and B=1 are sampled.
The property antecedent '!A' is evaluated to be true and at
that same time (overlapping operator) B==1. Hence the prop-
erty passes */

A=0;
B=1;
  #1;
  @ (posedge clk)

`else
 /* Following sequence of events will cause property 'ab' to
 FAIL. Here's the story. A=0 and B=1 change at the same time
 as posedge clk. This causes the sampled value of B to be
 equal to '0' and not '1' because the sampling edge (posedge
 clk) samples the variable values in the preponed region and
 B was equal to '0' in the preponed region. Note that A was
 equal to '0' in the preponed region because of its initial-
 ization in the 'initial' block above. So, now you have both
 'A' and 'B' == 0. Since A is 0, !A is true, and the property
 evaluation takes place. Property expects B==1 the same time
 (overlapping operator) that !A is true. However, 'B's sam-
 pled value is '0' and the property fails. */
```

```
    @ (posedge clk)
      A=0;
      B=1;
  `endif
   @ (negedge clk)
   $finish(2);
  end
  endmodule
```

Here are some detailed rules on how variables and expressions are sampled (at a clock edge). At this time, you may not understand them well. But once you better familiarize yourself with how sampled values of variables (local, automatic, normal) take place, this will be good reference.

The definition of a sampled value of an expression is based on the definition of a sampled value of a variable. The general rule for variable sampling is as follows:

- The sampled value of a variable in a time slot corresponding to time greater than 0 is the value of this variable in the *Preponed* region of this time slot.
- The sampled value of a variable in a time slot corresponding to time 0 is its default sampled value.

This rule has the following exceptions:

- Sampled values of automatic variables, local variables, and active free checker variables (see Chap. 22 for checker variables) are their current values. However,
- When a past or a future value of an active free checker variable is referenced by a *sampled value function* (e.g., $past), this value is sampled in the *Postponed* region of the corresponding past or future clock tick;
- When a past or a future value of an automatic variable (automatic variable is a SystemVerilog construct and not discussed in this book) is referenced by a sampled value function, the current value of the automatic variable is taken instead.

The sampled value of an expression is defined as follows:

- The sampled value of an expression consisting of a single variable is the sampled value of this variable.
- The sampled value of the triggered event property and the sequence methods . triggered and .matched (see Sect. 13.2) is defined as the current value returned by the event property or sequence method.
- The sampled value of any other expression is defined recursively using the values of its arguments. For example, the sampled value of an expression e1 & e2, where e1 and e2 are expressions, is the bitwise AND of the *sampled* values of e1 and e2. In particular, if an expression contains a function call, to evaluate the sampled value of this expression, the function is called on the sampled values of its arguments at the time of the expression evaluation.

6.4 Default Clocking Block

For a long chain of properties and sequences in the file, you can also use the default clocking block as explained in the Fig. 6.10. The figure explains the different ways in which clocking block can be declared and the scope in which it is effective.

The scope of a default clocking declaration is the entire module, interface, program, or checker in which it appears, including nested declarations of modules, interfaces, or checkers. A nested module, interface, or checker may, however, have its own default clocking declaration, which overrides a default from outside.

The scope of a default clocking declaration does not descend into instances of modules, interfaces, or checkers.

"clocking" is a keyword. The top block of the figure shows declaration of "default clocking cb1" which is then inherited by the properties "checkReqGnt" and "checkBusGrant" that follow. This default clocking block will be in effect until another default clocking block is defined. The bottom part of the figure is interesting. Here

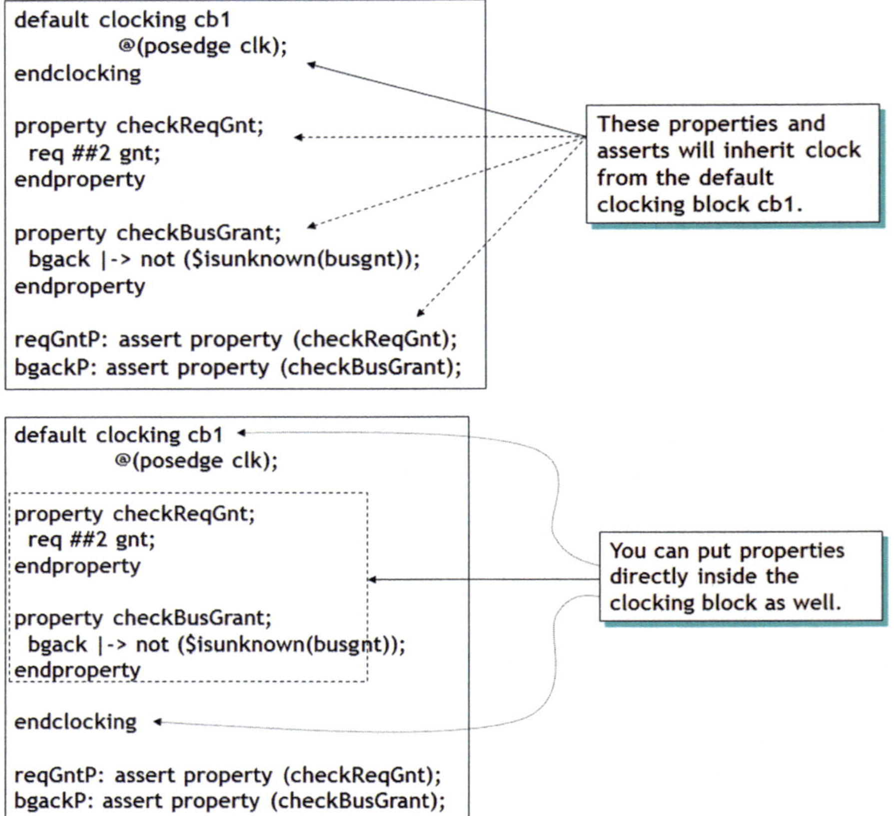

Fig. 6.10 Default Clocking block

the properties are directly embedded in the default clocking block. I don't recommend doing that though. The clocking block should only contain clock specifications, which will keep it modular and reusable. Use your judgment wisely on such issues.

Figure 6.11 declares two clocking blocks, namely "cb1" and "cb2" in a standalone Verilog module called "design_clocks." This is a great way to organize your clocking strategy in one module. Once defined, you can use any of the clocking block that is required simply by referring to it by its hierarchical instance name as shown in the figure.

Here's some food for thought. I have outlined a couple of pros and cons of using a default-clocking block. It is mostly advantageous but there are some caveats.

Pros: The argument towards default block is reusability. You may change the clocking relation in the default block, and it will be applicable to all the following blocks. You do not have to individually change clocking scheme in each property. This is indeed a true advantage and if you plan to change the clocking scheme in the default block without affecting the properties that follow, do use the default block by all means.

Cons: Readability/debuggability: When you see a property without any sampling edge, you have to scroll back to "someplace" to see what sampling edge is being used. You have to find the very preceding clocking block (in case of default clocking

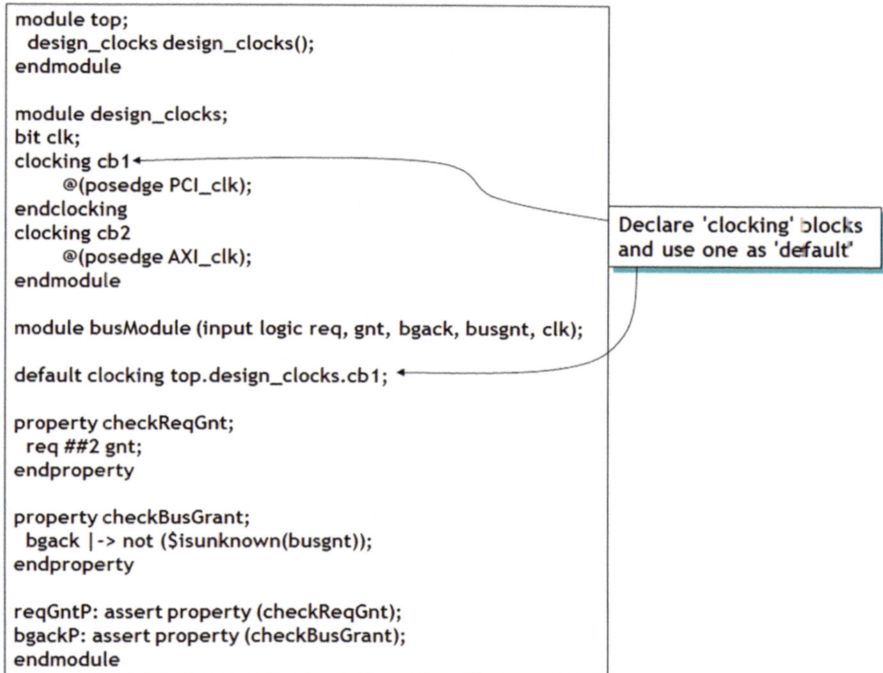

Fig. 6.11 "clocking" and "default clocking"

block in the same file as the property) and can't just go to the top of the file. I like properties that are mostly self-contained with the sampling edge. Sure, it's a bit more typing but a lot more readable.

Here's an example of how you can use the "default" clocking block but also override it with explicit (non-default) clocking.

```
module default_explicit_clocking;

default clocking negedgeClock @(negedge clk1); endclocking
clocking posedgeClock @(posedge clk2); endclocking

d2: assert property (x |=> y); //will inherit default clock -
negedgeClock
d3: assert property (z [=2] |-> a); //will inherit default clock -
negedgeClock
nd1: assert property (@posedgeClock b |=> c); //will use non-
default clocking posedgeClock

endmodule
```

Obviously, you don't *have* to declare the "clocking" block as shown above. You can simply use @ (posedge clk2) directly in the property assertion, as shown below.

```
module default_explicit_clocking;

default clocking negedgeClock @(negedge clk1); endclocking

d2: assert property (x |=> y); //will inherit default clock -
negedgeClock
d3: assert property (z [=2] |-> a); //will inherit default clock -
negedgeClock
nd1: assert property (@(posedge clk2) b |=> c); //explicit
declaration of clock - clk2

endmodule
```

Or you can model the same clocking structure as follows, using a property with its own explicit clock.

```
module default_explicit_clocking;

default clocking negedgeClock @(negedge clk1); endclocking

property nClk; @(posedge clk2) b |=> c; endproperty
```

```
d2: assert property (x |=> y); //will inherit default clock -
negedgeClock
d3: assert property (z[=2] |-> a); //will inherit default clock -
negedgeClock
nd1: assert property (nClk); //explicit declaration of clock -
clk2

endmodule
```

Note the following rules that apply to a clocking block:

1. Multi-clocked sequences and properties (Chap. 10) are not allowed within the clocking block.
2. If a named sequence or property that is declared outside the clocking block is instantiated within the clocking block, the instance is singly clocked, and its clocking event is identical to that of the clocking block.
3. An explicitly specified leading clocking event in a concurrent assertion statement supersedes a default clocking event.

Note the following example that shows the application of above rules and points to Legal and Illegal cases (courtesy SystemVerilog LRM).

```
property q1;
      $rose(a) |-> ##[1:5] b;
endproperty

property q2;
      @(posedge clk) q1;
endproperty

default clocking posedge_clk @(posedge clk);
      property q3;
           $fell(c) |=> q1;  // legal: q1 has no clocking event
      endproperty

      property q4;
           $fell(c) |=> q2;    // legal: q2 has clocking event
           identical to that of the clocking block
      endproperty

      sequence s1;
           @(posedge clk) b[*3]; // illegal: explicit clocking
           event in clocking block
      endsequence
endclocking
```

Following may not be too intuitive at this stage since we haven't seen concurrent assertions in practice. But these are important legal and illegal conditions which will help avoid unnecessary debugging. Examples are for properties that does not have a default clocking block.

```
module examples_NO_default (input logic a, b, c, clk);
     property q1;
          $rose(a) |-> ##[1:5] b;
     endproperty

     property q5;
          @(negedge clk) b[*3] |=> !b;
     endproperty

     property q6;
          q1 and q5;
     endproperty

     a5: assert property (q6); // illegal: no leading clocking
     event
     a6: assert property ($fell(c) |=> q6); // illegal: no
     leading clocking event

     sequence s2;
          $rose(a) ##[1:5] b;
     endsequence

     c1: cover property (s2); // illegal: no leading clocking event
     c2: cover property (@(negedge clk) s2); // legal: explicit
     leading clocking event,
                                        // @(negedge clk)

     sequence s3;
          @(negedge clk) s2;
     endsequence

     c3: cover property (s3); // legal: leading clocking event,
     @(negedge clk), determined from
                                   // declaration of s3

endmodule
```

```
assign clkstart = clk && cGate;

sequence sr1;
  req ##2 gnt;
endsequence

property pr1;
  @(posedge clkstart) cStart |-> sr1;
endproperty

reqGnt: assert property (pr1);
```

Gated Clock.

Fig. 6.12 Gated clock

6.5 Gated Clk

Figure 6.12 shows an interesting modeling application of using a gated clk as the sampling edge for a property. Note that "assign" is out of the scope of assertion. But its assigned value "clkstart" can indeed be used in the property. In general, any variable declared in a given scope in which the property/sequence is defined is available to the assertion. If the assertions are declared out of the module but bound to the module using "bind" method, the same rule applies. More on "bind" statement coming up soon.

In this example, the sampling edge will be the posedge of (clk && cGate). In the preponed region of this sampling edge, the variables in property and sequence will be sampled. This is also an example of an asynchronous clocking event.

6.6 Concurrent Assertions Are Multi-Threaded

This is about the most important concept you need to grasp when it comes to concurrent assertions. We all know SystemVerilog is a concurrent language but is it multi-threaded (except when automatic variables are used)? SVA by default is concurrent and multi-threaded.

In Fig. 6.13, we have declared the same assertion that you have seen before, namely at posedge clk, if cStart is sampled high that sr1 will be triggered at the same posedge clk which will then look for "req" to be high at that clock and "gnt" to be sampled high two clocks later.

Now, let us say that cStart is *sampled* high (S1) at a posedge of clk and that "req" is also sampled high at that same edge. After this posedge clk, the sequence will wait for 2 clocks to see if "gnt" is high.

But before the two clocks are over, clk cStart goes low and then goes high (S2) exactly two clocks after it was sampled high. This is also the same edge when our

```
sequence sr1;
  req ##2 gnt;
endsequence

property pr1;
  @(posedge clk) cStart |-> sr1;
endproperty

reqGnt: assert property (pr1) $display($stime,,,"\t\t %m PASS");
else $display($stime,,,"\t\t %m FAIL");
```

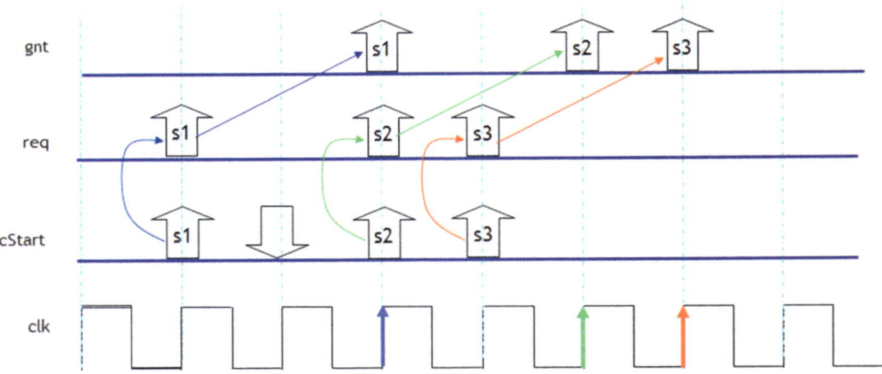

Fig. 6.13 Multi-threaded concurrent assertions

first trigger of assertion will look for gnt to be high (S1). So, what will the assertion do? Will it re-fire itself because it meets its antecedent condition (S2) and ignore "gnt" that it's been waiting for from the first trigger (S1)? Yes, it will re-fire but No, it will *not* ignore "gnt." It will sample "gnt" to be high (S1) and consider the first trigger (cStart (S1)) to PASS. So, what happens to the second trigger (cStart (S2))? It will start *another* thread. It will again wait for 2 clocks to check for "gnt." So far so good. We see one instance of SVA being threaded.

But life just got more interesting.

After S2, the very next clock cStart is sampled high again (S3). And "req" is high as well (req(S3)). Now what will the assertion do? Well, S3 will thread itself with S2. In other words, there are now two distinct threads of the same assertions waiting to sample "gnt" two clocks after their trigger. The figure perfectly (!) lines up "gnt" to be high two clocks after both S2 as well as after S3 and all 3 triggers of the same assertions will PASS.

This has many implications in terms of design of assertions and performance thereof. We will discuss this further when we discuss edge triggered antecedent. In other words, the way the property in our example is coded, it will drag your simulation performance because every time the property sees cStart to be high at posedge of clk, it will start a new thread. But if you want to evaluate the property only at the first rise of cStart and then ignore it if it stays high (unless it goes low and goes high again), then you have to use edge sensitive antecedent. More on this in Chap. 7. In addition, the concept of multi-threaded language gets much more interesting as you will see in Sect. 8.5.

Here's an explanation of further nuances of concurrent assertions.

It is important that the defined clock behavior be glitch free. Otherwise, wrong values can be sampled.

If a variable that appears in the expression for clock also appears in an expression with an assertion, the values of the two usages of the variable can be different. *The current value of the variable is used in the clock expression, while the sampled value of the variable (in preponed region) is used within the assertion.* This concept is especially important to understand if your "sampling edge" is not a synchronous clock but an "asynchronous edge." We will cover this concept via an explicit example in Chap. 19.

6.7 Formal Arguments

One of the key features of assertions is that they can be parameterized. In other words, assertions can be designed with formal arguments to keep them generic enough for use with different actual arguments.

Figure 6.14 is self-explanatory. Notice that the formal arguments can be specified in a sequence and in a property.

The application shows the advantage of formal arguments in reusability. Property "noChangeSig" has three formal arguments, namely pclk, refSig, and Sig. The property checks to see that if refSig is sampled low at posedge pclk, that the Sig is "stable." Once such a generic property is written you can invoke it with different clk, different refSig and Sig. CheckRd is a property that uses sysClk and OE_ and RdData to check for RdData to be stable while CheckWr uses WE_ and WrData to check for WrData to be "stable."

In any project, there are generic properties that can be reused multiple times by passing different actual arguments. This is reusable not only in the same project but also among projects.

Companies have created libraries of such pool of properties that projects look up and reuse according to their needs.

As shown in Fig. 6.15, properties can be both position based and name based. I highly recommend name based to make sure that actuals are connected to correct formals without ambiguity. This rule is the same as what we have been using in Verilog port connections.

Figure 6.16 describes the following points:

1. Default values can be assigned to the formal arguments.
   (a) If actual and formal both specify a "default" value, the actual will overwrite the formal default value.
   (b) You may leave passing an actual to a formal if the formal has a default value. Please refer to Fig. 6.16.

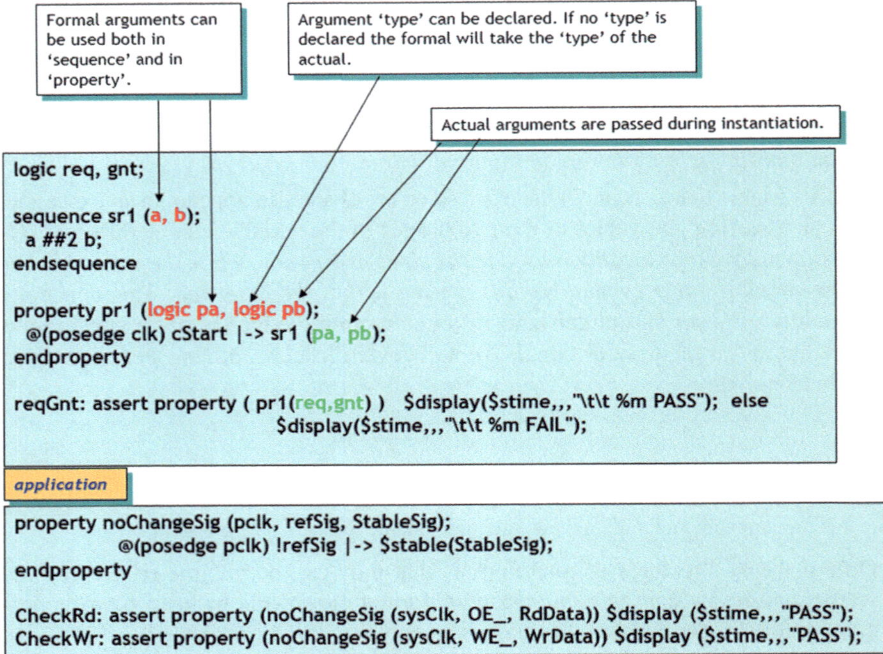

Fig. 6.14 Formal and actual arguments

default value assigned to a formal

```
logic req, gnt;

property pr1 (enb=1'b1, logic pa, logic pb);
  @(posedge clk) enb |-> pa ##2 pb;
endproperty

reqGnt: assert property ( pr1(cStart,req,gnt) );
```

position based connections...

```
reqGnt: assert property ( pr1 ( .enb(cStart), .pa(req), .pb(gnt)) );
```

name based connections...

Fig. 6.15 Formal and actual arguments—default value and name based connection

default value assigned to a formal

```
logic req, gnt;

property pr1 (enb=1'b1, logic pa, logic pb);
  @(posedge clk) enb |-> pa ##2 pb;
endproperty

reqGnt: assert property ( pr1(cStart,req,gnt) );
```

position based connections...

```
reqGnt: assert property ( pr1 (  .pa(req), .pb(gnt)) );
```

You may leave the formal with default value unconnected. Here we have not passed an 'actual' on the formal 'enb' since it has a default value assigned during it's formal declaration.

```
reqGnt: assert property ( pr1 (  ,req,gnt);
```

You may leave the formal with default value unconnected. *But (as in Verilog), for position based connections, you need to leave a "," for the unconnected formal.*

Fig. 6.16 Formal and actual arguments—default value and position based connection

This is a very interesting feature and very useful at that for reusability. A formal can be used for event control as well. A sampling edge can be passed as an actual to a formal and the actual can be used as a sampling edge in the property. We are passing "posedge clk" as an actual to the formal "csig." The property uses @ (csig) as its sampling edge. "@ (csig)" will change to "@ (posedge clk)" when the property "pr1" is called with "posedge clk" as the actual argument. Please refer to Fig. 6.17 for clarity on this point. Such properties can indeed be part of a common pool of properties that individual projects can reuse with their own sampling edge specification.

ILLEGAL :: Cannot use a 'formal' to size a local variable in a property. Size can only be a constant (or parameter) because it needs to be known at elaboration time.

```
property pr1 (int dSize, csig, enb=1'b1, logic pa, logic pb);

logic [dSize:0] Ldata;

 @(csig, Ldata=data) enb |-> pa ##2 pb;

endproperty

reqGnt: assert property ( pr1( 'd31, posedge clk, cStart, req, gnt) );
```

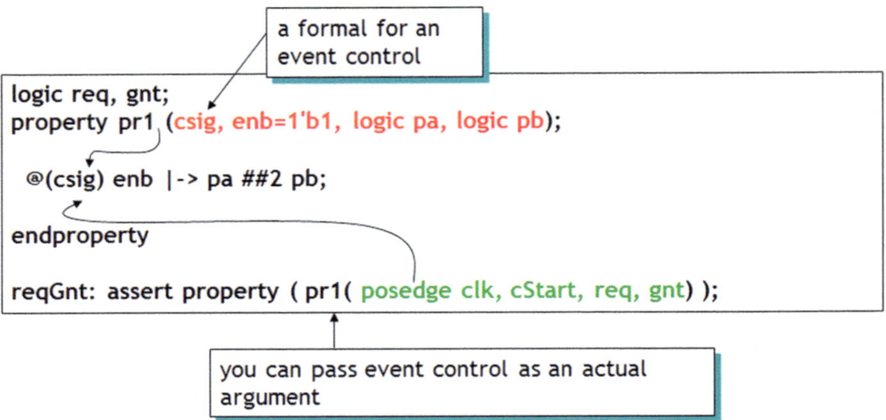

Fig. 6.17 Passing event control to a formal

Note also that you *cannot* pass a variable as an actual to size a local variable (see Chap. 11) in the property pr1. The size parameter needs to be a constant for sizing a local parameter.

Here are a some more rules governing binding between formal and actual.

A formal argument is said to be untyped if there is no type specified prior to its declaration in the port list. There is no default type for a formal argument.

The supported data types for sequence formal arguments are the types that are allowed for operands in assertion expressions and the keyword **untyped**.

Note that you can also use "$" as an actual argument. The terminal "$" may be an actual argument in an instance of a named sequence, either declared as a default actual argument or passed in the list of arguments of the instance. If "$" is an actual argument, then the corresponding formal argument must be untyped and each of its references will be an upper bound in a cycle_delay_const_range_expression.

6.8 Disable (Property) Operator—"disable iff"

Of course, you need a way to disable a property under conditions when the circuit is not stable (think Reset). That's exactly what "disable iff" operator does. It allows you to explicitly disable the property under a given condition. Note that "disable iff" reads as "disable if and only if." The example in Fig. 6.18 shows how you can disable an assertion during an active Reset. There is a good chance you will use this Reset based disable method in all your properties throughout the project.

So what happens if a property has started executing and the "disable iff" condition occurs in the middle of its execution?

The property in Fig. 6.18 checks to see that sdack_ falls (i.e., contained) within soe_ (don't worry; we'll see how such properties work in later chapters—see

```
sequence sdack_;
  !dack_[*7];
endsequence

sequence soe_;
  $fell (oe_) ##1 (!oe_[*8]) ;
endsequence

property pwin;
  @(posedge clk) disable iff (!reset) $fell(bMode) |=>
sdack_ within soe_;
endproperty
```

disable assertion testing if 'reset' is low.

Allowed only in a 'property' before 'antecedent'

NOT allowed in a 'sequence'

if disable condition is detected mid-way through checking of the assertion, the entire check is discarded.

```
# run -all
0  clk=1 reset=1 bMode=1 oe_=1 dack_=1
10  clk=1 reset=1 bMode=1 oe_=1 dack_=1
20  clk=1 reset=1 bMode=0 oe_=1 dack_=1
30  clk=1 reset=1 bMode=0 oe_=0 dack_=1
40  clk=1 reset=1 bMode=0 oe_=0 dack_=1
50  clk=1 reset=1 bMode=0 oe_=0 dack_=0
60  clk=1 reset=1 bMode=0 oe_=0 dack_=0
70  clk=1 reset=1 bMode=0 oe_=0 dack_=0
80  clk=1 reset=1 bMode=0 oe_=0 dack_=0
90  clk=1 reset=1 bMode=0 oe_=0 dack_=0
100  clk=1 reset=1 bMode=0 oe_=0 dack_=0
110  clk=1 reset=1 bMode=1 oe_=0 dack_=0
120    property pdiff PASS
120  clk=1 reset=1 bMode=1 oe_=0 dack_=1
130  clk=1 reset=1 bMode=1 oe_=1 dack_=1
140  clk=1 reset=1 bMode=1 oe_=1 dack_=1
```

```
# run -all
0  clk=1 reset=1 bMode=1 oe_=1 dack_=1
10  clk=1 reset=1 bMode=1 oe_=1 dack_=1
20  clk=1 reset=1 bMode=0 oe_=1 dack_=1
30  clk=1 reset=1 bMode=0 oe_=0 dack_=1
40  clk=1 reset=1 bMode=0 oe_=0 dack_=0
50  clk=1 reset=1 bMode=0 oe_=0 dack_=0
60  clk=1 reset=1 bMode=0 oe_=0 dack_=0
70  clk=1 reset=1 bMode=0 oe_=0 dack_=0
80  clk=1 reset=0 bMode=0 oe_=0 dack_=0
90  clk=1 reset=0 bMode=0 oe_=0 dack_=0
100  clk=1 reset=0 bMode=0 oe_=0 dack_=0
110  clk=1 reset=0 bMode=1 oe_=0 dack_=0
120  clk=1 reset=0 bMode=1 oe_=0 dack_=1
130  clk=1 reset=0 bMode=1 oe_=1 dack_=1
140  clk=1 reset=0 bMode=1 oe_=1 dack_=1
```

Fig. 6.18 "disable iff" operator

Sect. 8.18). It also has the "disable iff (! reset)" condition. Disable this property if and only if reset is asserted (active low).

Let us examine the simulations logs.

In the LHS simulation log, reset is never asserted, and the assertion completes (and passes in this case).

In the RHS simulation block, reset is asserted in the middle of check "sdack_ within soe" and the entire assertion is discarded. You will not see pass/fail for this assertion because it has been discarded. Entire assertion is disabled if the "disable iff" condition occurs in the middle of an executing assertion. Some folks mistake such discard as a failure, which is incorrect.

Once an assertion has been disabled with "disable iff" construct, it will re-start only after the "disable iff" condition is not true anymore.

Here are the rules that govern "disable iff"

1. "disable iff" can be used only in a property—not in a sequence
2. "disable iff" can only be used before the declaration of the antecedent condition.
3. *"disable iff" expression is not sampled at a clock edge* as with other expressions in the concurrent assertion. The expressions in a disable condition are evaluated using the *current* values of variables (not sampled in preponed region). "disable iff" expression can be thought of as asynchronous; it can trigger in between clock events or at clock event. This is an important point because we will be discussing a lot about sampled values and this here is an exception. In our example, we have disable iff (!reset). Here the "reset" signal is not sampled. The disable iff condition will trigger as soon as "reset" goes low.
4. Nesting of "disable iff" clauses, explicitly or through property instantiations, is not allowed.
5. The operator "disable iff" cannot be used in the declaration of a recursive property.
6. We haven't discussed .triggered or .matched methods but here's the rule for your reference later. "disable iff" may contain the sequence Boolean method .triggered. But it cannot contain any reference to local variables or to the sequence method .matched .

6.9 Default disable iff

If you have properties within a module, interface or program block that use the same "disable iff" condition, you may declare a "default disable iff" clause. It provides a default disable condition to all concurrent assertions in the scope and subs copes of the default disable iff declaration. Furthermore, the default extends to any nested module, interface, or program declarations, and to nested generate blocks. However, if a nested module, interface, or program declaration, or a generate block itself has a default "disable iff" declaration, then that default "disable iff" applies within the nested declaration or generate block and overrides any "default disable iff" from outside.

Here's an example.

```
module M1;
bit RST;

default disable iff RST;
     a1: assert property (@(posedge clk) propertyP1; //prop-
     ertyP1 is defined elsewhere
     a2: assert property (@negedge clk) propertyP2; //proper-
     tyP2 is defined elsewhere

endmodule
```

In this example, the "assert" statements will use the "default disable iff" RST as the default disable condition during the assertion. No need to explicitly call "disable iff" with each assertion.

A few rules governing the "default disable iff."

(a) If an assertion has a disable iff clause, then the disable condition specified in this clause will be used and any default disable iff declaration ignored for this assertion.
(b) If an assertion does *not* contain a disable iff clause, but the assertion is within the scope of a default disable iff declaration, then the disable condition for the assertion is inferred from the default disable iff declaration.

Here's another simple example.

```
module examples (input logic a, b, clk, rst, rst1);

default disable iff rst;  //Default disable condition is based
on the signal 'rst'
property p1;
     disable iff (rst1) a |=> b;   //explicit declaration of
     disable iff in property p1
endproperty

// Disable condition is rst1 - explicitly specified within a1
a1 : assert property (@(posedge clk) disable iff (rst1) a |=> b);

// Disable condition is 'rst1' - explicitly specified within p1
a2 : assert property (@(posedge clk) p1);

// Disable condition is 'rst' - no explicit specification, inferred from
// default disable iff declaration
a3 : assert property (@(posedge clk) a |=> b);

endmodule
```

6.10 Nested *disable iff*—ILLEGAL

Note that you cannot nest *disable iff* conditions. That is ILLEGAL. Here's an example,

```
module m_illegal_disable_nesting(logic reset, a, b, clk);
default clocking PCLK @(posedge clk); endclocking
     property PCI_disable;
          disable iff (reset) IRDY_ |=> FRAME_;
```

```
    endproperty
    Nest_disable: assert property (
         disable iff (reset1) PCI_disable;   //ILLEGAL
    ) else $error("FAIL");
endmodule
```

In this example, property PCI_disable defines a *disable iff* condition. Property Nest_disable also defines a *disable iff* condition. But Nest_disable then calls property PCI_disable. Since, PCI_disable has its own *disable iff* condition, we are in essence nesting the *disable iff* conditions. That is ILLEGAL.

6.11 Severity Levels (for Both Concurrent and Immediate Assertions)

Assertions also allow error reporting with different severity levels. $fatal, $error (default), $warning and $info. Figure 6.19 explains meaning of each.

$error is default, meaning if no failure clause is specified in the assert statement, $error will kick in and provide a simulator generated error message. If you have specified a label (and you should have) to the assertion, that will be (most likely) displayed in the $error message. I say most likely because the SystemVerilog LRM does not specify exact format of $error. It is simulator vendor specific. $warning and $info are self-explanatory as described in Fig. 6.19.

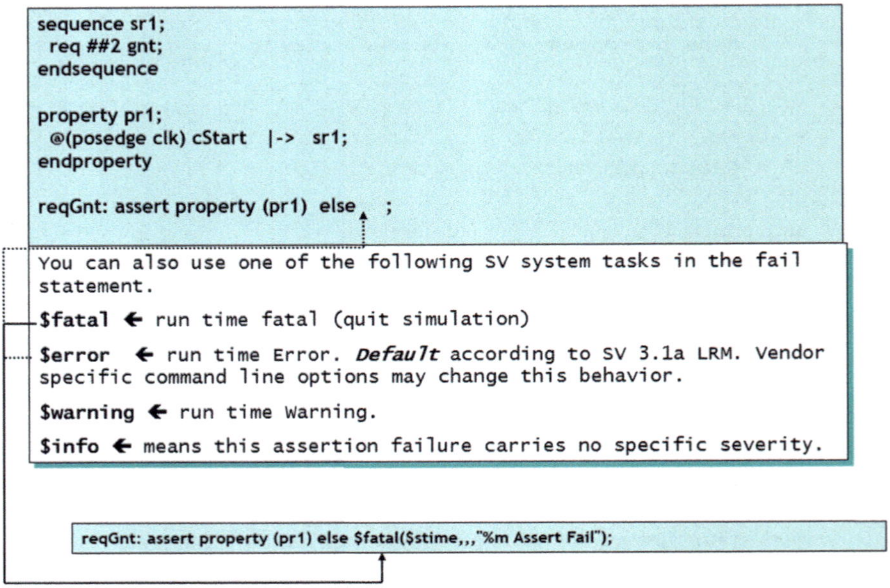

```
sequence sr1;
  req ##2 gnt;
endsequence

property pr1;
  @(posedge clk) cStart  |->  sr1;
endproperty

reqGnt: assert property (pr1) else  ;
```

You can also use one of the following SV system tasks in the fail statement.

$fatal ← run time fatal (quit simulation)

$error ← run time Error. *Default* according to SV 3.1a LRM. Vendor specific command line options may change this behavior.

$warning ← run time Warning.

$info ← means this assertion failure carries no specific severity.

```
reqGnt: assert property (pr1) else $fatal($stime,,,"%m Assert Fail");
```

Fig. 6.19 Severity levels for concurrent and immediate assertions

6.12 Binding Properties

"bind" allows us to keep design logic separate from the assertion logic. Design managers do not like to see anything in RTL that is not going to be synthesized. "bind" helps in that direction.

There are three modules in Fig. 6.20. The "designModule" contains the design. The "propertyModule" contains the assertions / properties that operate on the logic

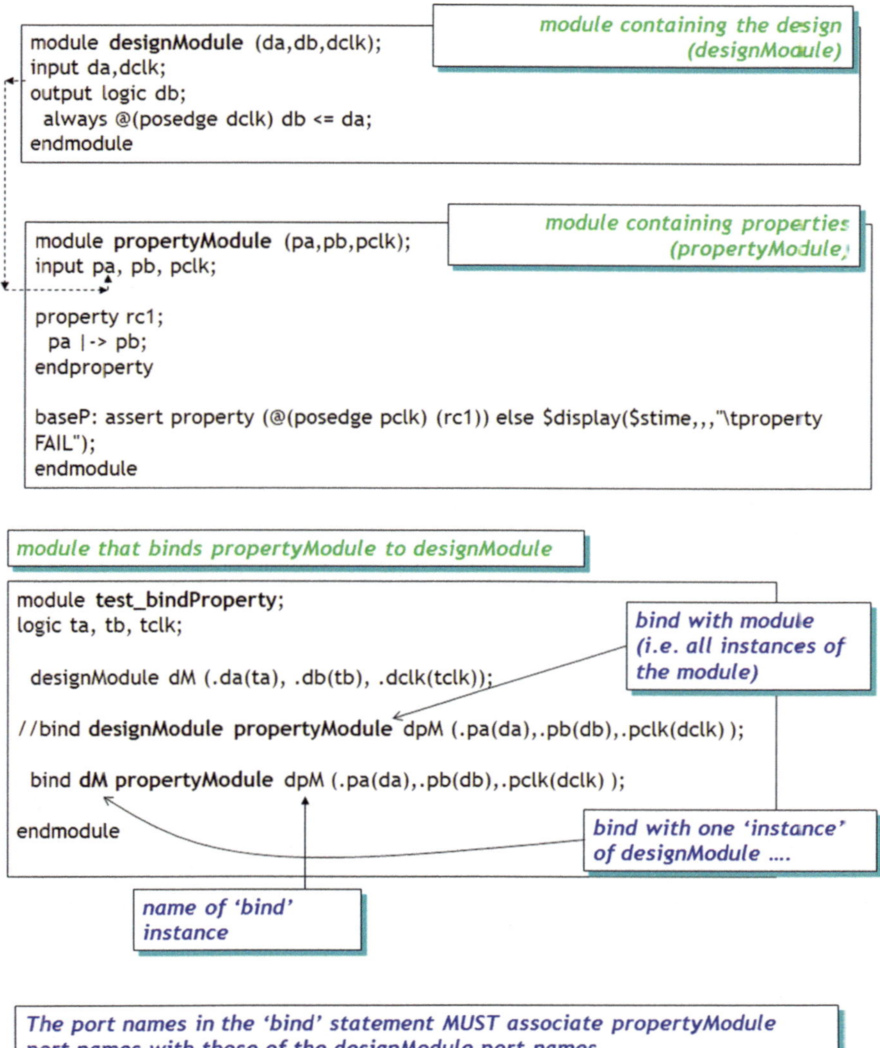

Fig. 6.20 Binding properties

in “designModule.” And the “test_bindProperty” module binds the propertyModule to the designModule. By doing so, we have kept the properties of the “property-Module” separate from the “designModule.” That is the idea behind “bind.” You do not have to place properties in the same module as the design module. As mentioned before, you should keep your design void of all constructs that are non-synthesizable. In addition, keeping assertions and design in separate modules allow both the design and the DV engineers work in parallel without restrictions of a database management system where a file cannot be modified by two engineers at the same time.

In order for “bind” to work, you have to declare either the instance name or the module name of the designModule in the “bind” statement. You need the design module/instance name, property module name, and the “bind” instance name for “bind” to work. In our case the design module name is designModule, its instance name is “dM” and the property module name is propertyModule.

The (uncommented) “bind” statement uses the module instance “dM” and binds it to the property module “propertyModule” and gives this “bind” an instance name “dpM.” It connects the ports of propertyModule with those of the designModule. With this the “property rc1” in propertyModule will act on design-Module ports as connected.

The commented “bind” statement uses the module name “designModule” to bind to the “propertyModule” whereby all instances of the “designModule” will be bound to the “propertyModule.”

In essence, we have kept the properties/assertions of the design and the logic of the design separate. This is the recommended methodology. You could achieve the same results by putting properties in the same module as the design module but that is highly non-modular and intrusive methodology. In addition, as noted above, keeping them separate allows both the DV and the Design engineer to work in parallel.

6.13 Binding Properties (Scope Visibility)

But what if you want to bind the assertions of the propertyModule to internal signals of the designModule? That is quite doable.

As shown in Fig. 6.21, “rda” and “rdb” are signals internal to designModule. These are the signals that you want to use in your assertions in the “propertyModule.” Hence, you need to make “rda” and “rdb” visible to the “propertyModule.” However, you do not want to bring “designModule” internal variables to external ports in order to make them visible to the “propertyModule.” You want to keep the “designModule” completely untouched. To do that, you need to add input ports to the “propertyModule” and bind those to the internal signals of the “designModule” as shown in Fig. 6.21. Note that in our example we bind the propertyModule ports “pa” and “pb” to the designModule internal registers “rda” and “rdb.” In other words, you can directly refer to the internal signals of designModule during “bind.” “bind” has complete scope visibility into the bound module “designModule.” Note

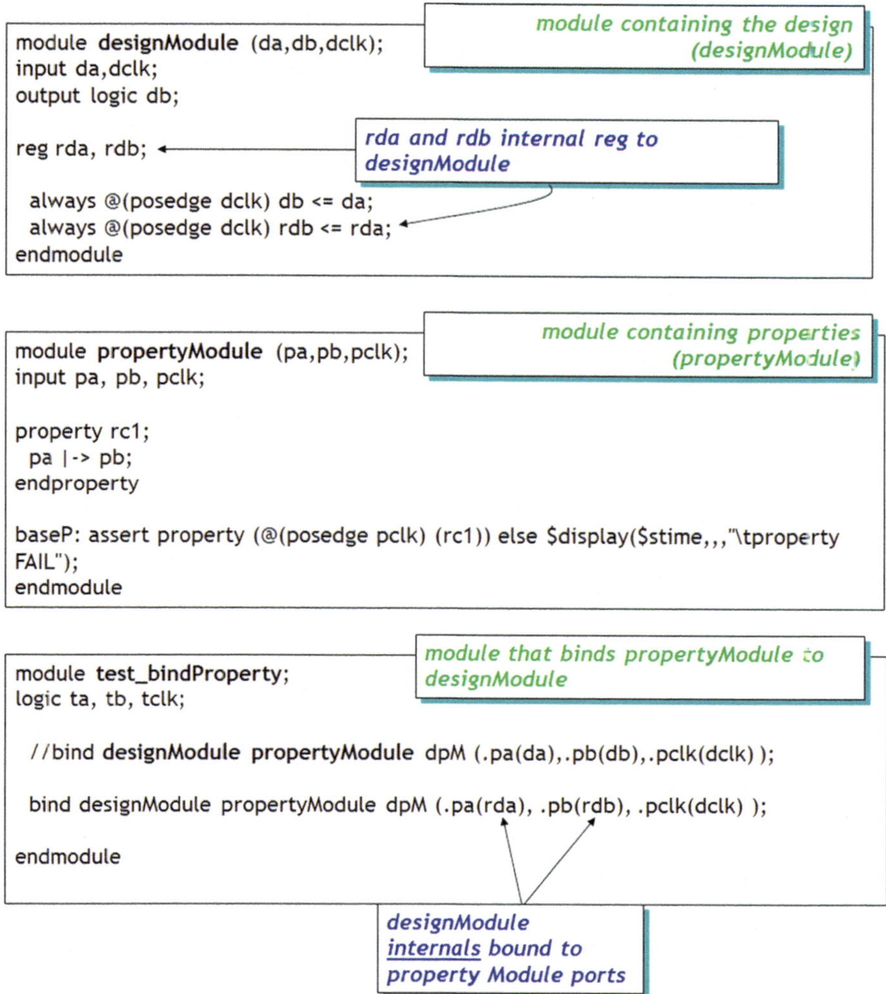

Fig. 6.21 Binding properties to design "module" internal signals (scope visibility)

that with this method you do not have to provide the entire hierarchical instance name when binding to "propertyModule" input ports.

6.14 VHDL DUT Binding with SystemVerilog Assertions

Yes, binding is independent of the language. The IEEE standard does not specify exactly how to accomplish it but EDA vendors have made it plenty simple and implemented it to suit their tools/methodology.

VHDL DUT

```
entity cpu is
port (
a : in std_logic;
b : in std_logic;
...);

architecture rtl of cpu is
...
signal c : std_logic;
begin
...
end rtl;
```

SVA

```
module cpu_props(input d,e,f);
...
assert property (@(posedge d) e |-> ##[1:2] f );
...
endmodule
```

'bind' SVA to VHDL DUT

```
module sva_wrapper;

bind cpu cpu_props cpu_sva_bind
        (.d(a),.e(b),.f(c));
                    // Connect SystemVerilog ports to VHDL ports (a and b)
                    // and to the internal signals (c)
endmodule
```

```
vlib work
vlog *.*sv
vcom *.vhd
vsim top sva_wrapper
```

Fig. 6.22 Binding VHDL DUT to SystemVerilog Assertions

Here's an example of how binding a VHDL entity to SystemVerilog Assertions module works. This is based on Mentor's Questa simulator.

The example in Fig. 6.22 shows a simple VHDL entity called "cpu" and a systemverilog assertions module called "cpu_props." The binding takes place in systemverilog module called "sva_wrapper." The binding mechanism is identical to that we saw in previous sections. There is no difference in binding a Verilog module to a Verilog module and binding a VHDL entity to a Verilog module.

When it comes to simulation, each EDA vendor has its own way to compiling VHDL and Verilog together to accomplish the "bind." The example in Fig. 6.22 shows the way Mentor's Questa simulator accomplishes this.

6.15 Assertion Adoption in Existing Design

Figure 6.23 shows that if you have an existing design, you can effectively use the "bind" construct to write assertions outside of the design scope and bind them. This can be very useful, if you are bringing in legacy blocks in your new SoC and want to make sure that the legacy blocks work well in your new design. This figure is a methodology component. Upfront in your project, determine your "bind" methodology. See that all the assertions are outside RTL and not a messy mix of some in RTL and some bound with external properties file.

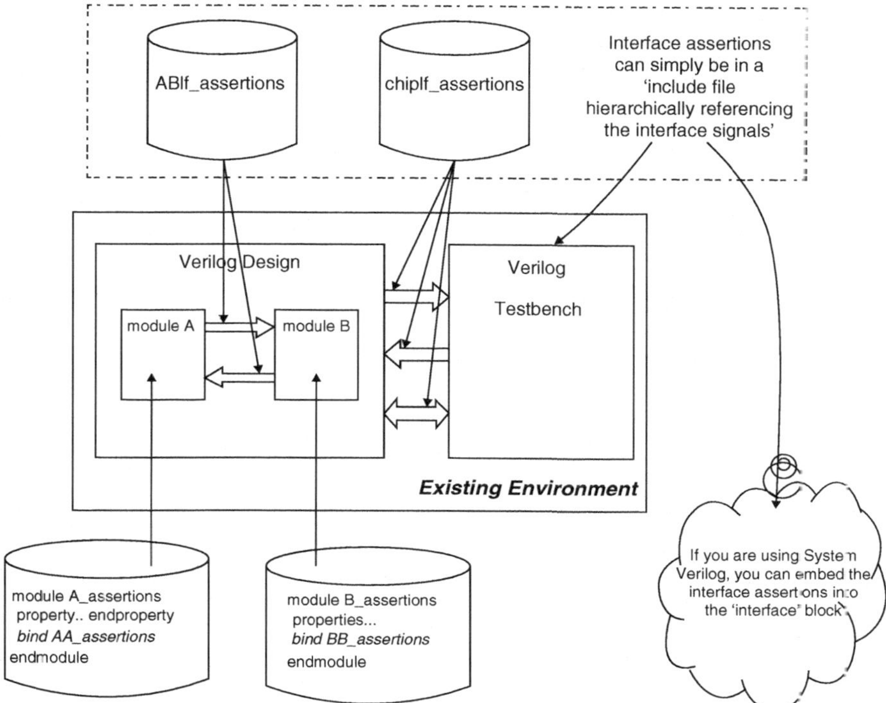

Fig. 6.23 Binding properties to an existing design. Assertions adoption in existing design

Other advantage of keeping assertions in a separate file is that they can be independently verified without the need to have control of RTL files. A big advantage when you want to make sure that both the design and verification progress in parallel.

6.16 Difference Between "sequence" and "property"

Now that we have seen assertions using sequences and properties, it is good to recap and clearly understand the differences between the two.

- **"sequence"**
  - A sequence is a building block. Think of it as a macro or a subroutine where you can define a specific relationship for a given set of signals.
  - A sequence on its own does not trigger. It must be "assert"ed. or "cover"ed.
  - A named sequence may be instantiated by referencing its name. The reference may be a hierarchical name.

  - ***A sequence does not allow implication operator***. Simply allows temporal (or combinatorial) domain relationship between signals.
  - A sequence can have optional formal arguments.
  - A clocking event can be used in a sequence.
  - A sequence can be declared in a module, an interface, a program, a clocking block, a package, a compilation-unit scope, a checker, and a generate block (***but –not- in a*** "***class***").

- "**property**"
  - A property also does not trigger by itself until "assert"ed. (or "cover"ed. or "assume"d).
  - ***Properties have implication operator*** that imply the relationship between an antecedent and a consequent.
  - Sequences can be used as building blocks of complex properties.
  - Clocking event can be specified in a property, in a sequence, or in both.
  - The formal and actual arguments can also be "property expressions"—meaning you can pass a "property" as an actual to a formal of type "property."
  - Local variable arguments (see Chap. 11) can only be "local input."
  - A property can be declared in a module, an interface, a program, a clocking block, a package, a compilation-unit scope, a checker and a generate block (***but—not—in a*** "***class***").

Chapter 7
Sampled Value Functions

Introduction: This chapter introduces and provides applications for Sampled Value Functions $rose, $fell, $past, and $stable. Note that there are also quite a few new sampled value functions introduced in 2009/2012 LRM (e.g., $changed, $rose_gclk, $sampled). These are covered in Chap. 20 which is solely devoted to the entire 2009/2012 LRM feature set.

A. B. Mehta, *SystemVerilog Assertions and Functional Coverage*,
https://doi.org/10.1007/978-3-030-24737-9_7

These sampled value functions (Fig. 7.1) allow for antecedent and/or the consequent to be edge triggered. $rose means that the least significant bit of the expression (in $rose(expression)) was sampled to be "0" at the previous clk edge (previous meaning the immediately preceding clk from current clk) where it was sampled "1." For $fell, just the opposite need to take place. Preceding value should be sampled "1" and current sampled value "0." As explained with examples below, one needs to understand the difference between level sensitive sample vs. edge sensitive sample.

But why do we call these functions "sampled value"? That's because they are triggered only when the sampled value of the expression in the preponed region differ at two successive clock edges as described above. In other words, $rose(abc) does *not* mean "posedge abc" as in Verilog. $rose(abc) does not evaluate to true as soon as abc goes from 0 to 1. $rose(abc) simply means that abc was sampled "1" at the current clock edge (in preponed region) and that it was *not* sampled a "1" at the immediately preceding clock edge.

Note also that both $rose and $fell work only on the Least Significant Bit of the expression. You will soon see what happens if you use a bus (vector) in these two sampled value functions.

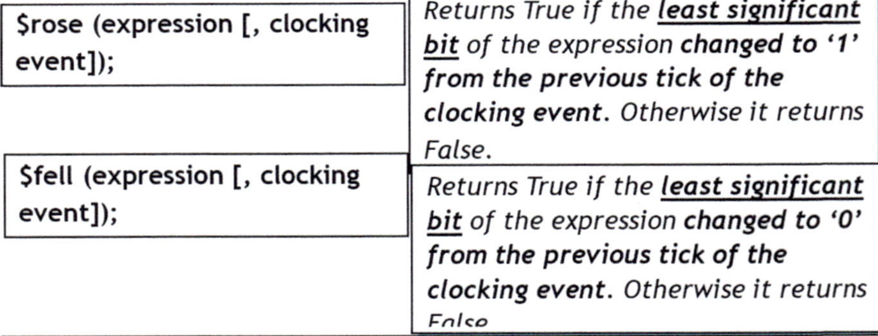

Notes:

- *The [, clocking event] is optional and usually derived from the clocking event of the assertion or from the inferred clock of the procedural block where the function is used*
- *When these functions are called at or before the simulation time step in which the first clocking event occurs, the results are computed by comparing the sampled value of the expression with its default sampled value.*
- *These functions can be used in property/sequence as well as in procedural code as expressions*

Fig. 7.1 Sampled value functions $rose, $fell—basics

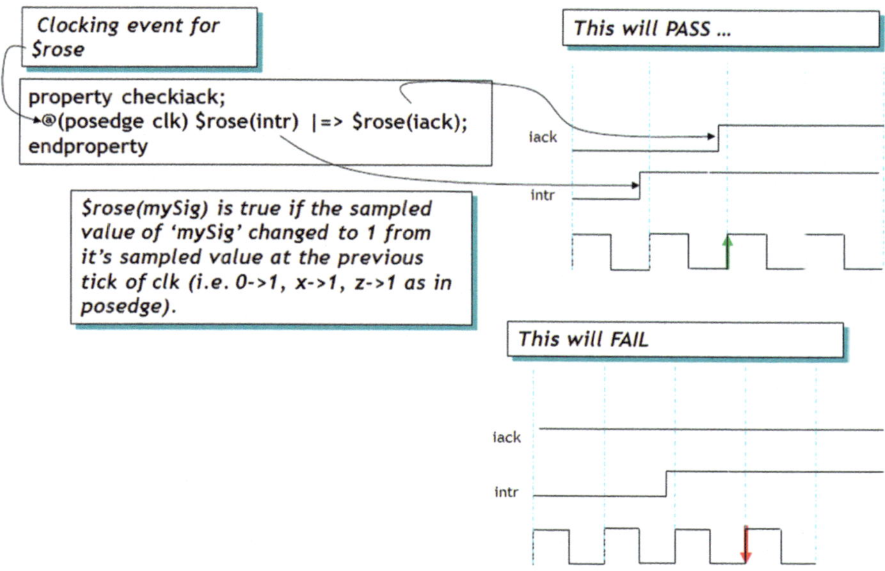

Fig. 7.2 $rose—basics

7.1 $rose—Edge Detection in Property/Sequence

property "checkiack" in the top logic/timing diagram will (Fig. 7.2) PASS because both the "inter" and "iack" signals meet the required behavior of $rose (value at two successive clks are different and are "0" followed by "1"). However, the logic in the bottom diagram fails because while $rose(intr) meets the requirement of $rose, but "iack" does not. "iack" does not change from "0" to "1" between the two clk edges.

Important Note: To reiterate the points made above. $rose does *not* mean posedge and $fell does *not* mean negedge (as in Verilog). In other words, the assertion won't consider $rose(intr) to be true as soon as a posedge on "intr" is detected. The $rose () / $fell () behavior is derived by "sampling" the expression at two successive clk edges and see if the values are opposite and in compliance with $rose () or $fell ().

In other words, the fundamentals of concurrent assertions specify that everything must be sampled at the sampling edge in the preponed region. Behavior is based on sampled value and not the current value.

7.2 Edge Detection is Useful Because …

This is a very important example. It explains the difference between use of level sensitive sampled values vs. edge sensitive. Both are correct to use, except that you need to know which to use when. As shown in Fig. 7.3, level sensitive evaluation is a superset of edge sensitive evaluation. But when you use level sensitive sample,

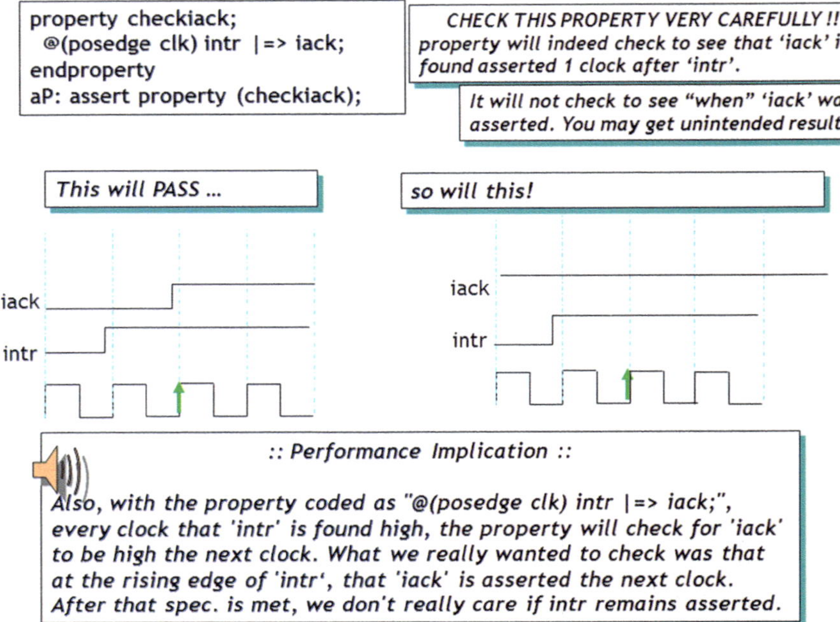

Fig. 7.3 Usefulness of "edge" detection and performance implication

you will degrade simulation performance, if in fact you meant for an edge sensitive evaluation.

In the above property, the following would be more appropriate if all you wanted to do was to check for iack to go from inactive "0" state to active state "1" once edge-sensitive intr is asserted. After that, the state of intr does not matter. Following is a better way to write the property if your intention was to check for rising edge of iack one clock after rising edge of intr.

```
property checkiack;
    @ (posedge clk) $rose(intr) |=> $rose(iack);
endproperty
```

But what if you decide to do the following (as shown in Fig. 7.4)? Now you are courting real trouble. As Fig. 7.4 explains, since "intr" is level sensitive sample, when it is sampled high it will look for the edge sensitive "iack." BUT since "intr" is level sensitive and high the very next clock, it will start a new thread and check for \$rose(iack) every clock. Since iack did not go from "0" to "1" on this second thread, the property fail. There is very good chance you did not want to see this failure.

In short, as simple as these functions look, you have to be careful in their usage and also keep in mind performance implications.

```
property checkiack;
  @(posedge clk) intr |=>
$rose(iack);
endproperty
aP: assert property (checkiack);
```

So, you decide to use $rose(iack) to make sure that iack indeed was low before it was detected to be high.

But.... intr is still level sensitive...

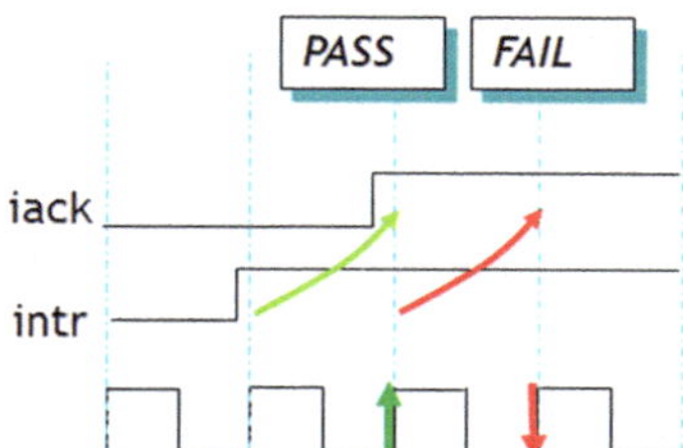

Fig. 7.4 $rose—finer points

7.3 $fell—Edge Detection in Property/Sequence (Fig. 7.5)

7.4 $rose, $fell—in Procedural

$rose and $fell are useful not only in concurrent assertions but also in sequential procedural blocks. They work exactly the same way as in concurrent assertions. Please see the examples in Fig. 7.6.

Since every assertion requires a clocking event (i.e., sampling edge), when you use a concurrent assertion in a procedural block without an explicit clocking event associated with them, the simulator looks for a clocking event in the code that precedes the concurrent assertion. We will discuss more on the use of concurrent assertions in procedural code, under the Advanced Topics Chap. 16.

In Fig. 7.6, $(posedge clk) is the preceding clocking event and acts as the sampling edge for $rose(iStreamDone) (example at the top of the figure).

Note the use of $rose and $fell in continuous assignment statement. Since continuous assign cannot have an edge behavior, you have to explicitly embed a clocking event with $rose () and/or $fell (). This is the same rule that applies to other sampled value functions.

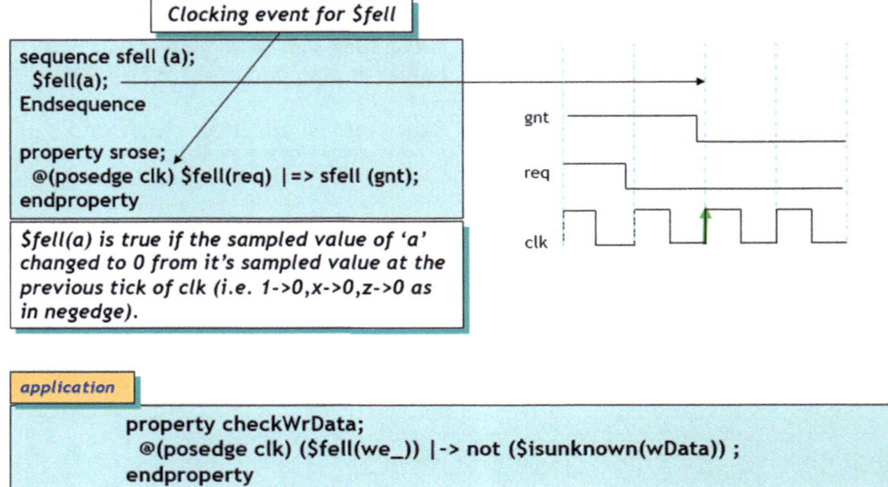

Fig. 7.5 $fell—basics

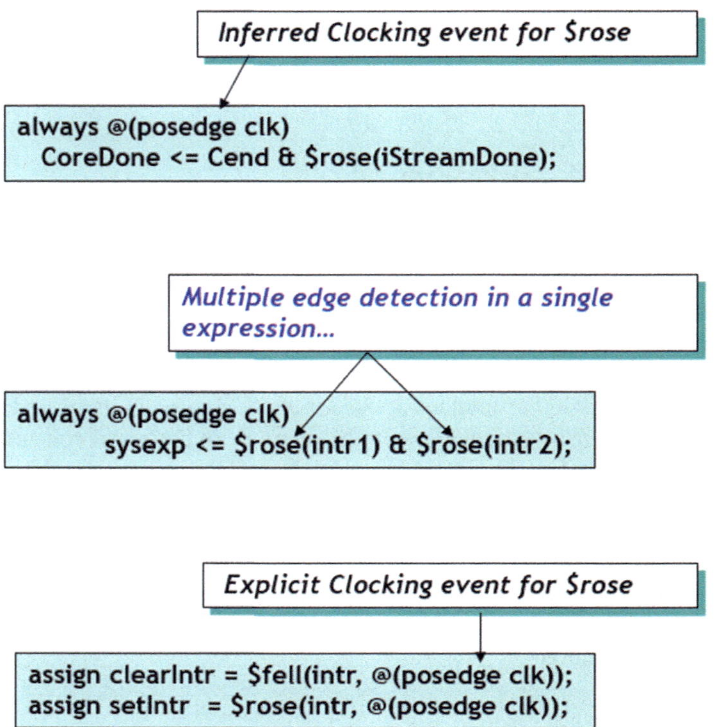

Fig. 7.6 $rose and $fell in procedural block and continuous assignment

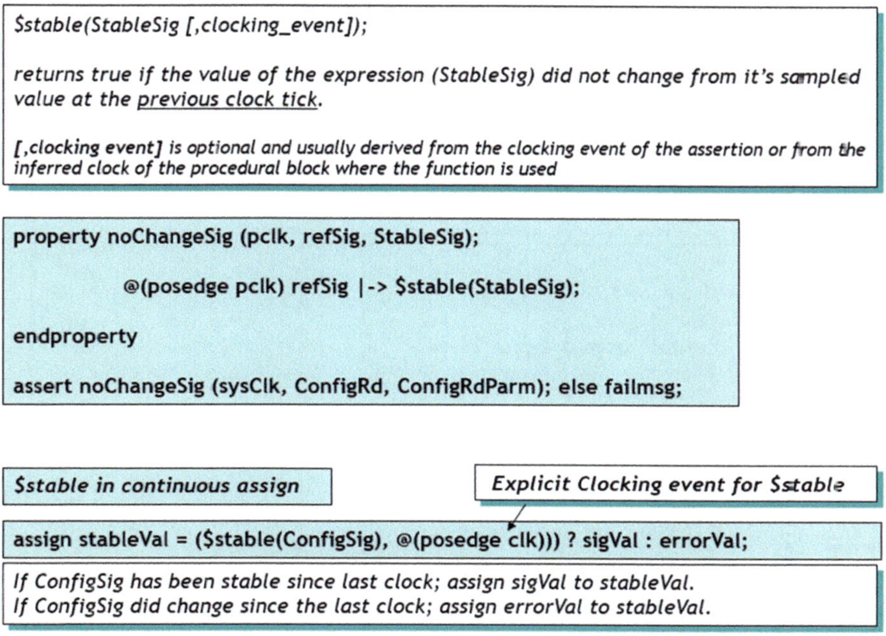

$stable(StableSig [,clocking_event]);

returns true if the value of the expression (StableSig) did not change from it's sampled value at the previous clock tick.

[,clocking event] is optional and usually derived from the clocking event of the assertion or from the inferred clock of the procedural block where the function is used

```
property noChangeSig (pclk, refSig, StableSig);

        @(posedge pclk) refSig |-> $stable(StableSig);

endproperty

assert noChangeSig (sysClk, ConfigRd, ConfigRdParm); else failmsg;
```

$stable in continuous assign

Explicit Clocking event for $stable

```
assign stableVal = ($stable(ConfigSig), @(posedge clk))) ? sigVal : errorVal;
```

If ConfigSig has been stable since last clock; assign sigVal to stableVal.
If ConfigSig did change since the last clock; assign errorVal to stableVal.

Fig. 7.7 $stable—basics

7.5 $stable

$stable (), as the name implies, looks for its expression to be stable between two clock edges (i.e., two sampling edges). It evaluates the expression at current clock edge and compares it with the value sampled at the immediately preceding clock edge. If both values are same, the check passes (Fig. 7.7).

Note the use of $stable in continuous assignment statement. Since continuous assign cannot have an edge behavior, you have to explicitly embed a clocking event with $stable. This is the same rule that applies to other sampled value functions.

But what if you want to check if the signal has been stable for more than one clock? Read on... $past () will solve that problem.

7.6 $stable in Procedural Block

As in $rose () and $fell (), $stable can also be embedded in the procedural code and works the same way as in a property or a sequence. As shown in the example above, $stable samples the value of expression ("a" and "b" in this example) at the current and the immediately preceding edge (posedge clk in this example) to see if the value of the expression did not change. At time 15, "b" has been stable, at time 25 "a" has been stable, and so on (Fig. 7.8).

```
always @(posedge clk)
begin
  if ($stable(a)) $display ($stime,,,"\t 'a' stable from previous clock");
  if ($stable(b)) $display ($stime,,,"\t 'b' stable from previous clock");

  if ($stable(a) && $stable(b))
    $display ($stime,,,"\t 'a' AND 'b' Stable this clock");
end
```

```
# run -all
#        5  clk=1 a=1 b=0
#       15  clk=1 a=0 b=0
#       15     'b' stable from previous clock

#       25  clk=1 a=0 b=1
#       25     'a' stable from previous clock

#       35  clk=1 a=1 b=0
#       45  clk=1 a=1 b=1
#       45     'a' stable from previous clock

#       55  clk=1 a=1 b=1
#       55     'a' stable from previous clock
#       55     'b' stable from previous clock
#       55     'a' AND 'b' Stable this clock
```

Fig. 7.8 $stable in procedural block

7.7 $past

$past () is an interesting function. It allows you to go into past as many clocks as you wish to. You can check for an "expression1" to have a certain value, number of clocks (strictly prior time ticks) in the past. Note that number of ticks is optional. If you do not specify it, the default will be to look for "expression1" one clock in the past.

Another caveat that you need to be aware of is when you call $past in the initial time ticks of simulation and there are not enough clocks to go in the "past." For example, you specify "a |-> $past (b)" and the antecedent "a" is a "1" at time 0. There isn't a clock tick to go in the past. In that case, the assertion will use the "initial" value of "b." What is the initial value of "b"? It's not the one in the "initial" block, it's the value with which the variable "b" was declared (as in "logic b = 1'b1;"). In our case, if "b" was not initialized in its declaration, the assertion will fail. If it was declared with an initial value of 1'b1, the assertion will pass.

You can also "gate" this check with expression2. The example in Fig. 7.10 shows how $past works. We are using a gating expression in this example. It is not a requirement as noted in Fig. 7.9, but it will most likely be required in your application. If expression2 is not specified, no clock gating is assumed.

If you understand the use of $past with a gating expression, then its use without one will be straightforward to understand.

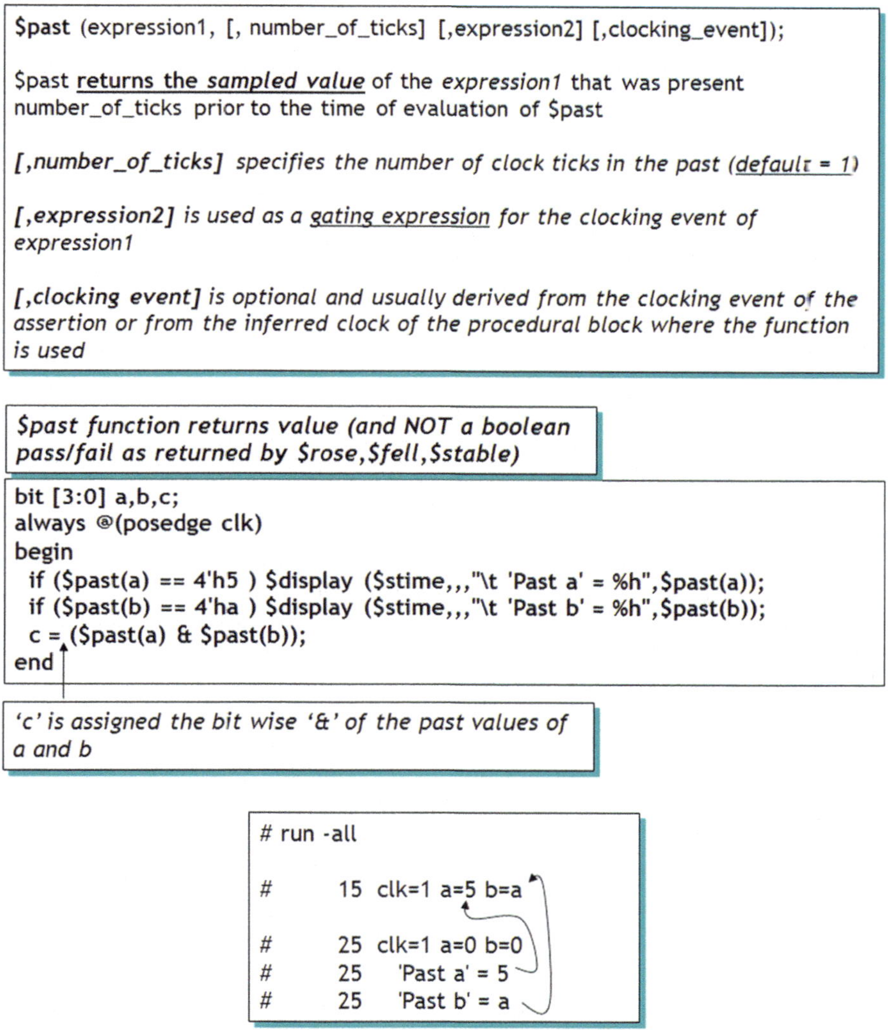

$past (expression1, [, number_of_ticks] [,expression2] [,clocking_event]);

$past returns the sampled value of the *expression1* that was present number_of_ticks prior to the time of evaluation of $past

[,number_of_ticks] specifies the number of clock ticks in the past (default = 1)

[,expression2] is used as a gating expression for the clocking event of expression1

[,clocking event] is optional and usually derived from the clocking event of the assertion or from the inferred clock of the procedural block where the function is used

$past function returns value (and NOT a boolean pass/fail as returned by $rose,$fell,$stable)

```
bit [3:0] a,b,c;
always @(posedge clk)
begin
  if ($past(a) == 4'h5 ) $display ($stime,,,"\t 'Past a' = %h",$past(a));
  if ($past(b) == 4'ha ) $display ($stime,,,"\t 'Past b' = %h",$past(b));
  c = ($past(a) & $past(b));
end
```

'c' is assigned the bit wise '&' of the past values of a and b

```
# run -all

#       15  clk=1 a=5 b=a

#       25  clk=1 a=0 b=0
#       25     'Past a' = 5
#       25     'Past b' = a
```

Fig. 7.9 $past—basics

Figure 7.10 asserts the following property.

```
assert property (@ posedge clk) done |-> lV (mySig,2, enb,
lastVal) $display(....);
```

And the property lV (lV stands for last Value) models the following.

```
property lV(Sig, numClocks, enb, lastV);
      (lastV == $past (Sig, numClocks, enb));
endproperty
```

```
(lastV == $past(Sig, numClocks, enb) );
          checks for the 'lastV' on Sig, numClocks in the past, gated by 'enb'
```

```
property lV(Sig,numClocks,enb,lastV);
  (lastV == $past(Sig, numClocks, enb) );
endproperty

assert property (@(posedge clk) done |-> lV(mySig, 2, enb, lastVal)) else
      $display($stime,,,"FAIL Expected lastVal=%h\n",lastVal);

cover property (@(posedge clk) done |-> lV(mySig, 2, enb, lastVal))
      $display($stime,,,"PASS Expected lastVal=%h\n",lastVal);

always @(posedge clk)
  $display($stime,,,"clk=%b mySig=%h past=%h   enb=%h done=%b", clk, mySig,
                      $past(mySig, 2, enb), enb, done);
```

```
'enb' in  (lastV == $past(Sig,numClocks,enb) );  means ::
          sampling of 'Sig' is performed based on it's clock gated by 'enb'.

In other words, $past evaluates 'Sig' iff 'enb' is true.
```

Fig. 7.10 $past—gating expression

The property says check on Sig, numClocks in the past and see if it has the value "lastV" and do this check *if and only if* "enb" (the gating expression) is high when you *start* the check (i.e., when antecedent done = 1 in the assert statement). *Let me re-emphasize that the gating expression is checked when* you *start* the check. The gating expression needs to be true when $past is called. Many seem to miss this point.

Let us look at the simulation log which will explain this concept. The example from previous page is repeated here with lastV = 'ha when we assert/cover the property "lV" for easy reference to the simulation log. Pay attention to the failure at time 80 (Fig. 7.11).

Let us examine the simulation log carefully to see how $past works. At time 30, done = 1 for the first time which means that the antecedent of the property is true implying that the property lV be executed. lV has formal arguments which are replaced by the actual arguments from the assert statement.

So, at time 30, the property first checks to see if the gating signal ("enb") is true. Since it is indeed true, the property calls/invokes past to evaluate the value of the actual "mySig" two clocks in the past. It sees that it is indeed h'a (at time 10 in the simulation log) and the property passes.

At time 50, done==1 and enb==1 but two clocks in the past (at time 30), mySig was not equal to h'a and the property fails.

At time 80, done = 1, but the gating signal "enb" is a "0." The interesting thing to note here is that the property FAILs even though the value of mySig is indeed h'a two clocks in the past at time 60. That's because the gating expression enb==0 at the current clock tick (when antecedent "done" is true). Since enb==0, the $past () does *not* evaluate itself. In other words, $past () did not look for mySig two clocks in the

```
property lV(Sig, numClocks, enb, lastV);
  (lastV == $past(Sig, numClocks, enb) );
endproperty

assert property (@(posedge clk) done |-> lV(mySig, 2, enb, 'ha)) else
      $display($stime,,,"FAIL Expected lastVal=%h\n",lastVal);

cover property (@(posedge clk) done |-> lV(mySig, 2, enb, 'ha))
      $display($stime,,,"PASS Expected lastVal=%h\n",lastVal);

always @(posedge clk)
  $display($stime,,,"clk=%b mySig=%h past=%h   enb=%h done=%b", clk, mySig,

$past(mySig, 2, enb), enb, done);
```

```
# run -all
#       10 clk=1 mySig=a past=0  enb=1 done=0
#       20 clk=1 mySig=5 past=0  enb=1 done=0
#       30 clk=1 mySig=5 past=a  enb=1 done=1
#       30 PASS Expected lastVal=a
#
#       40 clk=1 mySig=5 past=5  enb=1 done=0
#       50 clk=1 mySig=a past=5  enb=1 done=1
#       50 FAIL Expected lastVal=a
#
#       60 clk=1 mySig=a past=5  enb=1 done=0
#       70 clk=1 mySig=5 past=5  enb=0 done=0
#       80 clk=1 mySig=5 past=5  enb=0 done=1
#       80 FAIL Expected lastVal=a
#
```

A time 80 :

Even though mySig's $past(2) value (at time 60) is "a", && enb=1, the property fails at 80 when evaluated (with done=1) because enb=0 at the current clock tick and the lastV retains the previously sampled value of "5" and the comparison with "a" fails.

Fig. 7.11 Spast—gating expression—simulation log

past, instead returned its *previously* evaluated value h'5. The property compares this value of h'5 with expected value of h'a and fails.

Without a gating signal, the property will always evaluate whenever the antecedent is true and look for the required expression value N number of clocks in the past.

Useful Note on Debug: Note the use of $past in the $display statement. This way of using an expression in a $display is an excellent way to debug your design. You can always display what happened in the past to debug the current state of design.

Simulation efficiency tip: $past(..., n, ...) is not efficient for big delays "n." It is also recommended that you minimize the number of calls of $past in an assertion and avoid calling it whenever it is not essential.

For example, the following is inefficient:

```
a1: assert property (@(posedge clk) ##1 check |-> $past(c) ==
$past(a) + $past(b));
```

The same assertion can be written as follows, which is more efficient.

```
a2: assert property (@(posedge clk) ##1 check |-> $past(c) ==
$past (a + b);
```

7.8 Application: $past ()

Figure 7.12 is self-explaining. Note that $past is used in consequent in the application at the top of the figure and used as an antecedent in the bottom application.

Application: One more application using $past with consecutive range operator. Please refer to the consecutive repetition operator described in Sect. 8.8.

application

Specification:

If current 'state' is cacheRead, the past state cannot be cacheInv (you can never Read from an invalid line)

```
property rdPtrIncr;
        @(posedge clk) (state == cacheRead) |-> ($past(state) != cacheInv);
endproperty
```

application

Specification:

If pipe stall is asserted and data was ready to be sent the last clock that the current state must be data hold.

```
property dHoldCheck;
        @(posedge clk)  (     (pipeStall)
                              &&
                              ($past(State,2)==dataSend)
                        )
                        |->
                              (State == dataHold);
endproperty
```

Fig. 7.12 $past application

7.8.1 Specification

For a burst cycle, when the "read" signal is high until burst read is complete that during this period, the rd_addr is incremented by one in every clock cycle.

7.8.2 Solution

```
sequence check_rd_adr;
      ((rd_addr == $past (rd_addr+1)) && read) [*0:$] ##1 $fell
      (read);
endsequence

sequence read_cycle;
      ($rose(read) && reset_);
endsequence

property burst_check;
      @(posedge clk) read_cycle |-> check_rd_adr;
endproperty
```

This property reads as follows.

Sequence "check_rd_adr" checks to see that current rd_addr is past rd_addr+1. That is done using the $past sampled value function. We then continue to check this behavior "consecutively" until $fell(read) occurs (which is the end of the read cycle). So, every clock, the entire expression "(rd_addr == $past (rd_addr+1)) && read" is consecutively checked until "read" goes low. Please refer to the consecutive repetition operator described in Sect. 8.8.

Sequence read_cycle is simple enough in that it checks for the start of the read cycle.

And property "burst_check" connects the two sequences to check that the consecutive rd_addr is in increment of 1 until the read cycle ends.

7.9 $past Rescues $fell!

Figure 7.13 shows the difference between $rose/$fell with $past. Recall that $fell (or $rose) samples only the LSB of the expression/signal whose value we are evaluating. In contrast, $past evaluates the entire expression. Hence, for example, if you want to check the value of an entire "dBus," you have to use $past. As shown in the figure, $fell will give an incorrect evaluation of the 32-bit bus "dBus" if in fact you wanted to check how the entire bus evaluated at some number of clocks in the past. The figure explains how you can use $past to solve this problem which $fell could not.

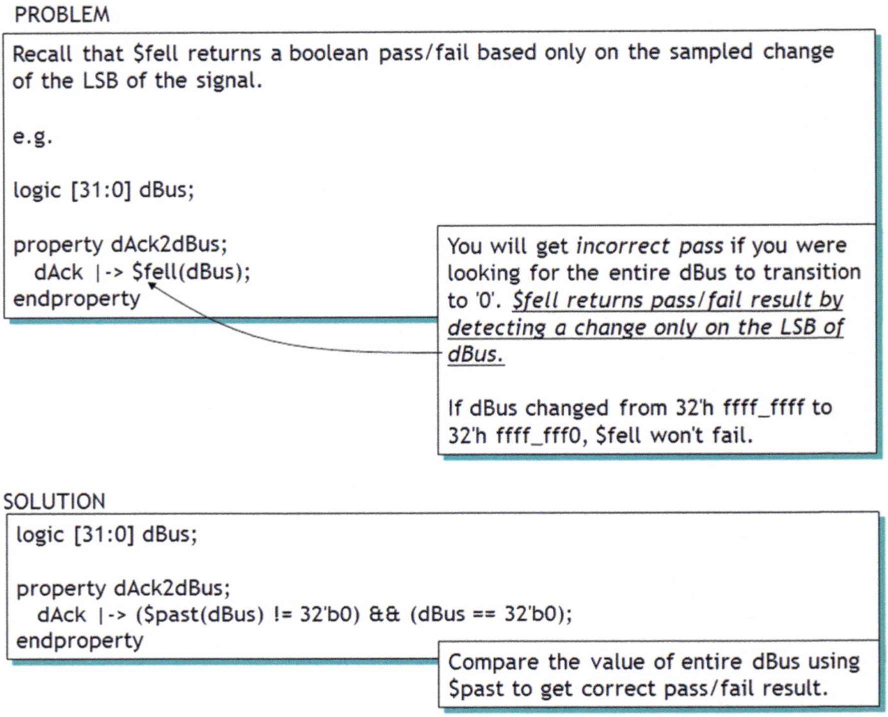

PROBLEM

```
Recall that $fell returns a boolean pass/fail based only on the sampled change
of the LSB of the signal.

e.g.

logic [31:0] dBus;

property dAck2dBus;
  dAck |-> $fell(dBus);
endproperty
```

You will get *incorrect pass* if you were looking for the entire dBus to transition to '0'. *Sfell returns pass/fail result by detecting a change only on the LSB of dBus.*

If dBus changed from 32'h ffff_ffff to 32'h ffff_fff0, $fell won't fail.

SOLUTION

```
logic [31:0] dBus;

property dAck2dBus;
  dAck |-> ($past(dBus) != 32'b0) && (dBus == 32'b0);
endproperty
```

Compare the value of entire dBus using $past to get correct pass/fail result.

Fig. 7.13 $past rescues $fell

7.10 $past() in Procedural Block

$past () sampled value function can also be used in procedural blocks *outside* of a sequence or a property.

Here's an example:

```
always @(posedge clk)
      iACK <= req & $past(gnt);
```

Another example:

```
always @(posedge clk)
      for (int i = 0; i < 4; i ++)
            if (cond[i])
            reg1[i] <= $past(b[i]);
......
```

In this example, $past () returns at each loop iteration the past value of the i-th bit of "b."

Chapter 8
Operators

Introduction: Chapter 8 is the big one! This chapter describes all the operators offered by the language (both for a "sequence" and for a "property") including Clock Delay with and without range, Consecutive repetition with and without range, non-consecutive repetition with and without range, "throughout," "within," "and," "or," "intersect," "first_match," "if...else," etc. Each of the operator description is immediately followed by examples and applications to solidify the concept.

A. B. Mehta, *SystemVerilog Assertions and Functional Coverage*,
https://doi.org/10.1007/978-3-030-24737-9_8

Following lists all the operators offered by the language (IEEE-1800, 2005). We will discuss operators of IEEE-1800- 2009/2012 LRM in a separate chapter (see Chap. 20). We will examine each operator in detail since these operators are the stronghold of the language (Tables 8.1 and 8.2).

8.1 ##m: Clock Delay (Fig. 8.1)

Clock delay is about the most basic of all the operators and probably the one you will use most often! First of all, note that ##m means a delay of "m" number of sampling edges (or clocks). In this example, the sampling edge is a "posedge clk," hence ##m means "m" number of posedge clks.

Table 8.1 Concurrent assertion operators

Operator	Description
##m ##[m:n]	Clock delay
[*m] [*m:n]	Repetition—Consecutive
[=m] [=m:n]	Repetition—Nonconsecutive
[− > m] [− > m:n]	GoTo Repetition—Nonconsecutive
sig1 throughout seq1	Signal sig1 must be true throughout sequence seq1
seq1 within seq2	sequence seq1 must be contained within sequence s2
seq1 intersect seq2	"intersect" of two sequences; same as "and" but both sequences must also "end" at the same time.
seq1 and seq2	"and" of two sequences. Both sequences must start at the same time but may end at different times
seq1 or seq2	"or" of two sequences. It succeeds if either sequence succeeds.
first_match complex_seq1	matches only the first of possibly multiple matches

Table 8.2 Concurrent assertions operators—contd

Operator	Description
not <property_expr>	If <property_expr> evaluates to true, then not <property_expr> evaluates to false; and vice-versa.
if (expression) property_expr1 else property_expr2	If…else within a property
|->	Overlapping implication operator
|=>	Nonoverlapping implication operator
always, s_always, eventually, s_eventually, until, s_until, until_with, s_until_with, nexttime, s_nexttime, followed by #-#, # = #	temporal operators
iff, implies	Equivalence operators
#-#, # = #	Followed by operators

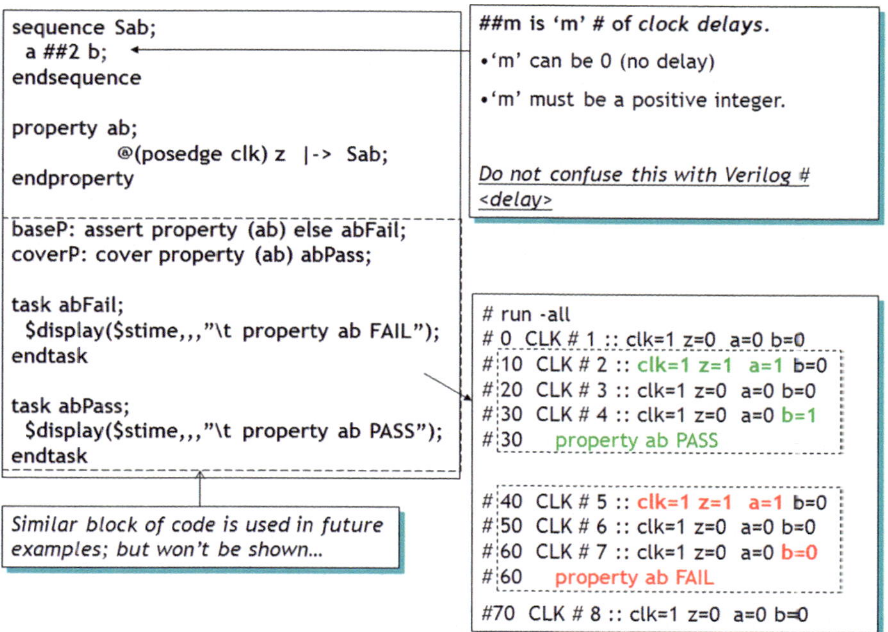

Fig. 8.1 ##m Clock Delay—basics

The property evaluates antecedent "z" to be true at posedge clk and implies the sequence "Sab." "Sab" looks for "a" to be true at that same clock edge (because of the overlapping operator used in the property) and if that is true, waits for two posedge clks and then looks for "b" to be true.

In the simulation log, we see that at time 10, posedge of clk, $z = 1$ and $a = 1$. Hence, the sequence evaluation continues. Two clks later (at time 30), it checks to see if $b = 1$, which it finds to be true and the property passes.

Similar scenario unfolds starting time 40. But this time, b is not equal to 1 at time 60 (two clks after time 40) and the property fails.

We can see that ##m is absolute delay. Can you have "m" to be a variable? Short answer is No. But there is an interesting way to make it variable using a "counter" technique. Please see Sect. 17.7.

Now, let us look at what happens if $m = 0$. That would mean ##m is equal to ##0... hmmm, no delay!

8.2 Clock Delay operator: ##m where $m = 0$ (Fig. 8.2)

We examine the property as before but with $m = 0$. As expected, the sequence "Sab" looks for "a" to be true and then at the same clock looks for "b" to be true. In addition, in this property we are using overlapping implication operator, which means when "z" is true, "a" should be true and so should be "b"—all at the same time. This

```
sequence Sab;
  a ##0 b;
endsequence

property ab;
          @(posedge clk) z  |->  Sab;
endproperty
```

```
# run -all
#        0 CLK # 1 :: clk=1 z=0  a=0 b=0
#       10 CLK # 2 :: clk=1 z=1  a=1 b=1
#       10    property ab PASS

#       20 CLK # 3 :: clk=1 z=0  a=0 b=0
#       30 CLK # 4 :: clk=1 z=1  a=1 b=0
#       30    property ab FAIL

#       40 CLK # 5 :: clk=1 z=0  a=0 b=1
#       50 CLK # 6 :: clk=1 z=0  a=0 b=1
```

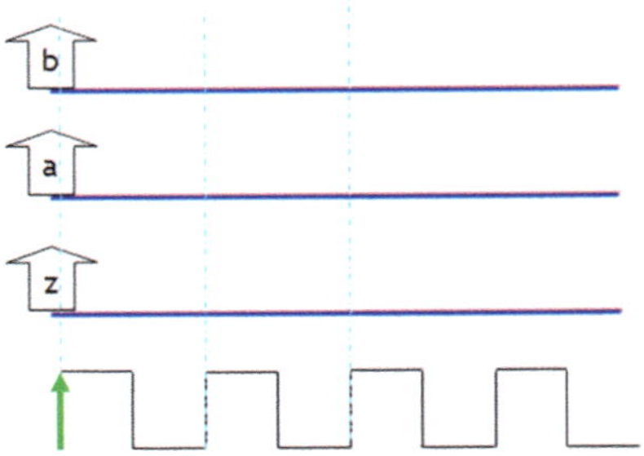

##0 acts as overlapping delay.

In this example, 'a' and 'b' must be '1' at the same edge of clk.

Fig. 8.2 ##m Clock Delay with $m = 0$

is one of the ways you can check for multiple expressions to be true at the same sampling edge (or "clk" edge).

Here's a good application where, in a complex sequence, you can effectively use ##0. *Note that you could have also used "&&" in place of ##0, obviously.*

8.3 Application: Clock Delay Operator:: ##m ($m = 0$) (Fig. 8.3)

8.4 ##[m:n]—Clock Delay Range

Since it is quite necessary for a signal or expression to be true in a given *range* of clocks (as opposed to fix number of clocks), we need an operator that does just the same.

##[m:n] allows a range of sampling edges (clock edges) in which to check for the expression that follows it. Figure 8.4 explains the rules governing the operator. Note that here also, m and n need to be constants. They cannot be variables.

The property "ab" in the figure says that if at the first posedge of clk that "z" is true that sequence "Sab" be triggered. Sequence "Sab" evaluates "a" to be true the

application

Application

##0 can be used as a overlapping delay operator, when within a complex sequence you need to guarantee that two events take place on the same clock.

For example, if tagError is detected that tErrorBit is Set the next clock and mCheck is asserted on the same clock.

```
@(posedge clk) $rose(tagError) |=> $rose(tErrorBit) ##0 $rose(mCheck);
```

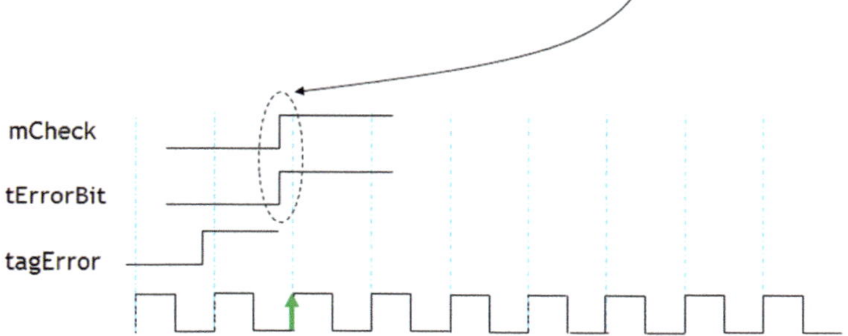

Fig. 8.3 ##0—application

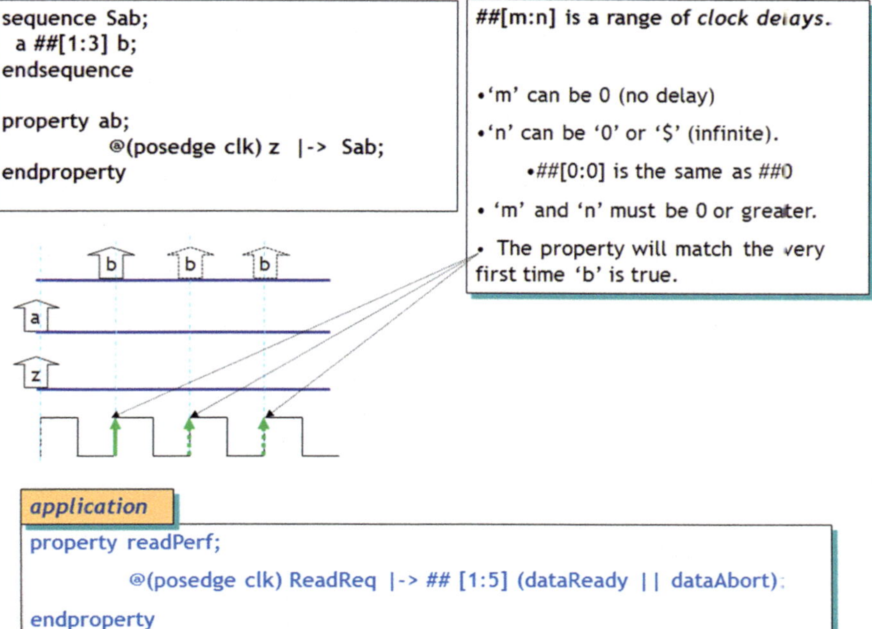

Fig. 8.4 ##[m:n] Clock delay range

same clock that "z" is true and then looks for "b" to be true delayed by either 1 clk or 2 clks or 3 clks. The very—first—instance that "b" is found to be true within the 3 clocks, the property will pass. If "b" is not asserted within 3 clks, the property will fail.

Note that in the figure you see 3 passes. That simply means that whenever "b" is true the first time within 3 clks that the property will pass. It does not mean that the property will be evaluated and pass 3 times. To reiterate, the property will pass as soon as (i.e., the first time) that "b" is true.

8.5 Clock Delay Range Operator: ##[m:n]: Multiple Threads

Back to multiple threads! But this is a very interesting behavior of multi-threaded assertions. This is something you really need to understand.

At s1, "rdy" is high and the antecedent is true. That implies that "rdyAck" be true within the next five clks. s1 thread starts looking for "rdyAck" to be true. The very next clock, rdyAck is not yet true but luck has it that "rdy" is indeed true at this next clk (s2). This will fork off another thread that will also wait for "rdyAck" to be true within the next 5 clks. The "rdyAck" comes along within 5 clks from s1 and that thread is satisfied and will pass.

But the second thread will also pass at the same time, because it also got its "rdyAck" within the 5 clks that it was waiting for.

This is a—very—important point to understand. *The range operator can cause multiple threads to complete at the same time.* This is in contrast to what we saw earlier with ##m constant delay where each thread will always complete only after the fixed ##m clock delays. There is a separate end to each separate thread. With the range delay operator multiple threads can end at the same time.

IMPORTANT: Let us further explore this concept since *it can indeed lead to false positive*. How would you know if rdyAck that satisfies both "rdy"s is for which "rdy"? If you did not receive "rdyAck" for the second "rdy," the property will *not* fail and you will get a false positive.

One hint is to keep the antecedent an edge sensitive function. For example, in the above example, instead of "@ (posedge clk) rdy" we could have used "@ (posedge clk) $rose(rdy)" which would have triggered the antecedent only once and there won't be any confusion of multiple threads ending at the same time. This is a performance hint as well. Use edge sensitive sampled value functions whenever possible. Level sensitive antecedent can fork off unintended multiple threads affecting simulation performance.

But a better solution is to use Local Variables to ID each "rdy" and "rdyAck." This will indeed make sure that you received a "rdyAck" for each "rdy" and also that each "rdyAck" is associated with the correct "rdy."

Now, I haven't explained Local Variables yet! Please refer to the chapter on Local Variables (see Chap. 11) to get familiar with it. That chapter is an in-depth study of Local Variables. But you don't need to understand that entire chapter to understand the following example. Just scan through the initial pages and you will understand that Local Variables are *dynamic* variables with each instance of the

```
property rdyProtocol;
  @(posedge clk) rdy |-> ##[1:5] rdyAck;
endproperty
```

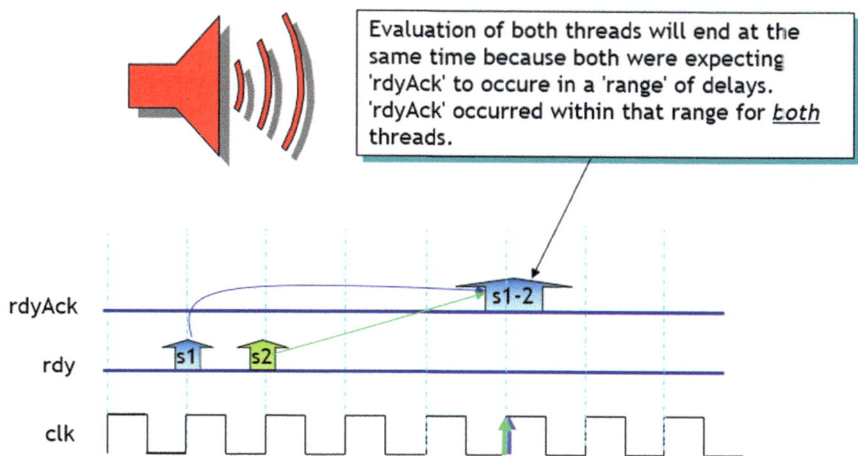

Fig. 8.5 ##[m:n]—multiple threads

sequence forking off an independent thread. Very powerful feature. You don't need to keep track of the pipelined behavior. The local variable does it for you. Having understood that, you should be able to follow the following example.

Problem Statement:

Please refer to Fig. 8.5: ##[m:n]—multiple threads. Two "rdy" signals are asserted on consecutive clocks with a range of clocks during which a "rdyack" must arrive. "rdyack" arrives that satisfies the time range requirements for *both* "rdy"s and the property passes. We have no idea whether a "rdyack" arrived for each "rdy." The PASS of the assertion does not guarantee that. In this example, I am also using the concept of attaching subroutines (subroutine being $display in this example), which is not covered so far but it is quite intuitive and you should be able to understand attachment of Sdisplay statements in the example.

This example simulates the property and shows that both instances of "rdy" will PASS with a single "rdyAck."

```
module range_problem;
logic clk, rdy, rdyAck;

initial
begin
  clk=1'b0; rdy=0; rdyAck=0;
  #500 $finish(2);
end
```

```
always begin
        #10 clk=!clk;
end

initial
begin
        repeat (5) begin @(posedge clk) rdy=~rdy; end
end

initial $monitor($stime,,,"clk=",clk,,,"rdy=",rdy,,,"rdyAck=",
rdyAck);


initial
begin
        repeat (4) begin @(posedge clk); end
        rdyAck=1;
end

sequence rdyAckCheck;
     (1'b1,$display($stime,,,"ENTER SEQUENCE rdy ARRIVES"))
     ##[1:5]
((rdyAck),$display($stime,,,"rdyAck ARRIVES"));
endsequence

gcheck: assert property (@(posedge clk) $rose (rdy) |->
rdyAckCheck)
begin $display($stime,,,"PASS"); end
        else begin $display($stime,,,"FAIL"); end

endmodule
```

```
/* Simulation log
     0  clk=0  rdy=0  rdyAck=0
#    10  clk=1  rdy=1  rdyAck=0
#    20  clk=0  rdy=1  rdyAck=0
#    30  clk=1  rdy=0  rdyAck=0
#    30  ENTER SEQUENCE rdy ARRIVES ← First 'rdy' is detected
#    40  clk=0  rdy=0  rdyAck=0
#    50  clk=1  rdy=1  rdyAck=0
#    60  clk=0  rdy=1  rdyAck=0
#    70  clk=1  rdy=0  rdyAck=1
#    70  ENTER SEQUENCE rdy ARRIVES ← Second 'rdy' is detected
#    80  clk=0  rdy=0  rdyAck=1
```

```
#      90   clk=1   rdy=0   rdyAck=1
#      90   rdyAck ARRIVES ← 'rdyack' detected for both 'rdy's and
            the property PASSes.
#      90   PASS
*/
```

Solution:

```
module range_solution;
logic clk, rdy, rdyAck;
byte rdyNum, rdyAckNum;

initial
begin
  clk=1'b0; rdy=0; rdyNum=0; rdyAck=0; rdyAckNum=0;
  #500 $finish(2);
end

always
begin
    #10 clk=!clk;
end

initial
begin
        repeat (4)
        begin
         @(posedge clk) rdy=0;
         @(posedge clk) rdy=1; rdyNum=rdyNum+1;
        end
end

initial
$monitor($stime,,,"clk=",clk,,,"rdy=",rdy,,,"rdyNum=",rdyNum,,,
"rdyAckNum",rdyAckNum,,,"rdyAck=",rdyAck);

always
begin
        repeat (4)
        begin
           @(posedge clk); @(posedge clk); @(posedge clk);
        rdyAck=1; rdyAckNum=rdyAckNum+1;
        @(posedge clk) rdyAck=0;
        end
end
```

```
sequence rdyAckCheck;
byte localData;
/* local variable 'localData' declaration. Note this is a
dynamic variable. For every entry into the sequence it will
create a new instance of localData and trigger an independent
thread. */

/* Store rdyNum in the local variable 'localData' */
(1'b1,localData=rdyNum,$display ($stime,,,"rdy ARRIVES: ",,,
"LOCAL rdyNum=",localData))

##[1:5]

/* Compare rdyAckNum with the rdyNum stored in 'localData' */
((rdyAck && rdyAckNum==localData),
$display($stime,,,"rdyAck  ARRIVES  ",,,"LOCAL",,,"rdyNum=",
localData,,, "rdyAck=",rdyAckNum));

endsequence

gcheck: assert property (@(posedge clk) (rdy) |-> rdyAckCheck)
begin $display($stime,,,"PASS"); end
        else begin
$display($stime,,,"FAIL",,,"rdyNum=",rdyNum,,,"rdyAckNum=",
rdyAckNum); end

endmodule

/*
#    0  clk=0  rdy=0  rdyNum=   0  rdyAckNum   0  rdyAck=0
#    10  clk=1  rdy=0  rdyNum=   0  rdyAckNum   0  rdyAck=0
#    20  clk=0  rdy=0  rdyNum=   0  rdyAckNum   0  rdyAck=0
#    30  clk=1  rdy=1  rdyNum=   1  rdyAckNum   0  rdyAck=0
#    40  clk=0  rdy=1  rdyNum=   1  rdyAckNum   0  rdyAck=0
#    50  clk=1  rdy=1  rdyNum=   1  rdyAckNum   1  rdyAck=1
#    50  rdy ARRIVES:  LOCAL rdyNum=   1 ← First 'rdy' arrives.
     A 'rdyNum' (generated in your testbench as shown above) is
     assigned to 'localData'. This 'rdyNum' is a unique number
     for each invocation of the sequence and arrival of 'rdy'.
#    60  clk=0  rdy=0  rdyNum=   1  rdyAckNum   1  rdyAck=1
#    70  clk=1  rdy=1  rdyNum=   2  rdyAckNum   1  rdyAck=0
#    70  rdyAck ARRIVES  LOCAL  rdyNum=  1  rdyAck=  1  //First
     'rdyAck' arrives
```

```
#     70  PASS ← When 'rdyAck' arrives, the sequence checks to
      see that its 'rdyAckNum' (again, assigned in the testbench)
      corresponds to the first rdyAck. If the numbers do not match
      the property fails. Here they are indeed the same and the
      property PASSes.

#     80  clk=0  rdy=1  rdyNum=   2  rdyAckNum   1  rdyAck=0
#     90  clk=1  rdy=0  rdyNum=   2  rdyAckNum   1  rdyAck=0
#     90  rdy ARRIVES:  LOCAL rdyNum =   2 ← Second 'rdy' arrives.
      localData is assigned the second
             'rdyNum'. This redNum will not overwrite the first rdy-
      Num. Instead a second thread is forked off and 'localData'
      will maintain (store) the second 'rdyNum'.

#     100  clk=0  rdy=0  rdyNum=   2  rdyAckNum   1  rdyAck=0
#     110  clk=1  rdy=1  rdyNum=   3  rdyAckNum   1  rdyAck=0
#     120  clk=0  rdy=1  rdyNum=   3  rdyAckNum   1  rdyAck=0
#     130  rdy ARRIVES:  LOCAL rdyNum =   3 //Third 'rdy' arrives
#     130  clk=1  rdy=0  rdyNum=   3  rdyAckNum   2  rdyAck=1
#     140  clk=0  rdy=0  rdyNum=   3  rdyAckNum   2  rdyAck=1
#     150  clk=1  rdy=1  rdyNum=   4  rdyAckNum   2  rdyAck=0
```

150 **rdyAck ARRIVES LOCAL rdyNum= 2 rdyAck= 2 //Second "rdyAck" arrives**

150 PASS ← When "rdyAck" arrives, the sequence checks to see that its "rdyAckNum" (again, assigned in the test-bench) corresponds to the second rdyAck. If the numbers do not match the property fails. Here they are indeed the same and the property PASSes. This is what we mean by pipelined behavior, in that, the second invocation of the sequence maintains its own copy of "localData" and compares with the second "rdyAck." This way there is no question of which "rdy" was followed by which "rdyAck." No false positive. Rest of the simulation log follows the same chain of thought.

Can you figure out why the property fails at #270? Hint: Start counting clocks at time #170 when fourth "rdy" arrives. Did a "rdyAck" arrive for that "rdy"?

```
#     160  clk=0  rdy=1  rdyNum=   4  rdyAckNum   2  rdyAck=0
#     170  rdy ARRIVES:  LOCAL rdyNum =   4 //Fourth 'rdy' arrives
#     170  clk=1  rdy=1  rdyNum=   4  rdyAckNum   2  rdyAck=0
#     180  clk=0  rdy=1  rdyNum=   4  rdyAckNum   2  rdyAck=0
#     190  clk=1  rdy=1  rdyNum=   4  rdyAckNum   2  rdyAck=0
#     200  clk=0  rdy=1  rdyNum=   4  rdyAckNum   2  rdyAck=0
#     210  clk=1  rdy=1  rdyNum=   4  rdyAckNum   3  rdyAck=1
#     220  clk=0  rdy=1  rdyNum=   4  rdyAckNum   3  rdyAck=1
#     230  clk=1  rdy=1  rdyNum=   4  rdyAckNum   3  rdyAck=0
```

```
#       230  rdyAck ARRIVES   LOCAL  rdyNum=   3  rdyAck=   3 //Third
        'rdyAck' arrives
#       230  PASS

#       240  clk=0  rdy=1  rdyNum=   4  rdyAckNum   3  rdyAck=0
#       250  clk=1  rdy=1  rdyNum=   4  rdyAckNum   3  rdyAck=0
#       260  clk=0  rdy=1  rdyNum=   4  rdyAckNum   3  rdyAck=0
#       270  clk=1  rdy=1  rdyNum=   4  rdyAckNum   3  rdyAck=0
#       270  FAIL  rdyNum=   4  rdyAckNum=   3  //Why does the asser-
        tion fail here? 'rdyNum' and 'rdyAckNum' are not the same...

#       280  clk=0  rdy=1  rdyNum=   4  rdyAckNum   3  rdyAck=0
#       290  clk=1  rdy=1  rdyNum=   4  rdyAckNum   4  rdyAck=1
*/
```

Sequence “rdyAckCheck” is explained as follows.

Upon entry in the sequence, a copy of localData is created and a rdyNum is stored into it. While the sequence is waiting for #[1:5] for the rdyAck to arrive another “rdy” comes in and sequence “rdyAckCheck” is invoked again. And again, the localData is assigned the next rdyNum and stored. This is where dynamic variable concept comes into picture. The second store of rdyNum into localData does not clobber the first store. A second copy of the localData is created and its thread will also now wait for #[1:5]. This way we make sure that for each “rdy” we will indeed a unique “rdyAck.” Please carefully examine the simulation log to see how this works. I’ve placed comments in the simulation log to explain the operation.

8.6 Clock Delay Range Operator:: ##[m:n] ($m = 0; n = \$$)

In Fig. 8.6 we are going extreme at both ends of the range, from 0 to infinity (“$” means infinite delay). As explained above, the sequence “Sab” will look for “b” to be true the same time as “a” or expect it to be true any time until the simulator ends. It is fine and good if it finds “b” to be true before simulation ends. If not, the simulator will (should) give a Warning that this assertion remains incomplete.

This example is similar to what we saw earlier. But in this example, we expect “tErrorBit” to rise in a certain range of clock delays. Figure 8.7 explains how the assertion works.

Important: Here’s how you need to read this assertion. We are essentially saying that look for $rose(tErrorBit) from now to infinity *but only until $rose(mCheck) arrives*. This is the key to this assertion. We are not waiting for $rose(tErrorBit) forever. We are waiting for it to arrive *only until* $rose(mCheck) arrives. If $rose(mCheck) arrives but $rose(tErrorBit) does not arrive, the assertion fails. Once you understand this concept you will be able to use this seemingly simple operator in powerful ways.

```
sequence Sab;
  a ##[0:$] b;
endsequence

property ab;
          @(posedge clk) z  |-> Sab;
endproperty
```

##[0:$] means (*clock*) delay range from '0' (no delay) to infinite delay.

'a' being true requires that 'b' must be true anytime from the clock edge that 'a' is true until the end of simulation.

The property will match the very first time 'b' is true.

```
# run -all
#        0 CLK # 1 :: clk=1 z=0  a=0 b=0
#       10 CLK # 2 :: clk=1 z=1  a=1 b=1
#       10    property ab PASS
#       20 CLK # 3 :: clk=1 z=1  a=1 b=0
#       30 CLK # 4 :: clk=1 z=0  a=0 b=1
#       30    property ab PASS
#       40 CLK # 5 :: clk=1 z=1  a=1 b=0
#       50 CLK # 6 :: clk=1 z=0  a=0 b=0
#       60 CLK # 7 :: clk=1 z=0  a=0 b=0
#       70 CLK # 8 :: clk=1 z=0  a=0 b=1
#       70    property ab PASS
#       80 CLK # 9 :: clk=1 z=0  a=0 b=1
```

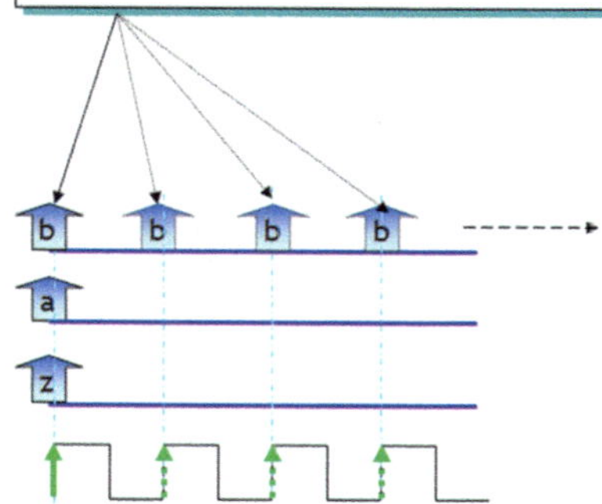


```
sequence Sab;
  a ##[0:$] b;
endsequence

property ab;
          @(posedge clk) z  |-> Sab;
endproperty
```

Simulator may report an Error if 'b' is never found asserted until the end of Simulation.

OR

Simulator may report this as an Incomplete assertion.

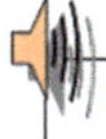

Note that we will discuss 'strong' properties under the chapter on 1800-2009 that determines what happens if we run out of simulation time before the property reaches its end.

Fig. 8.6 ##[m:n] Clock delay range with $m = 0$ and $n = \$$

Note also that you could use "&&" in place of ##0 to achieve the same results. Since assertions are mainly temporal domain, I prefer to tie everything with temporal domain constructs. But that's a matter of preference.

Note also the following two semantically equal statement but with different syntax. These are short forms. I am not a fan of such short forms. They reduce the readability of your code. But that's just me.

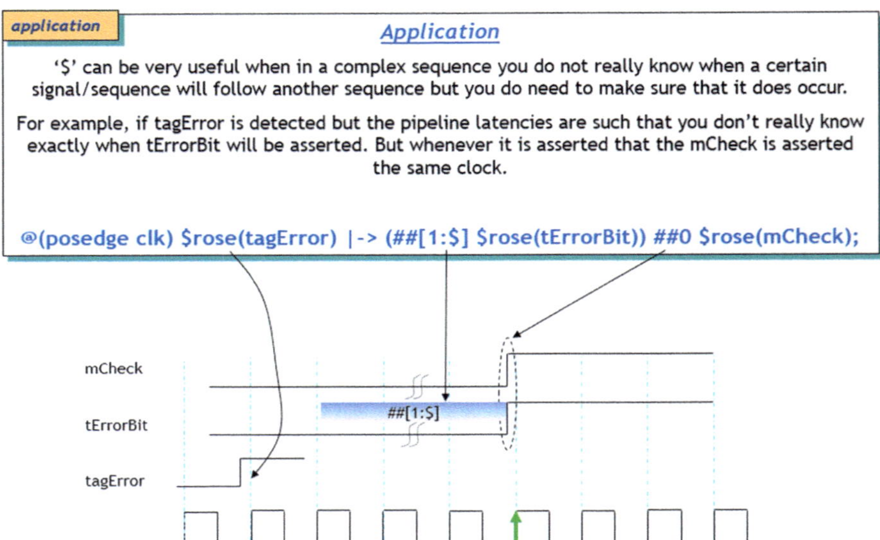

Fig. 8.7 ##[1:$] Delay range application

- `##[*]` is used as an equivalent representation of `##[0:$]`.
- `##[+]` is used as an equivalent representation of `##[1:$]`.

8.7 [*m]—Consecutive Repetition Operator

As depicted in Fig. 8.8 the consecutive repetition operator [*m] sees that the signal/expression associated with the operator stays true for "m" consecutive clocks. Note that "m" *cannot* be $ (infinite # of consecutive repetition).

The important thing to note for this operator is that it will match at the *end* of the last iterative match of the signal or expression.

The example in Fig. 8.8 shows that when "z" is true that at the next clock, sequence "Sc1" should start its evaluation. "Sc1" looks for "a" to be true and then waits for 1 clock before looking for 2 consecutive matches on "b." This is depicted in the simulation log. At time 10 "z" is high; at 20 "a" is high as expected (because of nonoverlapping operator in property); at time 30 and 40, "b" remains high matching the requirement b[*2]. At the end of the second high on "b" the property meets all its requirements and passes.

The very next part of the log shows that the property fails because "b" does not remain high for 2 consecutive clocks. Again, the comparison ends at the *last* clock where the consecutive repetition is supposed to end and then the property fails.

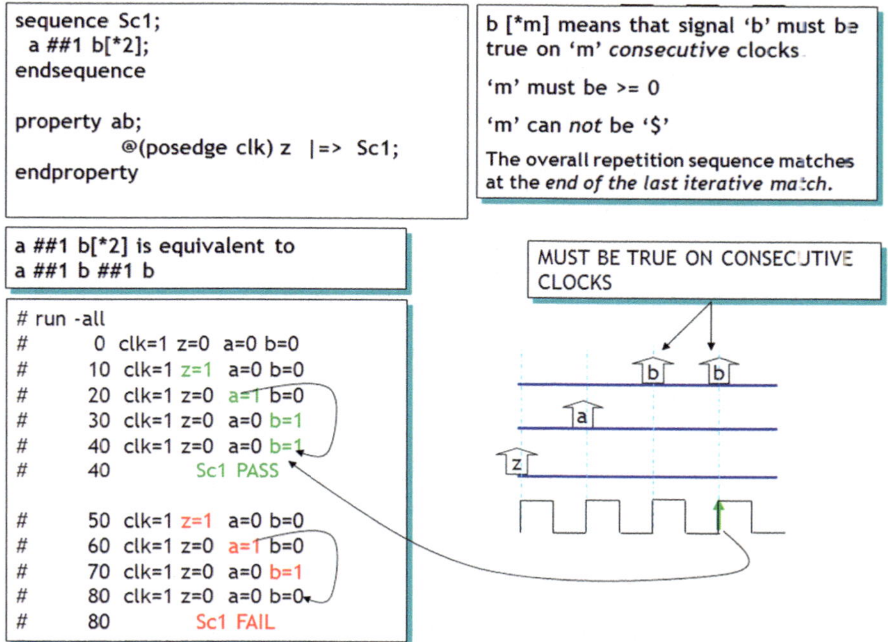

Fig. 8.8 [∗m]—Consecutive repetition operator: basics

Figure 8.9 shows an interesting application where we effectively use the "not" of the repetition operator. At posedge busClk, if ADS is high, then starting the same busClk (overlapping operator), ADS is checked to see if it remains high consecutively for 2 busClk(s). If it does, then we take a "not" of it to declare that the property has failed. This is a very useful property, as simple as it looks. In many protocols one needs to make sure that certain signals follow very strict protocol. This application models just such a protocol. Note also the use of parameterized property.

Interesting note is that if ADS is high for 3 consecutive clocks, the property will fail twice during those 3 clocks. Please see if you can figure out why. Hint, "Sig" is level sensitive.

What if you use [∗m] on the antecedent side? Here's an example that explains that.

```
property cons_on_antecedent;
    @(posedge clk) a[*2] |-> ((##[1:3] c) or (d |=> e));
endproperty
```

Property "cons_on_antecedent" says that if "a" holds for consecutive cycles, then either "c" must hold at some point one to three cycles after the current cycle or, if "d" holds in the current cycle, then "e" must hold one cycle later. Note that here

application

Specification: Verify that the address strobe (ADS) is not asserted for consecutive 2 clocks.

```
property checkConsecutive (clk,Sig,numClk);
   @(posedge clk) disable iff (rst) Sig |-> not (Sig[*numClk]);
endproperty
checkADS: assert property (checkConsecutive(busClk, ADS,2))
       else $display($stime,,," Error: ADS asserted consecutively for 2 Clocks");
```

```
#       5 busClk=1 ADS=1
#      15 busClk=1 ADS=1
#      15           Error: ADS asserted
consecutively for 2 Clocks
#      25 busClk=1 ADS=0
#      35 busClk=1 ADS=1
#      45 busClk=1 ADS=1
#      45           Error: ADS asserted
consecutively for 2 Clocks
#      55 busClk=1 ADS=1
#      55           Error: ADS asserted
consecutively for 2 Clocks
```

Fig. 8.9 [∗m] Consecutive repetition operator—application

a[∗2] means that the "a" should have been consecutively held asserted for 2 clocks *at* the posedge clk.

NOTE: *I also want you to be careful about nested implication operators as shown in this example. It may give inadvertent results. I've discussed this at length in Sect. 17.10 .*

8.8 [∗m:n]—Consecutive Repetition Range Operator

This is another important operator that you need to understand carefully, as benign as it appears to be.

Let us start with the basics. Sig[∗m:n] means that sig should remain true for minimum "m" number of consecutive clocks but no more than maximum "n" number of consecutive clocks. That is simple enough. But here's the first thing that differs from the sig[∗m] operator we just learnt. The consecutive range operator match ends at the *first* match of the sequence that meets the required condition. Note this point carefully. It ends at the *first* match of the range operator (in contrast the non-range operator [*m] which ends at the *last* match of the "m").

Figure 8.10 outlines the fact that b[*2:5] is essentially an OR of four different matches. When any one of these four sequences match that the property is considered to match and pass. In other words, the property first waits looking for two consecutive high on "b." If it finds that sequence, the property ends and passes. If it does *not* find the second "b" to be true, the property will fail. It does *not* wait for the third consecutive high on "b" because there isn't a third consecutive "b" if it wasn't consecutively high in the second clock. The chain was already broken. So, if "b" arrives in the second clock, the property will pass. If "b" was not true in the second clock, the property would fail. It will not wait for the max range. So, what's the use of the upper bound ":5"??

Back to the range b[*2:5]. If you think about it,:5 will *never* get executed!! If "b" is true for 2 consecutive clocks, the property matches and ends (because the property ends at first match). And if "b" wasn't true for 2 consecutive edges, the property will fail. Please study simulation log in Fig. 8.10 carefully.

So, why do we need the max range? When does the maximum range:5 come into picture? What does:5 really mean? See Fig. 8.11, it will explain what max range:5 means and how it gets used.

Note that we added "##1 c" in sequence Sc1. It means that there must be a "c" (high) 1 clock after the consecutive operator match is complete. Ok, simple enough.

Now let's look at the simulation log. Time 30 to 90 is straightforward. At time 30, $z = 1$ and $a = 1$, the next clock "b" = 1 and remains "1" for two consecutive clocks and then 1 clock later $c = 1$ as required and the property passes. But what if "c" was not equal to "1" at time 90? That is what the second set of events show.

$Z = 1$ and $a = 1$ at time 110 and the sequence Sc1 continues. OK. $b = 1$ the next two clocks. Correct. But why doesn't the property end here? Isn't it supposed to end at the first match? Well, the reason the property does not end at 150 is because it needs to wait for $c = 1$ the next clock. So, it waits for $C = 1$ at 170. But it does not see a $c = 1$. Shouldn't the property now fail? NO. This is where the max range: 5 comes into picture. Since there is a range [*2:5], if the property does not see a $c = 1$ after the first two consecutive repetitions of "b," it waits for the next consecutive "b" (total 3 now) and then looks for "$c = 1$." If it does not see $c = 1$ it waits for the next consecutive $b = 1$ (total 4 now) and then looks for $c = 1$. Still no "c"? It finally waits for max range: 5, $b = 1$ and then the next clock looks for $c = 1$. If it finds one, the property ends and passes. If not, the property fails.

Continuing with the simulation log, the last part shows how the property would fail. One way it would fail is what I have described above. The other way is shown in the log file. I have repeated the log file here to help us concentrate only on that part of the log file.

```
sequence Sc1;
  a ##1 b[*2:5];
endsequence

property ab;
          @(posedge clk) z  |-> Sc1;
endproperty
```

b [*m:n] means that signal 'b' must be true on

minimum 'm' *consecutive* clocks and *maximum* 'n' *consecutive* cycles.

'm' must be >= 0;

'n' can be >=0 or $

The overall repetition sequence matches at the first match of the sequence that meets the required condition.

```
a ##1 b[*2:5] is equivalent to

a ##1 b ##1 b                    ||
a ##1 b ##1 b ##1 b              ||
a ##1 b ##1 b ##1 b ##1 b        ||
a ##1 b ##1 b ##1 b ##1 b ##1 b
```

IMPORTANT POINT ::
The 'max' value (:5) in this example has meaning only if there is a qualifying event -*after*- b[*2:5].

as in, a ##1 b[*2:5] ##1 c;

In other words, if there isn't a "##1 c", the sequence will simply wait for the first 2 Consecutive 'b' and it will pass if it found them or fail if it didn't.

It would never wait for the max :5, because this is an OR.

So how does ":5" work???

```
# run -all
#      90  clk=1 z=0  a=0 b=0
#      110 clk=1 z=1  a=1 b=0
#      130 clk=1 z=0  a=0 b=1
#      150 clk=1 z=0  a=0 b=1
#      150   Sc1 PASS
#      170 clk=1 z=0  a=0 b=1
#      190 clk=1 z=0  a=0 b=1
#      210 clk=1 z=0  a=0 b=1
#      230 clk=1 z=0  a=0 b=0
#      250 clk=1 z=1  a=1 b=0
#      270 clk=1 z=0  a=0 b=1
#      290 clk=1 z=0  a=0 b=0
#      290 Sc1 FAIL
```

Fig. 8.10 [*m:n] Consecutive repetition range—basics

```
#     250   clk=1  z=1   a=1  b=0  c=0
#     270   clk=1  z=0   a=0  b=1  c=0
#     290   clk=1  z=0   a=0  b=1  c=1
#     310   clk=1  z=0   a=0  b=0  c=0
#     310   Sc1  FAIL
```

At time 250, $z = 1$ and $a = 1$ so the sequence evaluation continues to consecutive operator. "b" is equal to 1 for the next two consecutive clocks. Good. But at time 310, $b = 0$ *and* –also- $c = 0$. Hence the property fails. After two consecutive "b," there should be either a third "b" or a "$c = 1$." Neither of them is present and the property fails. If $C = 1$ at time 310, the property would pass. If $b = 1$ and $c = 0$ at time 310, the property would continue to evaluate until it sees five consecutive "b" or a $c = 1$ before five consecutive "b" are encountered. Or after five consecutive "b" that there is a $c = 1$ as shown in the previous part of the simulation log file.

```
sequence Sc1;
  a ##1 b[*2:5] ##1 c;
endsequence

property ab;
          @(posedge clk) z  |-> Sc1;
endproperty
```

```
b [*m:n] means that signal 'b' must
be true on

minimum 'm' consecutive clocks and
maximum 'n' consecutive cycles.

'm' must be >= 0;

'n' can be >=0 or $

The overall repetition sequence
matches at the first match of the
sequence that meets the required
condition.
```

```
a ##1 b[*2:5] is equivalent to

a ##1 b ##1 b           ##1 c      ||
a ##1 b ##1 b ##1 b ##1 c          ||
a ##1 b ##1 b ##1 b ##1 b ##1 c ||
a ##1 b ##1 b ##1 b ##1 b ##1 b ##1 c
```

```
# run -all
#     10  clk=1 z=0  a=0 b=0 c=0
#     30  clk=1 z=1  a=1 b=0 c=0
#     50  clk=1 z=0  a=0 b=1 c=0
#     70  clk=1 z=0  a=0 b=1 c=0
#     90  clk=1 z=0  a=0 b=0 c=1
#     90     Sc1 PASS
#    110  clk=1 z=1  a=1 b=0 c=0
#    130  clk=1 z=0  a=0 b=1 c=0
#    150  clk=1 z=0  a=0 b=1 c=0
#    170  clk=1 z=0  a=0 b=1 c=0
#    190  clk=1 z=0  a=0 b=1 c=0
#    210  clk=1 z=0  a=0 b=1 c=0
#    230  clk=1 z=0  a=0 b=0 c=1
#    230     Sc1 PASS
#    250  clk=1 z=1  a=1 b=0 c=0
#    270  clk=1 z=0  a=0 b=1 c=0
#    290  clk=1 z=0  a=0 b=1 c=1
#    310  clk=1 z=0  a=0 b=0 c=0
#    310  Sc1 FAIL
```

```
Requirement: After at least 2
consecutive High on 'b', if 'b' goes
Low, that 'c' must go High the next
clock.

But 'c' is low at #310 and the property
fails.
```

Fig. 8.11 [∗m:n] Consecutive repetition range—example

Important: Here's how you need to read this assertion. Just as in clock delay range, here also (what I call) the qualifying event, namely "##1 C" plays an important role. We are essentially saying that look for "##1 C" after 2 clocks or maximum 5 clocks *but only until "##1 C" arrives*. We are looking for "b[∗2:5] " to occur but only until "C" arrives within those [∗2:5] clocks or at the max after 5 clocks. So, it's the end event aka qualifying event *after* the consecutive range is satisfied (whether after 2 clocks or 3 clocks or 4 clocks or maximum 5 clocks) that the qualifying event ends the assertion. Read this carefully until you understand what's going on.

Once you understand this concept you will be able to use this seemingly simple operator in powerful ways.

Confusing? That could be the case at first. However, please see the next few applications and this concept will be clear. *This is one of the most useful operators in the language* and the better you understand it, the more productive you will be.

Note also following new shortcuts introduced in LRM 2009. I personally prefer explicit declaration and not the short cuts, unless you want to impress your boss!!!

```
[*]is an equivalent representation of [*0:$]
[+]is an equivalent representation of [*1:$]
```

8.9 Application: Consecutive Repetition Range Operator

This application is again on the same line that we have been following. Reason to carry on with the same example is to show how specification can change around seemingly similar logic.

Property in Fig. 8.12 says that at $rose(tagError), check for tErrorBit to remain asserted until mCheck is asserted. If tErrorBit does not remain asserted until mCheck gets asserted, the property should fail.

So, at $rose(tagError) and one clock later we check to see that $rose(tErrorBit) occurs. If it does, then we move forward at the same time (##0) with tErrorBit[*1:$]. This says that we check to see that tErrorBit remains asserted consecutively (i.e., at every posedge clk) *until the qualifying event* $rose(mCheck) arrives. In other words, the qualifying event is what makes consecutive range operator very meaningful as well as useful. *Think of the qualifying event as the one that ends the property*. This way, you can check for some expression to be true until the qualifying event occurs.

A PCI cycle starts when FRAME_ is asserted (goes Low) and the CMD is valid (Fig. 8.13). A CMD == 4'b0001 specifies the start of a PCI Special cycle. On the start of such a cycle (i.e., the antecedent being true), the consequent looks for DEVSEL_ to be high forever consecutively at every posedge clk *until* FRAME_ is de-asserted (goes High). This is by far the easiest way to check for an event / expression to remain true (and we do not know for how long) until another condition/ expression is true (i.e., *until* what I call the qualifying event or end event, is true).

Note also that you can mix edge sensitive and level sensitive expressions in a single logic expression. That is indeed impressive and useful.

Property in Fig. 8.14 states that if the currentState of the state machine is not IDLE and if the currentState remains stable consecutively for 32 clocks that the property should fail.

There are a couple of points to observe.

Note that the entire expression **((currentState**! = **IDLE) && $stable(currentState))** is checked for consecutive repetition of 32 times because we need to check at every clock for 32 clocks that the currentState is not IDLE and that it remains "! = IDLE" for 32 consecutive clocks. ,In other words, you have to make sure that within these 32 clocks, the current State does not go back to IDLE.

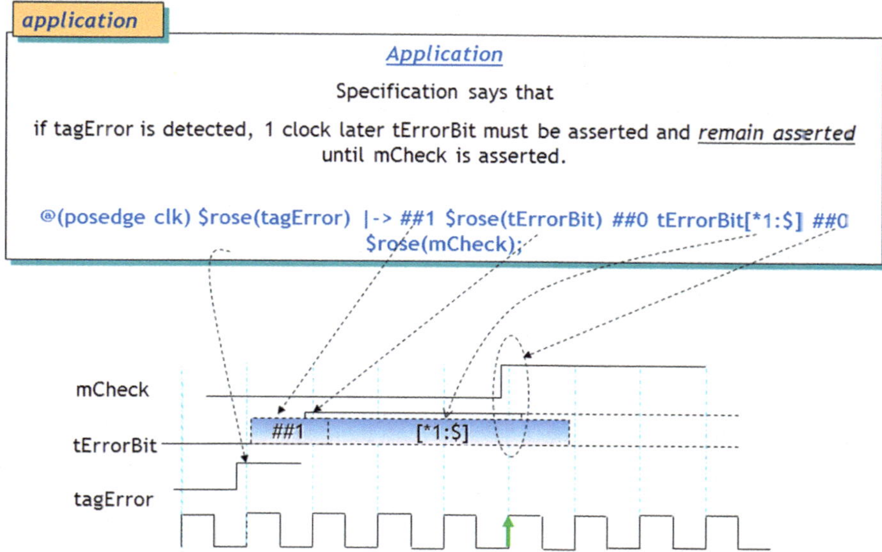

Fig. 8.12 [∗m:n] Consecutive repetition range—application

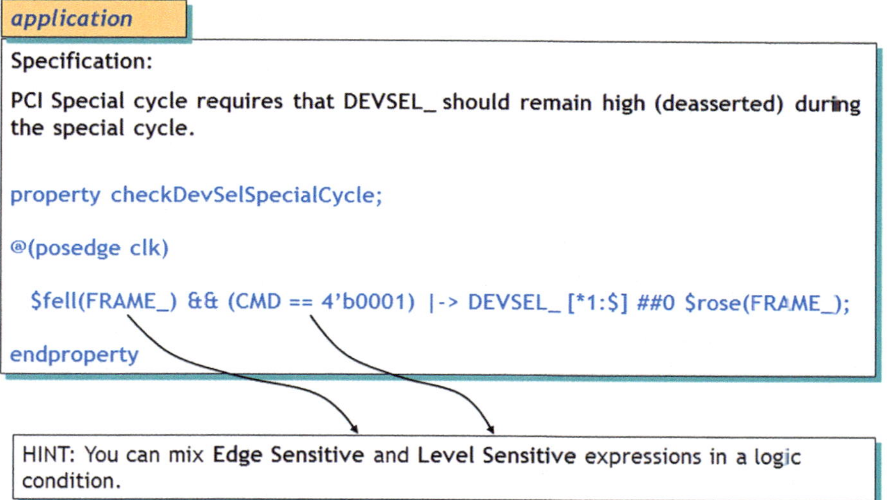

Fig. 8.13 [∗m:n] Consecutive repetition range—application

application

Specification:

Make sure that the state machine does not get stuck in current state except 'IDLE'.

```
property StuckState;
@(posedge clk) disable iff (rst)
        ((currentState != IDLE) && $stable(currentState))[*32] |=> 1'b0;
endproperty
```

HINT: You can simply delcare your consequent as a failure.

Fig. 8.14 [∗m:n] Consecutive repetition range—application

If it does, then the antecedent does not match, and it will start all over again to check for this condition to be true (i.e., the antecedent to be true).

Note that if the antecedent is indeed true, it would mean that the state machine is indeed stuck into the same state for 32 clocks. In that case, we want to assertion to fail. That is taken care of by a *hard failure* in the consequent. We simply program consequent to fail without any pre-requisite.

As you notice, this property is unique in that the condition is checked for in the antecedent. The consequent is simply used to declare a failure.

This application states that the state machine matches the state transition specification (Fig. 8.15). If we are in `readStart state that after 1 clock, the state machine should be in `readID state and stays in that state until the state machine reaches `readData state. It then is expected to stay in `readData state until `readEnd arrives. In short, we have made sure that the state machine does not stray and do an illegal transition until it reaches `readEnd.

Here's complete SystemVerilog code with built-in test-bench and simulation log that exemplifies above property. Code is simple and self-explanatory. I tactfully (!) crafted the test-bench such that the property passes.

Exercise: See if you can tweak the test-bench to make the property fail. That will further solidify your concepts.

Exercise: What happens if `readData arrives while you are waiting for a transition from `readStart to `readID? In such a case, the assertion should fail. Hint: Apply a Boolean condition to `readData. Hmmm, since I am a nice guy, I've provided the solution with the grayed outlines. See that you understand the solution.

Note the use of `define to establish temporal relationship between signals and states. This makes the code very readable. Below I am showing the actual code of the assertion property, the test-bench, and the simulation log.

```
Specification:
The state machine must follow specified state transitions.

`define readStart (read_enb ##1 readStartState)
`define readID (readStartState ##1 readIDState)
`define readData (readIDState ##1 readDataState)
`define readEnd (readDataState ##1 readEndState)

sequence checkReadStates;

  @(posedge clk)
  `readStart          ##1
  `readID      [*1:$] ##1 //`readID && !`readData
  `readData    [*1:$] ##1  //`readData && !`readEnd
  `readEnd     ;

endsequence
```

Fig. 8.15 [∗m:n] Consecutive repetition range—application

```
module state_transition;
int readStartState, readIDState, readDataState, readEndState;
logic clk, read_enb;

`define readStart (read_enb ##1 readStartState)
`define readID (readStartState ##1 readIDState)
`define readData (readIDState ##1 readDataState)
`define readEnd (readDataState ##1 readEndState)

property checkReadStates;
@(posedge clk)
    `readStart    ##1
    `readID     [*1:$]     ##1
    `readData[*1:$] ##1
    `readEnd;
endproperty

sCheck: assert property (checkReadStates) else $display ($stime,,,
"FAIL");
cCheck: cover property (checkReadStates) $display ($stime,,,
"PASS");
```

```
initial
begin
    read_enb=1; clk=0;
    @(posedge clk) readStartState=1;
    @(posedge clk) @(posedge clk); readIDState=1;
    @(posedge clk) @(posedge clk); readDataState=1;
    @(posedge clk) @(posedge clk); readEndState=1;

end

initial $monitor($stime,,,"clk=",clk,
                "read_enb=%0b",read_enb,,,
                "readStartState=%0b",readStartState,,
                "readIDState=%0b",readIDState,,
                "readDataState=%0b",readDataState,,
                "readEndState=%0b",readEndState);

always #10 clk=!clk;

endmodule

/*
#     0 clk=0read_enb=1 readStartState=0 readIDState=0
      readDataState=0 readEndState=0
#     10 clk=1read_enb=1 readStartState=1 readIDState=0
      readDataState=0 readEndState=0
#     20 clk=0read_enb=1 readStartState=1 readIDState=0
      readDataState=0 readEndState=0
#     30 clk=1read_enb=1 readStartState=1 readIDState=0
      readDataState=0 readEndState=0
#     40 clk=0read_enb=1 readStartState=1 readIDState=0
      readDataState=0 readEndState=0
#     50 clk=1read_enb=1 readStartState=1 readIDState=1
      readDataState=0 readEndState=0
#     60 clk=0read_enb=1 readStartState=1 readIDState=1
      readDataState=0 readEndState=0
#     70 clk=1read_enb=1 readStartState=1 readIDState=1
      readDataState=0 readEndState=0
#     80 clk=0read_enb=1 readStartState=1 readIDState=1
      readDataState=0 readEndState=0
#     90 clk=1read_enb=1 readStartState=1 readIDState=1
      readDataState=1 readEndState=0
#     100 clk=0read_enb=1 readStartState=1 readIDState=1
      readDataState=1 readEndState=0
```

```
#      110 clk=1read_enb=1 readStartState=1 readIDState=1
       readDataState=1 readEndState=0
#      120 clk=0read_enb=1 readStartState=1 readIDState=1
       readDataState=1 readEndState=0
#      130 clk=1read_enb=1 readStartState=1 readIDState=1
       readDataState=1 readEndState=1
#      140 clk=0read_enb=1 readStartState=1 readIDState=1
       readDataState=1 readEndState=1
#      150 clk=1read_enb=1 readStartState=1 readIDState=1
       readDataState=1 readEndState=1
#      150 PASS
*/
```

Let us examine one more application as follows (Fig. 8.16).

Simulation log explains the behavior (Fig. 8.17).

One more application using consecutive range operator.

Specification:

For a burst cycle, when the "read" signal is high until burst read is complete that during this period, the rd_addr is incremented by one in every clock cycle.

Solution:

```
sequence check_rd_adr;
    ( (rd_addr == $past (rd_addr+1) ) && read ) [*0:$] ##1
    $fell (read);
endsequence

sequence read_cycle;
    ($rose(read) && reset_);
endsequence

property burst_check;
    @(posedge clk) read_cycle |-> check_rd_adr;
endproperty
```

This property reads as follows.

Sequence "check_rd_adr" checks to see that current rd_addr is past rd_addr+1. That is done using the $past sampled value function. We then continue to check this behavior "consecutively" until $fell(read) occurs (which is the end of the read cycle). So, every clock, the entire expression " (rd_addr == $past (rd_addr+1)) &&read" is consecutively checked until "read" goes low. In other words, we are also making sure that the rd_addr is incremented every clock.

Sequence read_cycle is simple enough in that it checks for the start of the read cycle.

Specification:

- When request is asserted that grant is asserted the very next clock.
  - *grant must have been de-asserted prior to it's assertion.*
- grant must remain asserted as long as request is asserted.
- grant must de-assert the very next clock after request is de-asserted.

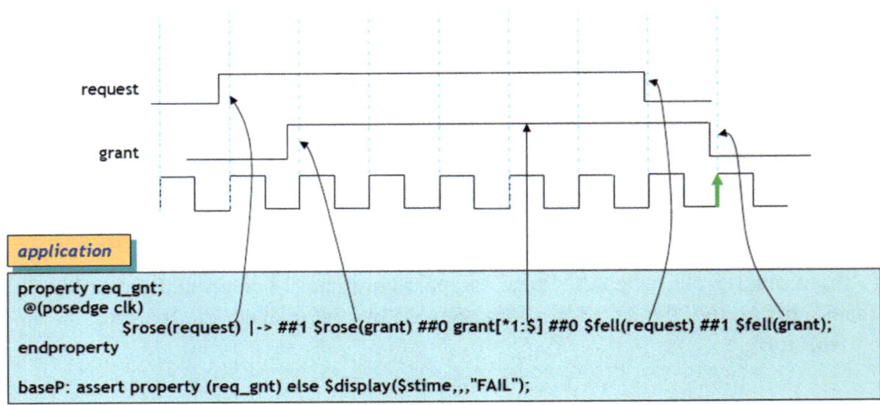

Fig. 8.16 Design application

```
# run -all
#      5  clk=1 request=0 grant=0
#     15  clk=1 request=0  grant=0
#     25  clk=1 request=1  grant=0
#     35  clk=1 request=1 grant=1
#     45  clk=1 request=1 grant=1
#     55  clk=1 request=1 grant=1
#     65  clk=1 request=0  grant=1
#     75  clk=1 request=0 grant=0
#     75  PASS
#     85  clk=1 request=0 grant=0
#     95  clk=1 request=1  grant=0
#    105  clk=1 request=1 grant=1
#    115  clk=1 request=1 grant=1
#    125  clk=1 request=1 grant=1
#    135  clk=1 request=1 grant=1
#    145  clk=1 request=1  grant=0
#    145  FAIL
```

grant falls before request. Hence the property fails.

Fig. 8.17 Design application—simulation log

And property "burst_check" connects the two sequences to check that the consecutive rd_addr is in increment of 1 until the read cycle ends.

8.10 [=m]: Non-consecutive Repetition

Non-consecutive repetition is another useful operator (just as the consecutive operator) and used very frequently. In many applications, we want to check that a signal remains asserted or de-asserted a number of times and that we need *not* know when exactly these transitions take place. For example, (as we will see in Fig. 8.21), if there is a non-burst READ of length 8, that you expect 8 RDACK. These RDACK may come in a consecutive sequence *or not* (based on read latency). But you must have 8 RDACK before read is done.

In Fig. 8.18, property "ab" says that if "a" is sampled high at the posedge of clock that starting next clock, "b" should occur twice not necessarily consecutively. They can occur any time after the assertion of "a." The interesting (and important) thing to note here is that even though the property uses nonoverlapping implication operator (i.e., the first "b" should occur 1 clock after "*a*" = 1), the first "b" can occur *any time after* 1 clock after "a" is found high. Not necessarily exactly 1 clock after "a"!!!

In the simulation log, the first part shows that "b" does occur 1 clock after "a" and then is asserted again a few clocks later. This meets the property requirements and the assertion passes.

But note that second part of the log. "b" does—not—occur 1 clock after "a," rather 2 clocks later. And then it occurs again a few clocks later. Even this behavior is considered to meet the property requirements and the assertion passes.

Exercise: Based on the description above, do you think this property will ever fail? Please experiment and see if you can come up with the answer. It will also further confirm your understanding. Hint: There is no qualifying (i.e., end) event after "b[=2] ."

Continuing with the same analogy, refer to the example below. Here again, just like in the consecutive operator, the qualifying event ("##1 C" in the example below) plays a significant role.

The example in Fig. 8.19 is identical to the previous except for the "##1 C" at the end of the sequence. The behavior of "a |= > b[=2] " is identical to what we have seen above. "##1 c" tells the property that after the last "b," "c" must occur once and that it can occur any time after one clock after the last "b." Note again that even though we have "##1 c," "c" does *not* necessarily need to occur 1 clock after the last "b." It can occur after any # of clks after 1clock after the last "b"—as long as—no other "b" occurs while we are waiting for "c." Confusing! Not really. Let us look at the simulation log in Fig. 8.19. That will clarify things.

In the log, $a = 1$ at time 5; $b = 1$ at time 25 and then at 45. So far so good. We are marching along just as the property expects. Then "c = 1" occurs at time 75. That also meets the property requirement that "c" occurs any time after last b = 1.

```
property ab;
            @(posedge clk) a |=> b [=2];
endproperty
```

```
a |=> b[=2]; is equivalent to
a ##1 ..... b ##1 ...... b
```

Will this property ever FAIL ??

```
# run -all
#        5  clk=1 a=1 b=0
#       15  clk=1 a=0 b=1
#       25  clk=1 a=0 b=0
#       35  clk=1 a=0 b=0
#       45  clk=1 a=0 b=1
#       45            property abc PASS
#       55  clk=1 a=0 b=0

#       65  clk=1 a=1 b=0
#       75  clk=1 a=0 b=0
#       85  clk=1 a=0 b=1
#       95  clk=1 a=0 b=0
#      105  clk=1 a=0 b=1
#      115            property abc PASS
```

b [=m]

means that signal 'b' must be true on 'm' clocks, _not necessarily consecutive_ (i.e. there can be a delay of 1 or more clocks between one match of the operand and the next successive match and no match strictly in between).

m must be >= 0 (cannot be "$")

The overall repetition sequence matches "*at or after the last iterative match*" of the operand

NON-CONSECUTIVE CLOCKS.
Note that the first occurrence of 'b' does not necessarily have to happen exactly 1 clock after 'a'.

b b a

Fig. 8.18 Repetition non-consecutive operator—basics

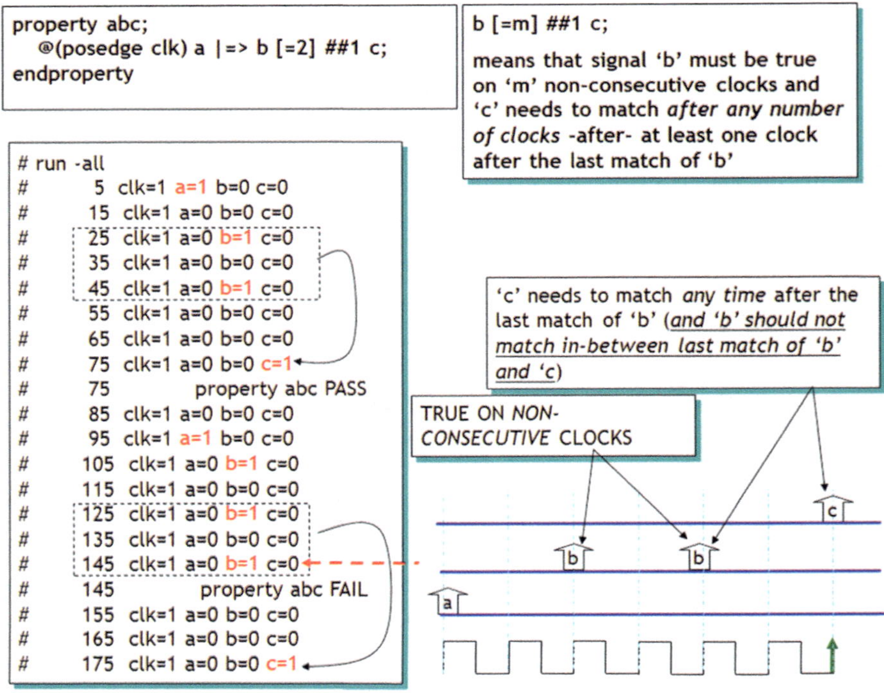

Fig. 8.19 Non-consecutive repetition operator—example

BUT note that *before* $c = 1$ arrived at time 75, "b" did not go to a "1" after its last occurrence at time 45. The property passes. Let us leave this at that for the moment. Now let us look at the second part of the log.

$a = 1$ at time 95; then $b = 1$ at 105 and 125; we are doing great. Now we wait for $c = 1$ to occur any time after last "*b*." $C = 1$ occurs at time 175. But the property fails before that!! What is going on? Note $b = 1$ at time 145. That is not allowed in this property. The property expects a $c = 1$ after the *last* occurrence of "b" but *before* any other $b = 1$ occurs. If another $b = 1$ occurs *before* $c = 1$ (as at time 145), then all bets are off. Property does not wait for the occurrence of $c = 1$ and fails as soon as it sees this extra $b = 1$. In other words, (what I call) the qualifying event "##1 c" encapsulates the property and strictly checks that b[=2] allows only two occurrences of "b" before "c" arrives.

8.11 [=m:n]: Non-consecutive Repetition Range

Property in Fig. 8.20 is analogs to the non-consecutive (non-range) property, except that this has a range. The range says (in the example above) that "b" must occur minimum two times or maximum five times after which "c" can occur one clock later any time and that no more than maximum of 5 occurrences of "b" occur between the last occurrence of $b = 1$ and $c = 1$.

First simulation log (Top left) shows that after $a = 1$ at time 5, b occurs twice (the minimum # of times) at time 15 and 45 and then $c = 1$ at time 75. Why didn't the property wait for five occurrences of $b = 1$? That is because after the second $b = 1$ at time 45, $c = 1$ arrives at time 75 and this $c = 1$ satisfies the property requirement of minimum of two $b = 1$ followed by a $c = 1$. The property passes and does not need to wait for any further $b = 1$. In other words, the property starts looking for the qualifying end event "$c = 1$" after the minimum required (2) "$b == 1$." Since it did find a "$c = 1$" after two "$b = 1$," the property ends there and passes.

Similarly, the simulation log on bottom left shows that "b" occurs 5 (max) times and then "c" occurs without any occurrence of b. The property passes. This is how that works. As explained above, after two "$b = 1$," the property started looking for "$c == 1$." But before the property detects "$c == 1$," it sees another "$b == 1$." That's OK because "b" can occur maximum of five times. So, after the third "$b == 1$," the property continues to look for either "$c == 1$" or "$b == 1$" until it has reached maximum of five "$b == 1$." This entire process continues until five "b"s are encountered. Then the property simply waits for a "c."

So, what happens if you don't get a "$c == 1$" after five "b"s? While waiting for a "c" at this stage, if a sixth "b" occurs, the property fails. This failure behavior is shown in simulation log in the bottom right corner of Fig. 8.20.

```
property abc;
    @(posedge clk) a |=> b [=2:5] ##1 c;
endproperty
```

b [=m:n] ##1 c;

means the property matches over an interval of clocks provided 'a' is true on the first clock tick, 'c' is true on the last clock tick and there are at least '*m*' and at most '*n*' not-necessarily consecutive clocks strictly in-between the first and the last on which 'b' is true (LRM :: SV 3.1a)

m must be >= 0 (cannot be "$")

n must be >= 0 (can be "$")

```
#       5 clk=1 a=1 b=0 c=0
#      15 clk=1 a=0 b=1 c=0
#      25 clk=1 a=0 b=0 c=0
#      35 clk=1 a=0 b=0 c=0
#      45 clk=1 a=0 b=1 c=0
#      55 clk=1 a=0 b=0 c=0
#      65 clk=1 a=0 b=0 c=0
#      75 clk=1 a=0 b=0 c=1
#      75   property abc PASS
```

```
# run -all
#       5 clk=1 a=1 b=0 c=0
#      15 clk=1 a=0 b=0 c=0
#      25 clk=1 a=0 b=1 c=0
#      35 clk=1 a=0 b=1 c=0
#      45 clk=1 a=0 b=0 c=0
#      55 clk=1 a=0 b=1 c=0
#      65 clk=1 a=0 b=0 c=0
#      75 clk=1 a=0 b=1 c=0
#      85 clk=1 a=0 b=0 c=0
#      95 clk=1 a=0 b=1 c=0
#     105 clk=1 a=0 b=0 c=0
#     115 clk=1 a=0 b=0 c=0
#     125 clk=1 a=0 b=0 c=1
#     125   property abc PASS
```

```
# run -all
#       5 clk=1 a=1 b=0 c=0
#      15 clk=1 a=0 b=1 c=0
#      25 clk=1 a=0 b=0 c=0
#      35 clk=1 a=0 b=1 c=0
#      45 clk=1 a=0 b=0 c=0
#      55 clk=1 a=0 b=1 c=0
#      65 clk=1 a=0 b=0 c=0
#      75 clk=1 a=0 b=1 c=0
#      85 clk=1 a=0 b=0 c=0
#      95 clk=1 a=0 b=1 c=0
#     105 clk=1 a=0 b=0 c=0
#     115 clk=1 a=0 b=1 c=0
#     115   property abc FAIL:: # of
posedge 'b' = 6
#     125 clk=1 a=0 b=0 c=0
#     135 clk=1 a=0 b=0 c=0
#     145 clk=1 a=0 b=0 c=1
```

Fig. 8.20 Repetition Non-consecutive range—basics

8.12 Application: Non-consecutive Repetition Operator

Here is a practical example of using non-consecutive operator. The specs are provided in Fig. 8.21.

The property RdAckCheck will wait for nBurstRead to be high at the posedge clk. Once that happens, it will start looking for 8 RdAck before ReadDone comes along. If ReadDone comes in after 8 RdAck (and not 9) the property will pass. If ReadDone comes in before 8 RdAck come in, the property will fail. Note that this will guarantee that the non-burst protocol is completely adhered to.

Following is an interesting example, just for fun... (Fig. 8.22).

The first case is interesting. At time 5 "a" is 1 (antecedent is true) which triggers the consequent. At time 15, $c = 1$ and the property passes. But there was no occur-

application

Specification:

- If nonBurst Read of length 8 is asserted that the RdAck must be asserted 8 times and ReadDone must be asserted anytime after the last Read and that there are no more RdAck's between the last RdAck and ReadDone.

```
property RdAckCheck (int length);

        @(posedge clk) nBurstRead |=> RdAck [=length] ##1 ReadDone;

endproperty

aP: assert property (RdAckCheck (8));
```

Fig. 8.21 Repetition non-consecutive range—application

```
property abc;
        @(posedge clk) a |=> b [=0:$] ##1 c;
endproperty
```

b [=0:$] ##1 c;

means that the signal 'b' should be true either at no time 1 clock (after 'a' is true in this example) or it can be true infinite times until 'c' is asserted.

Will this property ever fail??

```
# run -all
#     5  clk=1 a=1 b=0 c=0
#    15  clk=1 a=0 b=0 c=1
#    15   property abc PASS

#    35  clk=1 a=1 b=0 c=0
#    45  clk=1 a=0 b=0 c=0
#    55  clk=1 a=0 b=1 c=0
#    65  clk=1 a=0 b=0 c=1
#    65   property abc PASS

#    75  clk=1 a=1 b=0 c=0
#    85  clk=1 a=0 b=1 c=1
#    85   property abc PASS
#    95  clk=1 a=0 b=1 c=1
```

Fig. 8.22 Repetition non-consecutive range—[=0:$]

rence of "b." b[=0] is an *empty* sequence which states that "b" should *not* occur. We'll discuss empty sequences later in the book (see Sect. 17.20). The log from time 35 to 65 is quite straightforward. "*a* == 1" at time 35; "*b* == 1" at 55 and "*c* == 1" at time 65. Since the property states b[=0:$] and since "b" did occur once followed by a "c," the property passes.

But let us examine the log from time 75. At time 75, $a = 1$ so the consequent fires. At time 85, both $b = 1$ and $c = 1$ and property passes! How? Note again that b[=0] part of the b[=0:$] range states that "b" may never occur. That property is satisfied at time 75 (i.e., "b" does not occur) and 1 clock later "c" arrives, hence the property passes. Once you learn a bit more about the empty sequences (Sect. 17.20), you will better understand this example. But the point here is that you need to be careful using the minimum and maximum range in an operator. The results may not be that apparent.

8.13 [− > m] Non-consecutive GoTo Repetition Operator

This is the so-called non-consecutive goto operator! Very similar to [=m] non-consecutive operator. Note the symbol difference. The goto operator is [−>2].

In Fig. 8.23, b[−>2] acts exactly the same as b[=2]. So, why bother with another operator with the same behavior? It is the *qualifying event* that makes the difference. Recall that the qualifying event is the one that comes *after* the non-consecutive or

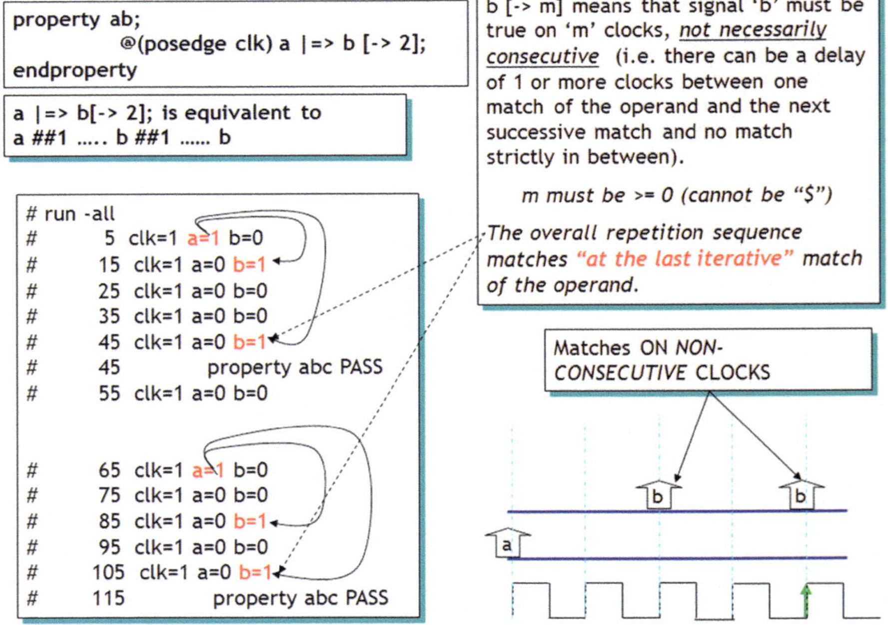

Fig. 8.23 GoTo non-consecutive repetition—basics

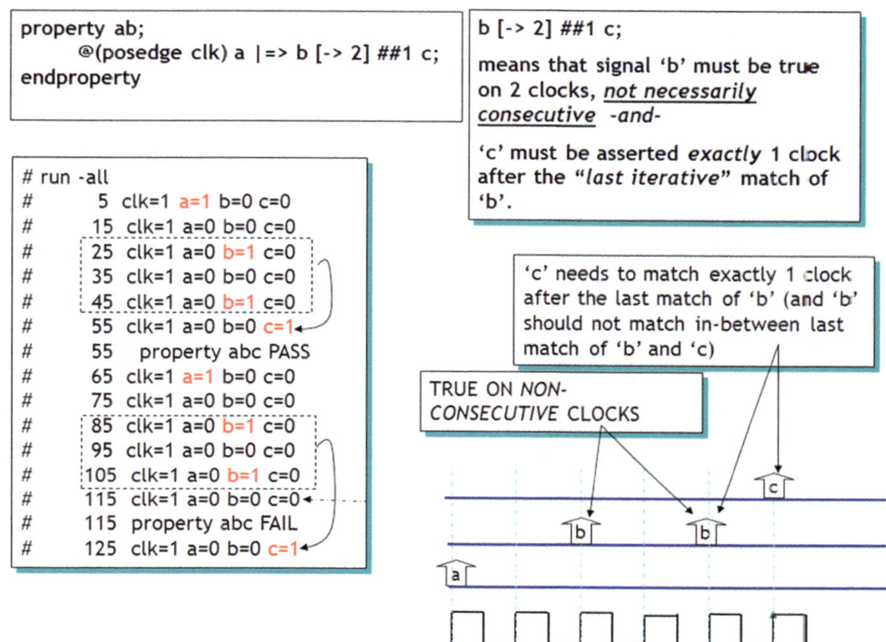

Fig. 8.24 Non-consecutive repetition—example

the "goto" non-consecutive operator. I call it qualifying because it is *the end event* that qualifies the sequence that precedes for final sequence matching.

In Fig. 8.24, we have the so-called qualifying event "##1 c." The property says that on finding $a = 1$ at posedge clk, b must be true 2 times (1 clock after $a = 1$) non-consecutively and "c" must occur *exactly* 1 clock after the last occurrence of "b." In contrast, with "b[=2] ##1 c", "c" could occur *any time* after 1 clock after the last occurrence of "c." That is the difference between [=m] and [–>m].

The simulation log in Fig. 8.24 shows a PASS and a FAIL scenario. PASS scenario is quite clear. At time 5, $a == 1$, then two non-consecutive "b" occur and then *exactly* 1 clock after the last "$b = 1$," "$c = 1$" occurs. Hence, the property passes. The FAIL scenario shows that after two occurrences of $b == 1$, $c == 1$ does *not* arrive *exactly* 1 clock after the last occurrence of $b = 1$. That is the reason the b[–>2] ##1 c check fails.

8.14 Difference Between [=m:n] and [– > m:n]

The simulation log in Fig. 8.25 is quite self-explaining. The left and the right-side properties are identical except that the LHS uses [=2:5] and RHS uses [–>2:5]. The LHS log, i.e., the one for b[=2:5] PASSes while the one for b[–>2:5] fails because

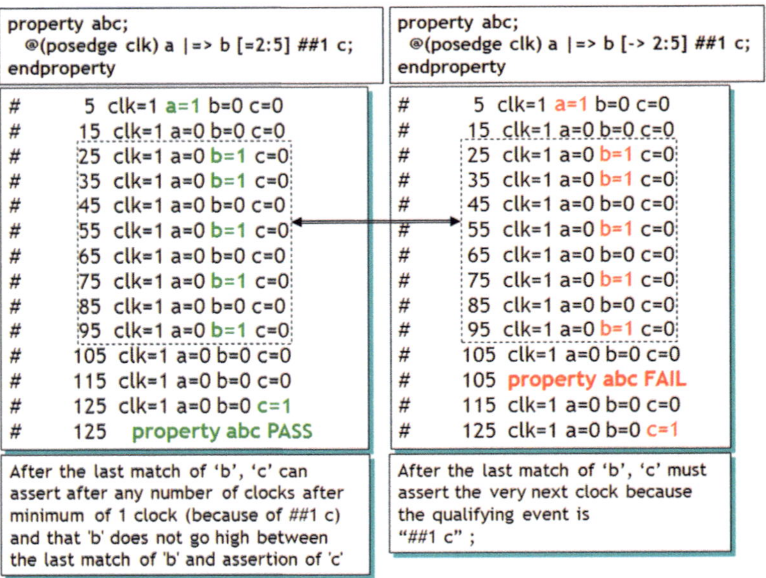

Fig. 8.25 Difference between [=m:n] and [−>m:n]

according to the semantics of "b[−>2:5] ##1 c," "c" must arrive exactly 1 clock after the last occurrence of "b."

Now here is a very important point. Note that "c" is expected to come in 1 clk after the last occurrence of $b = 1$ because of "##1 c." But what if you have "##2 c" in the property?

b[=m] ##2 C: This means that after "m" non-consecutive occurrence of "b," "c" can occur any time *after* 2 clocks. If "$c = 1$" arrives before 2 clocks, the property will fail.

b[− > m]##2 c: This means that after "m" non-consecutive occurrence of "b," "c" must occur *exactly* after 2 clocks. No more no less.

8.15 Application: GoTo repetition: Non-consecutive Operator

The application says that at the rising edge of "req," after 1 clock (because of non-overlapping operator) "ack" must occur once and that it must de-assert (go low) exactly 1 clock after its occurrence. If "ack" is not found de-asserted (low) exactly 1 clock after the assertion of "ack," the property will fail (Fig. 8.26).

application

Specification:

- For every 'req' you must get *at least* 1 'ack' and 'ack' must clear the next clock.

```
property ReqAckCheck;

          @(posedge clk) $rose(req) |=> ack[->1] ##1 !ack;

endproperty

aP: assert property (reqAckCheck);
```

Fig. 8.26 GoTo repetition: non-consecutive operator—application

8.16 sig1 Throughout seq1

The "throughout" operator (Fig. 8.27) makes it that much easier to test for condition (signal or expression) to be true throughout a sequence. Note that the LHS of "throughout" operator can only be a signal or an expression, but it cannot be a sequence (or subsequence). The RHS of "throughout" operator can be a sequence. So, what if you want a sequence on the LHS as well? That is accomplished with the "within" operator, discussed right after "throughout" operator.

Let us examine the application in Fig. 8.28 which will help us understand the throughout operator.

8.17 Application: sig1 Throughout seq1

In Fig. 8.28 the antecedent in property pbrule1 requires bMode (burst Mode) to fall. Once that is true, it first checks to see that (dack_ && oe_ == 0) within 2 clocks and if so, requires checkbMode to execute.

checkbMode makes sure that the bMode stays low "throughout" the data_transfer sequence. Note here that we are, in a sense, making sure that the antecedent remains true through the checkbMode sequence. If bMode goes high *before* data_transfer is over, the assertion will fail. The data_transfer sequence requires both dack_ and oe_ to be asserted (active low) and to remain asserted for 4 consecutive cycles. Throughout the data_transfer, burst mode (bMode) should remain low.

There are two simulation logs presented in Fig. 8.29. Both are for FAIL cases! FAIL cases are more interesting than the PASS cases, in this example! The first simulation log (left hand side) shows $fell(bMode) at time 20. Two clocks later at

‘T throughout Seq1’ matches along a finite interval of consecutive clock ticks provided Seq1 matches along the interval and T evaluates true at each clock tick of the interval

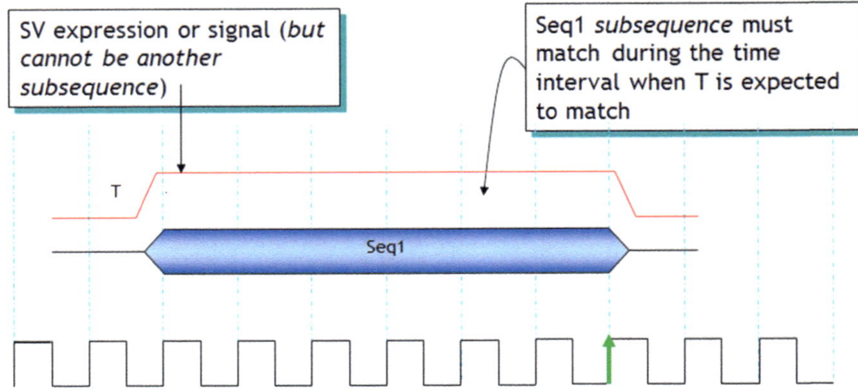

Useful when you want to describe that a logical condition must hold true (or not) throughout a transaction.

Fig. 8.27 sig1 *throughout* seq1

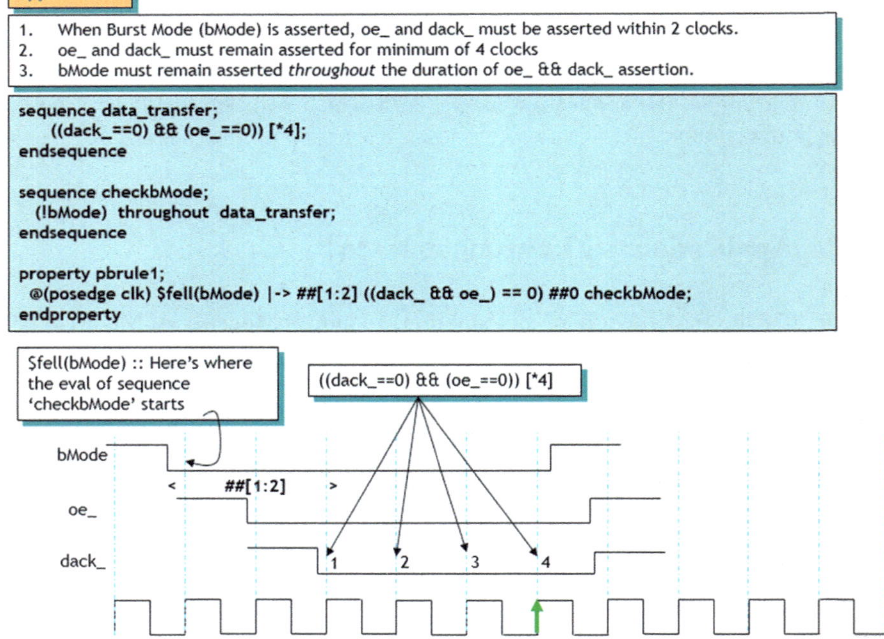

Fig. 8.28 sig1 *throughout* seq1—application

bMode goes High a clock too early.

```
# 0  CLK #1 :: clk=1 bMode=1 oe_=1 dack_=1
#10  CLK #2 :: clk=1 bMode=1 oe_=1 dack_=1
#20  CLK #3 :: clk=1 bMode=0 oe_=1 dack_=1
#30  CLK #4 :: clk=1 bMode=0 oe_=0 dack_=1
#40  CLK #5 :: clk=1 bMode=0 oe_=0 dack_=0
#50  CLK #6 :: clk=1 bMode=0 oe_=0 dack_=0
#60  CLK #7 :: clk=1 bMode=0 oe_=0 dack_=0
#70    property pbrule1 FAIL
#70 CLK #8 :: clk=1 bMode=1 oe_=0 dack_=0
```

(dack_==0 && oe_==0) does not hold for 4 clocks

```
#100  CLK #11 :: clk=1 bMode=1 oe_=1 dack_=1
#110  CLK #12 :: clk=1 bMode=0 oe_=1 dack_=1
#120  CLK #13 :: clk=1 bMode=0 oe_=0 dack_=1
#130  CLK #14 :: clk=1 bMode=0 oe_=0 dack_=0
#140  CLK #15 :: clk=1 bMode=0 oe_=0 dack_=0
#150  CLK #16 :: clk=1 bMode=0 oe_=0 dack_=0
#160    property pbrule1 FAIL
#160  CLK #17 :: clk=1 bMode=0 oe_=0 dack_=1
#170  CLK #18 :: clk=1 bMode=0 oe_=0 dack_=1
#180  CLK #19 :: clk=1 bMode=0 oe_=0 dack_=1
#190  CLK #20 :: clk=1 bMode=1 oe_=1 dack_=1
```

Fig. 8.29 sig1 throughout seq1—application simulation log

time 40, oe_ = 0 and dack_ = 0 are detected. So far so good. oe_ and dack_ retain their state for three clocks That's good too. But in the fourth cycle (time 70). bMode goes high. That's a violation because bMode is supposed to stay low throughout the data-transfer sequence, which is four clocks long.

The second simulation log (Right hand side) also follows the same sequence as above but after 3 consecutive clocks that the oe_ and dack_ remain low, dack_ goes high at time 160. That is a violation because data_transfer (oe_ = 0 and dack_ = 0) is supposed to stay low for four consecutive cycles.

This also points to a couple of other important points

1. Both sides of the *throughout* operator must meet their requirements. In other words, if either the LHS or the RHS of the throughout sequence fails, the assertion will fail. Many folks assume that since bMode is being checked to see that it stays low (in this case), only if bMode fails that the assertion will fail. Not true as we see from the two failure logs.
2. *Important Point*: In order to make it easier for the reader to understand this burst mode application, I broke it down into two distinct subsequences. But what if someone just gave you the timing diagram and asked you to write assertions for it?

Break down any complex assertion requirement into smaller chunks. This is probably the most important advice I can part to the reader. If you look at the entire AC protocol (the timing diagram) as one monolithic sequence, you will indeed make mistakes and spend more time debugging your own assertion then debugging the design under test.

Exercise: How would you model this application using only the consecutive repetition [*m] operator? Please experiment to solidify your concepts of both the *throughout* and the [*m] operators.

Couple more examples.

Specification:

1. If Frame_ is high then Frame_ must be low the very next cycle (1 clock pulse) and remain low until the next strictly subsequent cycle in which IRDY_ is high.

Solution:

```
1. Frame_to_IRDY: assert property (
        Frame_ |=> !Frame_ throughout IRDY_ [-> 1]
   );
```

Specification:

Once an Ethernet Frame starts transmitting, that exactly 16 non-consecutive packets are sent between “frame_start” and “frame_end.” “packet_sent” remains asserted every time a packet is sent.

Solution:

```
Frame_check:
assert property (frame_start) |=>
(!frame_end throughout (packet_sent [->16] ##1 !packet_sent))
##1 frame_end;
```

The property reads as: frame_end must remain de-asserted “throughout” the sequence where non-consecutive 16 packets are sent. Once the packets are sent, the frame_end must be de-asserted 1 clock after the packet send sequence is over. Note that the “throughout” operator in this example applies to the sequence “(packet_sent [− > 16] ##1!packet_sent)”. Both sequences, namely (!frame_end) and (packet_sent [−>16] ##1!packet_sent), are monitored in parallel. If either of them fails, the property fails.

8.18 seq1 Within seq2

Analogous to “throughout,” the “within” operator sees if one sequence is contained within or of the same length as another sequence. Note that the “throughout” operator allowed only a signal or an expression on the LHS of the operator. “within” operator allows a sequence on both the LHS and RHS of the operator.

'Seq1 within Seq2' matches along a finite interval of consecutive clocks ticks provided that Seq2 matches along the interval and Seq1 matches along some sub-interval of consecutive clock ticks.
Note that both Seq1 and Seq2 can be sequences.

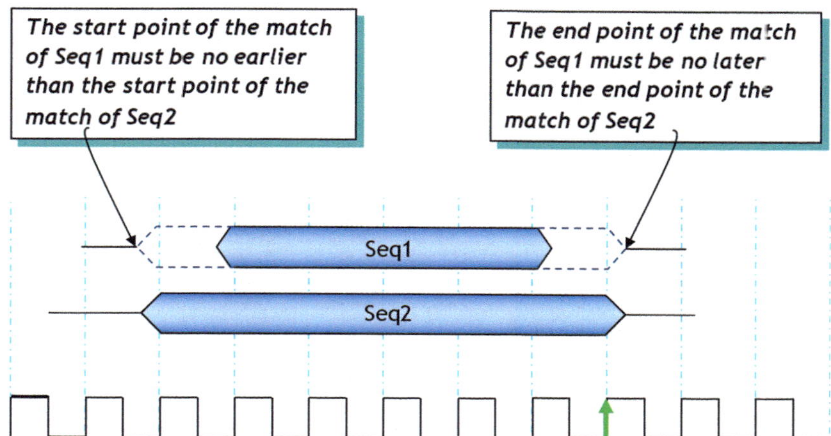

Fig. 8.30 seq1 within seq2

The property "within" ends when the larger of the two sequences end, as shown in the figure above (Fig. 8.30).

Let us understand "within" operator with the application in Fig. 8.31.

8.19 Application: seq1 *within* seq2

In Fig. 8.31, we again tackle the nasty protocol of burst mode! When burst mode is asserted, the master transmits (smtrx) must remain asserted for 9 clocks and the Target Ack (tack) must remain asserted for 7 clocks and that the "tack" sequence occurs within the "smtrx" sequence. This makes sense because from the protocol point of view, the target responds only after the master starts the request and the master completes the transaction after target is done.

In Fig. 8.31, the assertion of bMode ($fell(bMode)) implies that "stack" is valid "within" "smtrx." Now, carefully see the implication property "@ (posedge clk) $fell(bMode) |=>stack **within** smtrx;"

LHS and RHS sequences start executing once the consequence fires. "stack" will evaluate to see if its condition remains true and "smtrx" starts its own evaluation. At the same time, the "within" operator continually makes sure that "stack" is

Specification:

- Assertion of burst Mode (bMode) requires that Master Tx and Target Ack cycles follow the protocol below.
- Master Trx: mtrx must assert the clock after bMode and remains asserted for 9 clocks.
- Target Ack: tack must remain asserted for 7 clocks *within* mtrx transaction.

```
sequence stack;
  $fell(tack) ##0 !tack[*7];
endsequence

sequence smtrx;
  $fell (mtrx) ##0 (!mtrx[*9]) ;
endsequence

property pwin;
  @(posedge clk) $fell(bMode) |=> stack within smtrx;
endproperty
```

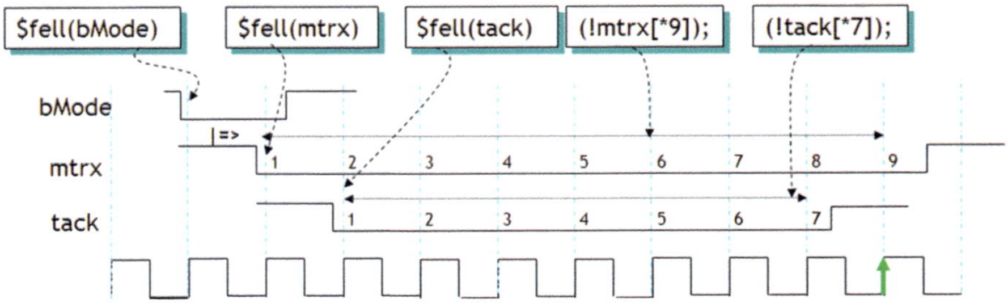

Fig. 8.31 seq1 *within* seq2—application

contained with “smtrx.” The annotations in Fig. 8.31 show how the property handles different parts of the protocol. Simulation logs are presented in Fig. 8.32.

8.19.1 “within” operator PASS Cases

On the left hand side of Fig. 8.32, bMode is “1” at time 0 (not shown) and at time 10, it goes to “0.” That satisfies $fell(bMode). After that the consequent starts execution. *Both “stack” and “smtrx” sequences start executing.* As shown in the left side simulation log, “mtrx” falls and stays low for 9 clocks, as required. “tack” falls after “mtrx” falls, stays low for 7 clocks and goes high the same time when “mtrx” goes high (i.e., both sequences end at the same time). In other words, “tack” is contained within “mtrx.” This satisfies the “within” operator requirement and the property passes. Note that the operator “within” can have either sequences start or end at

```
 10 CLK #2 :: clk=1 bMode=0 mtrx=1 tack=1
 20 CLK #3 :: clk=1 bMode=0 mtrx=0 tack=1
 30 CLK #4 :: clk=1 bMode=0 mtrx=0 tack=1
 40 CLK #5 :: clk=1 bMode=0 mtrx=0 tack=0
 50 CLK #6 :: clk=1 bMode=0 mtrx=0 tack=0
 60 CLK #7 :: clk=1 bMode=0 mtrx=0 tack=0
 70 CLK #8 :: clk=1 bMode=0 mtrx=0 tack=0
 80 CLK #9 :: clk=1 bMode=0 mtrx=0 tack=0
 90 CLK #10 :: clk=1 bMode=0 mtrx=0 tack=0
100 CLK #11 :: clk=1 bMode=0 mtrx=0 tack=0
          110 property pwin PASS
110 CLK #12:: clk=1 bMode=1 mtrx=1 tack=1
```

mtrx and tack deassert at the same. That's OK, as long as tack did remain asserted for required clocks within mtrx.

```
140 CLK #15 :: clk=1 bMode=1 mtrx=1 tack=1
150 CLK #16:: clk=1 bMode=0 mtrx=1 tack=1
160 CLK #17:: clk=1 bMode=0 mtrx=0 tack=0
170 CLK #18 :: clk=1 bMode=0 mtrx=0 tack=0
180 CLK #19 :: clk=1 bMode=0 mtrx=0 tack=0
190 CLK #20 :: clk=1 bMode=0 mtrx=0 tack=0
200 CLK #21 :: clk=1 bMode=0 mtrx=0 tack=0
210 CLK #22 :: clk=1 bMode=0 mtrx=0 tack=0
220 CLK #23 :: clk=1 bMode=0 mtrx=0 tack=0
230 CLK #24 :: clk=1 bMode=0 mtrx=0 tack=1
240 CLK #25 :: clk=1 bMode=0 mtrx=0 tack=1
       250        property pwin PASS
250 CLK #26 :: clk=1 bMode=1 mtrx=1 tack=1
```

mtrx *and tack assert at the same.* That's OK, as long as tack did remain asserted for required clocks within mtrx.

Fig. 8.32 *within* operator: simulation log example—PASS cases

the same time. Similarly, the right-side log shows that both sequences start at the same time and "tack" is contained within "mtrx" and the property passes. Now let us look at fail cases.

8.19.2 "within" operator: FAIL Cases

The simulation logs show different cases of failure. In the top log of Fig. 8.33: *within* operator—simulation log example—FAIL cases, "tack" is indeed contained within "mtrx," but "tack" does not remain asserted for required 7 clocks and the property fails.

In the bottom log, again "tack" is contained within "mtrx" but this time around, "mtrx" is asserted 1 clock too less.

The last log shows both "tack" and "mtrx" asserted for their required clks, but "tack" starts one clk before the falling edge of "mtrx," thus violating the "within" semantics. Sequences on either side of "within" can start at the same time or end at the same time but the sequence that is to be contained within the larger sequence cannot start earlier or end later than the larger sequence.

```
#     0  CLK #1 :: clk=1 bMode=1 mtrx=1 tack=1
#    10  CLK #2 :: clk=1 bMode=1 mtrx=1 tack=1
#    20  CLK #3 :: clk=1 bMode=0 mtrx=1 tack=1
#    30  CLK #4 :: clk=1 bMode=0 mtrx=0 tack=1
#    40  CLK #5 :: clk=1 bMode=0 mtrx=0 tack=0
#    50  CLK #6 :: clk=1 bMode=0 mtrx=0 tack=0
#    60  CLK #7 :: clk=1 bMode=0 mtrx=0 tack=0
#    70  CLK #8 :: clk=1 bMode=0 mtrx=0 tack=0
#    80  CLK #9 :: clk=1 bMode=0 mtrx=0 tack=0
#    90  CLK #10 :: clk=1 bMode=1 mtrx=0 tack=0
#   100  CLK #11 :: clk=1 bMode=1 mtrx=0 tack=1
#   110  CLK #12 :: clk=1 bMode=1 mtrx=0 tack=1
#       property pwin FAIL
```

tack is asserted for 1 clock too less.

```
#   280  CLK #29 :: clk=1 bMode=1 mtrx=1 tack=1
#   290  CLK #30 :: clk=1 bMode=1 mtrx=1 tack=1
#   300  CLK #31 :: clk=1 bMode=0 mtrx=1 tack=1
#   310  CLK #32 :: clk=1 bMode=0 mtrx=0 tack=1
#   320  CLK #33 :: clk=1 bMode=0 mtrx=0 tack=0
#   330  CLK #34 :: clk=1 bMode=0 mtrx=0 tack=0
#   340  CLK #35 :: clk=1 bMode=0 mtrx=0 tack=0
#   350  CLK #36 :: clk=1 bMode=0 mtrx=0 tack=0
#   360  CLK #37 :: clk=1 bMode=0 mtrx=0 tack=0
#   370  CLK #38 :: clk=1 bMode=1 mtrx=0 tack=0
#   380  CLK #39 :: clk=1 bMode=1 mtrx=0 tack=0
#   390  CLK #40 :: clk=1 bMode=1 mtrx=1 tack=1
#       property pwin FAIL
```

mtrx is asserted 1 clock too less ...

tack is asserted for 7 clocks but started a clock too early, so was asserted 1 clock earlier 'within' the mtrx sequence

```
#   140  CLK #15 :: clk=1 bMode=1 mtrx=1 tack=1
#   150  CLK #16 :: clk=1 bMode=1 mtrx=1 tack=1
#   160  CLK #17 :: clk=1 bMode=0 mtrx=1 tack=0
#   170  CLK #18 :: clk=1 bMode=0 mtrx=0 tack=0
#   180  CLK #19 :: clk=1 bMode=0 mtrx=0 tack=0
#   190  CLK #20 :: clk=1 bMode=0 mtrx=0 tack=0
#   200  CLK #21 :: clk=1 bMode=0 mtrx=0 tack=0
#   210  CLK #22 :: clk=1 bMode=0 mtrx=0 tack=0
#   220  CLK #23 :: clk=1 bMode=0 mtrx=0 tack=0
#   230  CLK #24 :: clk=1 bMode=1 mtrx=0 tack=1
#   240  CLK #25 :: clk=1 bMode=1 mtrx=0 tack=1
#   250  CLK #26 :: clk=1 bMode=1 mtrx=0 tack=1
     property pwin FAIL
```

Fig. 8.33 *within* operator: simulation log example—FAIL cases

Another **important point** to note from these simulation logs is that the *property ends when the larger of the two sequences end.* In our case, the property does not end as soon as there is a violation on "stack." It waits for "smtrx" to end to make a judgment call on pass/fail of the property "pwin."

8.20 seq1 *and* seq2

As the name suggests, "and" operator expects both the LHS and RHS side of the operator "and" to evaluate to true. It does not matter which sequence ends first as long as both sequences meet their requirements. The property ends when the longer of the two sequences ends. *But note that both the sequences must start at the same time* (Fig. 8.34).

The "and" operator is very useful, when you want to make sure that certain concurrent operations in your design, start at the same time and that they both complete/match satisfactorily. As an example, in the processor world, when a Read is issued to L2 cache, L2 will start a tag match and issue a DRAM Read as well, both at the same time, in anticipation that the tag may not match. If there is a match, it will abort the DRAM Read. So, one sequence is to start tag compare while other is to start a DRAM Read (ending in DRAM Read Complete or Abort). The DRAM Read sequence is designed such that it will abort as soon as there is a tag match. This way we have made sure that both sequences start at the same time and that both end. Let

'Seq1 *and* Seq2' match if

- **Both sequences start at the same time**
- **Both sequences match**
- **The end time of each sequence can be different.**

The end time is the end time of either Seq1 or Seq2, *whichever matches last*.

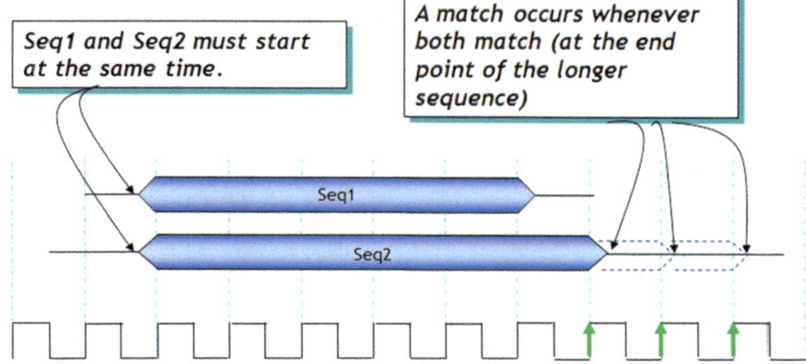

Fig. 8.34 seq1 *and* seq2—basics

us look at the following cases to clearly understand "and" semantics. The figures are self-explaining with annotation within the figures.

8.21 Application: "and" Operator (Figs. 8.35 and 8.36)

In Fig. 8.37, we "and" two expressions in a property. In other words, as noted before, an "and" operator allows a signal, an expression, or a sequence on both the LHS and RHS of the operator. The simulation log is annotated with pass/fail indication.

8.22 seq1 "*or*" seq2

"or" of two sequences means that when either of the two sequences match its requirements that the property will pass. Please refer to Fig. 8.38 and examples that follow to get a better understanding.

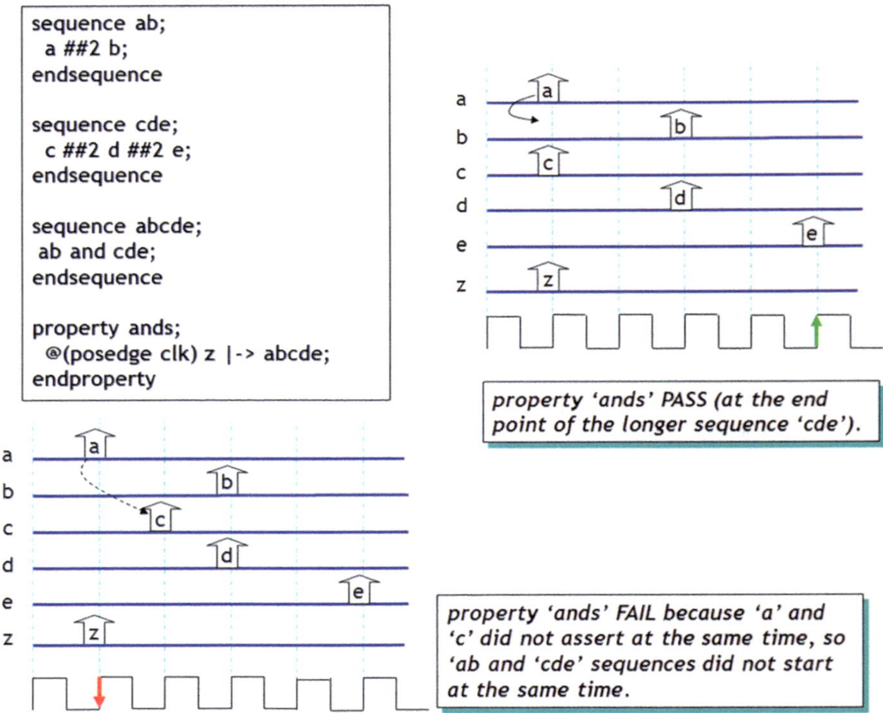

Fig. 8.35 *and* operator—application

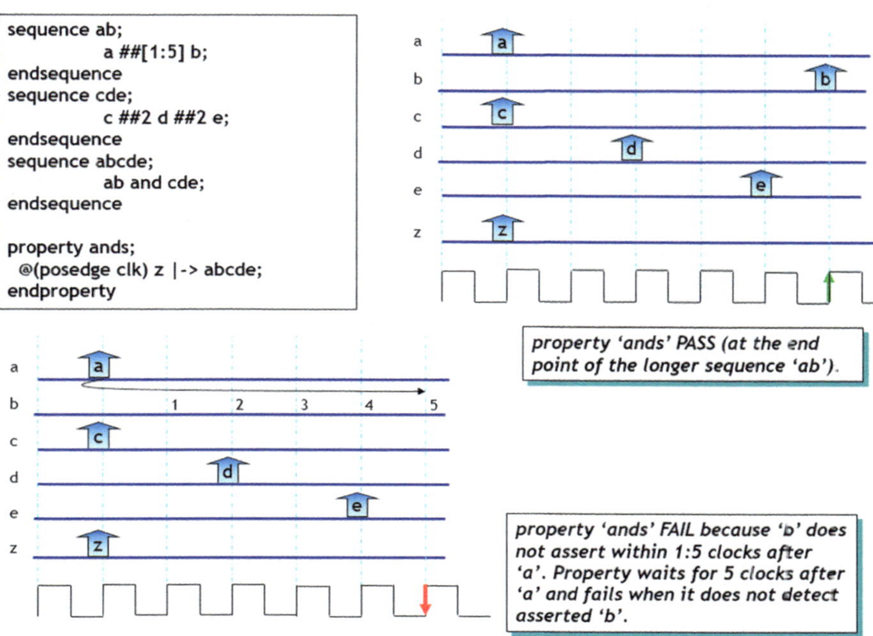

Fig. 8.36 *and* operator—application -II

```
property ands;
  @(posedge clk) z |-> (a==b) and (c==d);
endproperty
```

```
# run -all
#        5  CLK # 1 :: clk=1 z=0 a=0 b=0 c=0 d=0
#       15  CLK # 2 :: clk=1 z=0 a=0 b=0 c=0 d=0
#       25  CLK # 3 :: clk=1 z=1 a=1 b=1 c=0 d=0
#       25    property ands PASS

#       35  CLK # 4 :: clk=1 z=0 a=0 b=0 c=1 d=1
#       45  CLK # 5 :: clk=1 z=1 a=1 b=1 c=1 d=1
#       45    property ands PASS

#       55  CLK # 6 :: clk=1 z=0 a=0 b=1 c=0 d=1
#       65  CLK # 7 :: clk=1 z=1 a=1 b=0 c=1 d=1
#       65    property ands FAIL

#       75  CLK # 8 :: clk=1 z=1 a=1 b=1 c=0 d=1
#       75    property ands FAIL
```

Note that you can do an 'and' of sequences or expressions or a combination of the two.

Fig. 8.37 *and* of expressions

'Seq1 *or* Seq2' match if

- **operand 'or' is used when at least one of the two operand sequences is expected to match.**

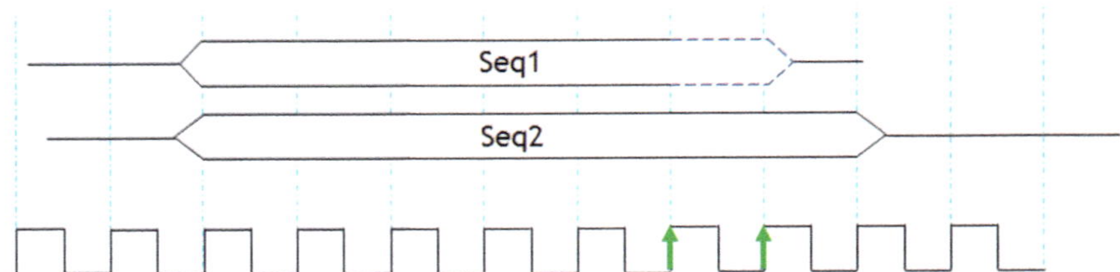

Fig. 8.38 seq1 or seq2—basics

The feature to note with "or" is that as soon as either of the LHS or RHS sequence meets its requirements that the property will end. This is in contrast to "and" where only after the longest sequence ends that the property is evaluated.

Note also that if the shorter of the two sides fails, the sequence will continue to look for a match on the longer sequence. Following examples make this clear.

Note that "seq1" and "seq2" need not start at the same time.

8.23 Application: or Operator

A simple property is presented in Fig. 8.39. Different cases of passing of the property are shown. On the top right of the figure, both "ab" and "cde" sequences start at the same time. Since this is an "or," as soon as "ab" completes, the property completes and passes. In other words, the property does not wait for "cde" to complete anymore.

On the bottom left corner, we see that "ab" sequence fails. However, since this is an "or" the property continues to look for "cde" to be true. Well, "cde" does turn out to be true and the property passes (Figs. 8.40, 8.41, and 8.42).

8.24 seq1 "intersect" seq2

So, with "throughout," "within," "and," and "or" operators who needs another operator that also seem to verify that sequences match?

"throughout" or "within" or "and" or "or" does *not* make sure that both the LHS and RHS sequences of the operator are of exactly the same length. They can be of the same length, but the operators do not care as long as the signal/expression or sequence meets their requirements. That's where "intersection" comes into picture

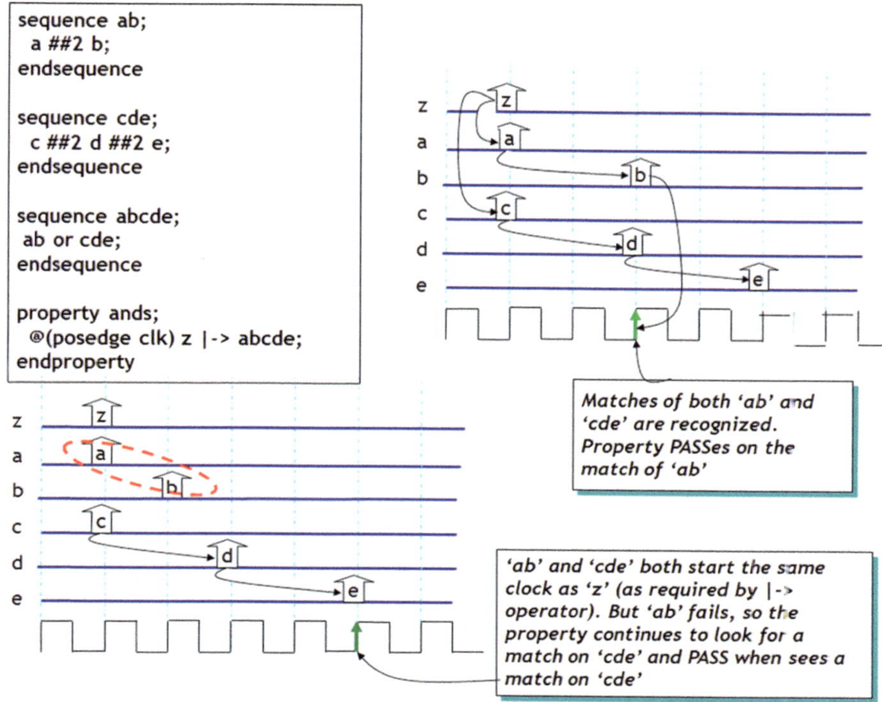

Fig. 8.39 *or* operator—application

(Fig. 8.43). It makes sure that the two sequences indeed *start at the same time and end at the same time* and satisfy their requirements. In other words, they intersect. Essentially, "intersect" is an "and" with length restriction.

As you can see, the difference between "and" and "intersect" is that "intersect" requires both sequences to be of the same length and that they both start at the same time and end at the same time, while "and" can have the two sequences of different lengths. I have shown that difference with timing diagrams further down the chapter. But first, some simple examples to understand "intersect" better.

8.25 Application: "intersect" Operator

Figure 8.44 shows two cased of failure with the "intersect" operator.

Property "isect" says that if "z" is sampled true at the posedge clk that sequence "abcde" should be executed and hold true. I have broken down the required sequence into two subsequences. Sequence "ab" requires "a" to be true at posedge clk and then "b" be true any time from 1 to 5 clks. Sequence "cde" is a fixed temporal domain sequence which requires c to be true at posedge clk, then d to be true 2 clocks later and "e" to be true 2 clocks after "d."

```
sequence ab;
 a ##2 b;
endsequence

sequence cde;
 c ##2 d ##2 e;
endsequence

sequence abcde;
 ab or cde;
endsequence

property ands;
 @(posedge clk) z |-> abcde;
endproperty
```

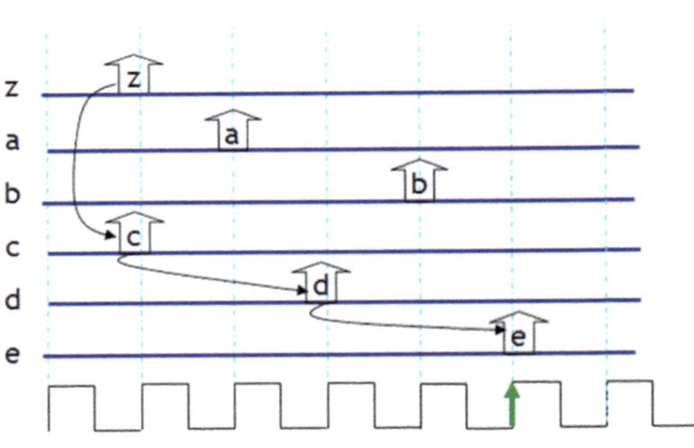

Here 'a' is asserted 1 clock later and 'ab' does satisfy it's requirement, but 'a' was not asserted the same time as 'z' (as required by overlap implication). however, 'c' was indeed asserted when 'z' was asserted, the property is looking for 'cde' to match. Since 'cde' does match, the property passes at the end of 'cde' (and not at the end of 'ab').

```
sequence ab;
 a ##2 b;
endsequence

sequence cde;
 c ##2 d ##2 e;
endsequence

sequence abcde;
 ab or cde;
endsequence

property ands;
 @(posedge clk) z |-> abcde;
endproperty
```

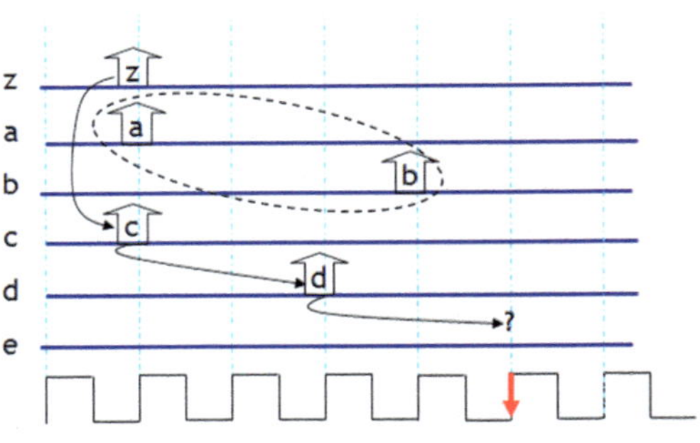

Here, 'ab' does not match; but property keeps looking to see if 'cde' matches. But when 'e' does not follow 2 clocks after d, 'cde' also fails and the property FAILs at that time.

Fig. 8.40 or operator—application II

Top Right timing diagram in Fig. 8.44 shows that both "ab" and "cde" meet their requirements but the property fails because they both do not end at the same time (even though they start at the same time). Similarly, the bottom left timing diagram shows that both "cde" and "ab" meet their requirements but don't end at the same time and hence the assertion fails.

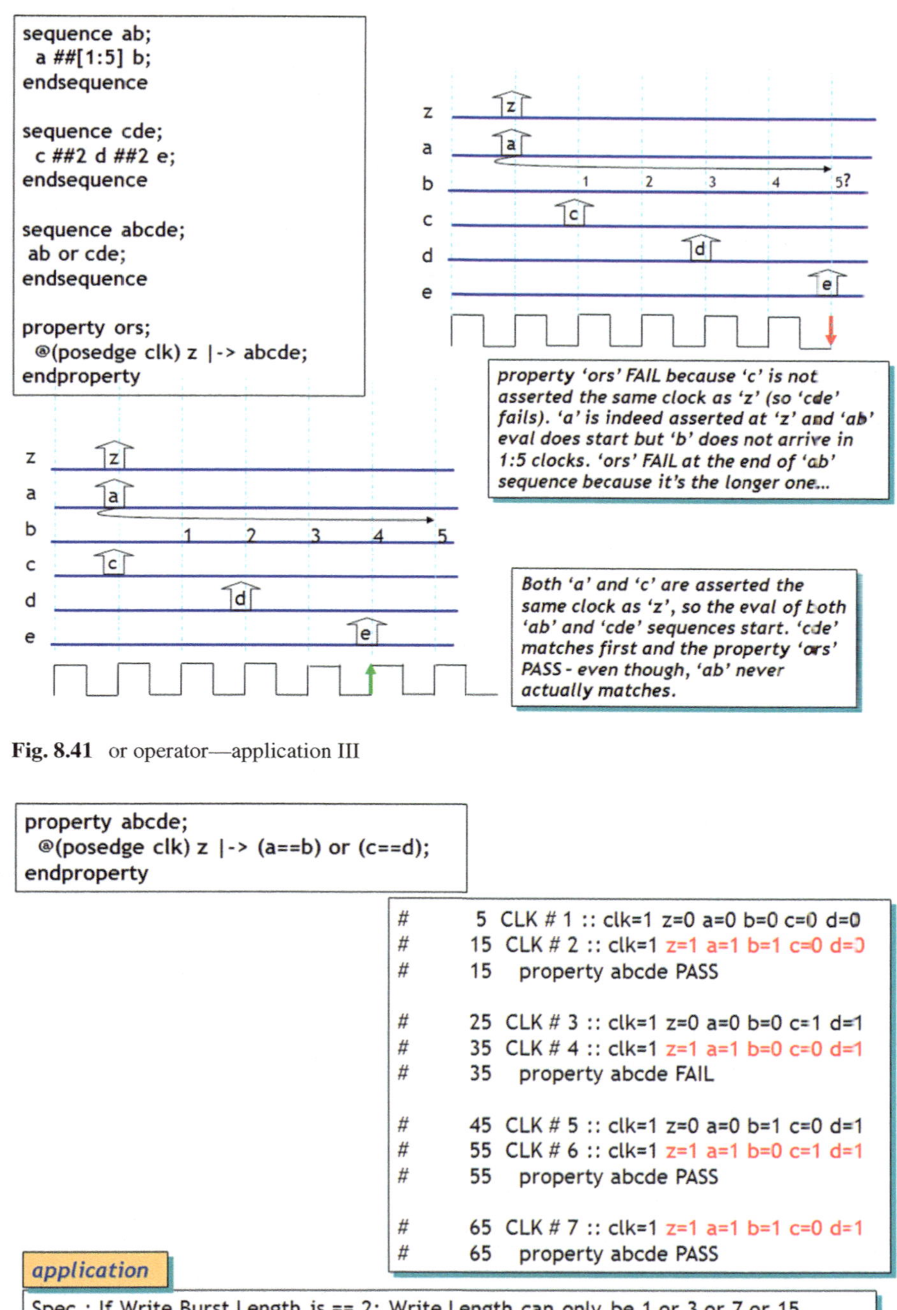

Fig. 8.41 or operator—application III

application

Spec : If Write Burst Length is == 2; Write Length can only be 1 or 3 or 7 or 15

```
property BurstLengthRestrict;
   @(posedge clk) disable iff (!rst)
     ( (bLength==2) |->
           (rwlen==1) or (rwlen==3) or (rwlen==7) or (rwlen==15)) ;
endproperty
aP: assert property(BurstLengthRestrict);
```

Fig. 8.42 *or* of expressions

‘Seq1 *intersect* Seq2’ match if

- Both sequences start at the same time
- Both sequences must match
- *The lengths of the two matches of the operand sequences must be the same.*

The end time is when both sequences match and end at the same time.

The main difference between ‘and and ‘intersect’ is the requirement on the length of the two sequences. For ‘and’ each sequence can be of any length. For ‘intersect’ they must be of the same length.

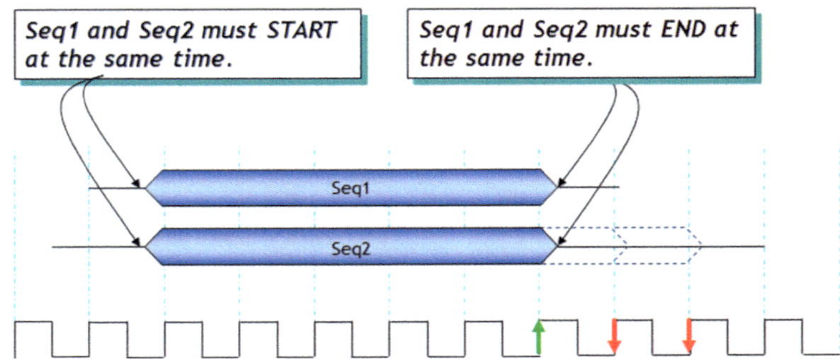

Fig. 8.43 seq1 intersect seq2

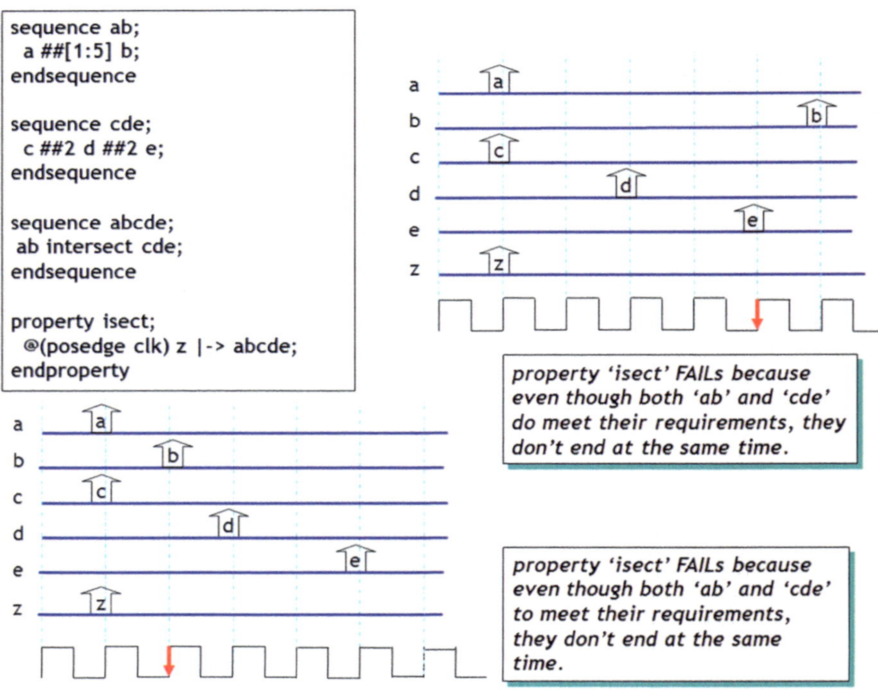

Fig. 8.44 seq1 “intersect” seq2—application

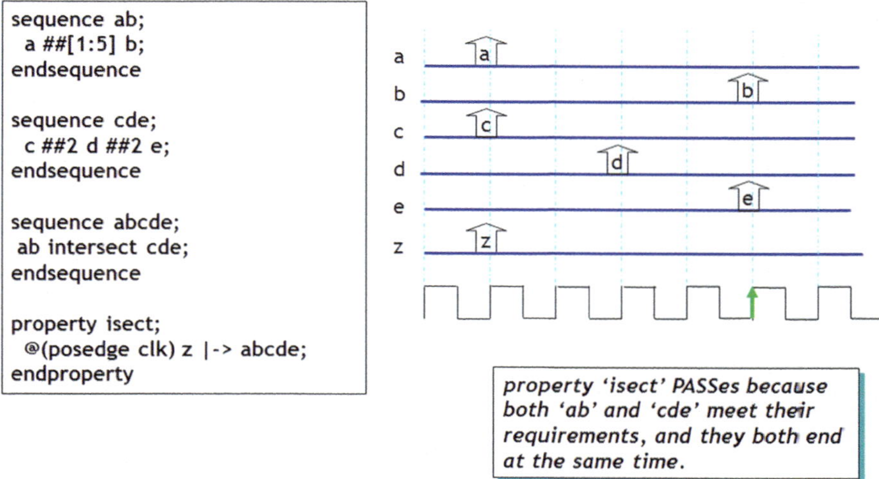

Fig. 8.45 seq1 intersect seq2—application II

In Fig. 8.45 we show a PASS case of the same property (repeated here for the sake of convenience). Both "ab" and "cde" meet their requirements and end at the same time. Hence the assertion passes.

Now let's look at the example in Fig. 8.46. This one does not use a range operator in sequence "ab" (as in the above example). It is obvious that without the range operator that the two sequences with fixed lengths and ending at different times, the intersect property will never pass. *Hence, it makes sense to use subsequences with a range while using an "intersect" operator.*

8.26 Application: *intersect* Operator *(Interesting Application)* (Fig. 8.47)

OK, I admit this property could have been written simply as.

```
@ (posedge clk) $rose(Retry) |-> ##[1:4] $rose(dataRead);
```

So, why are we making it complicated? I just want to highlight an interesting way to use "intersect."

When $rose(Retry) is true, the consequent starts execution. The consequent uses "intersect" between `true[*1:4] and (Retry ##[1:$] $rose(dataRead). The LHS of "intersect" says that it will be True for consecutive 4 cycles. The RHS says that $rose(dataRead) should occur anytime (##[1:$]) after "Retry" has been asserted. Now, recall that "intersect" requires both the LHS and RHS to be of "same" length.

```
sequence ab;
  a ##2 b;
endsequence

sequence cde;
  c ##2 d ##2 e;
endsequence

sequence abcde;
 ab intersect cde;
endsequence

property isect1;
  @(posedge clk) z |-> abcde;
endproperty
```

Will this property ever pass ???

'a ##2 b' matches; but 'cde' did not end on the match of 'a ##2 b'; so the property FAILs

```
#75  CLK # 8 :: clk=1 z=1; a=1 b=0 c=1 d=0 e=0
#85  CLK # 9 :: clk=1 z=0; a=0 b=0 c=0 d=0 e=0
#95  CLK # 10 :: clk=1 z=0; a=0 b=1 c=0 d=1 e=0
#95    property ab intersect cde FAIL
#105  CLK # 11 :: clk=1 z=0; a=0 b=0 c=0 d=0 e=0
#115  CLK # 12:: clk=1 z=0; a=0 b=0 c=0 d=0 e=1
```

This property will never pass because both sequences are of fixed length and end at different times.

Hence, it makes sense to use 'intersect' with subsequences with 'ranges'.

Fig. 8.46 *intersect* makes sense with subsequences with range

Specification:

See that dataRead is asserted within 4 clocks after a rising edge on Retry.

```
`define true 1'b1

property retryCheck;

  @(posedge clk)  $rose(Retry) |->     `true[*1:4]        intersect

                              (Retry ##[1:$] $rose(dataRead)) ;

endproperty
```

If the subsequence "(Retry ##[1:$] $rose(dataRead))" does not match within 4 clocks, the sequence "`true[*1:4]" will end and the property will Fail.

Fig. 8.47 intersect operator: interesting Application

If \$rose(dataRead) does *not* arrive in 4 cycles, the RHS will continue to execute beyond 4 clocks. But since `true[∗1:4] has now completed and since "intersect" requires both sides to complete at the same time, the assertion will fail.

If \$rose(dataRead) does occur within 4 clks, the property will PASS. Why? Let us say \$rose(dataRead) occurs on the third clock. That sequence will end and at the same time `true[*1:4] will end as well, since it is `true anytime within the four clocks. This satisfies the requirements of "intersect" and the property passes.

So, what's the practical use of such a property. Any time you want to contain a large sequence to occur within a certain period, it is very easy to use the above technique. A large sequence may have many time domains and temporal complexities, but with the above method, you can simply superimpose `true construct with "intersect" to achieve the desired result.

Some more examples.

Specification:

In PCI protocol, once frame_ is asserted (going low), irdy_ must be asserted (going low) within 8 clocks.

Solution: This solution is similar to what we just discussed above.

```
`define true 1'b1
 property check_irdy;
      @(posedge clk) `true[*1:8] intersect ($fell(frame_)
      ##[1:$] $fell(irdy_));
 endproperty
 iCheck: assert property (check_ irdy);
```

In this property, we check for the sequence $fell(frame_) ##[1:$] $fell(irdy_) to occur (intersect) with `true[*1:8], meaning if the sequence does not occur within 8 clocks, the property will fail. Both the `true[*1:8] and ($fell (frame_) ##[1:$] $fell(irdy_)) start executing at the same time and end within 8 clocks. If irdy_ is asserted within 8 clocks, then both sequences end at the same time and the property passes. If they don't end at the same time and since both sequences must end at the same time, the property fails.

Specification:

See that once the CPU starts a cycle (for a certain instruction), that two READs are issued in order and one WRITE is issued. Condition is that the WRITE must complete at the end of the second READ. In other words, the WRITE completion and the second READ completion must happen simultaneously.

Solution:

```
property checkRW;
@(posedge clk)
    $rose(CPU_Start) |=> WRITE_complete [-> 1] intersect
    READ_complete [-> 2];
endproperty
```

This property will start consequent evaluation at $rose(CPU_Start) and check to see that WRITE_complete occurs at least once and the READ_complete occurs at least twice. And that the completion of the WRITE *intersects* (i.e., ends at the same time) with the completion of second READ.

Continuing with above example, here's another variation of it.

Specification:

Between CPU_Start and CPU_End commands, there must be at least 3 READs and 2 WRITEs and that READs and WRITEs complete the same time as the rising edge of CPU_End.

Solution:

```
property checkNumRW;
@(posedge clk)
    $rose(CPU_start) |-> (READ_complete[=3] intersect WRITE_
    complete[=2]) ##0 $rose(CPU_End);
endproperty
```

8.27 "*intersect*" and "*and*":: What's the Difference? (Fig. 8.48)

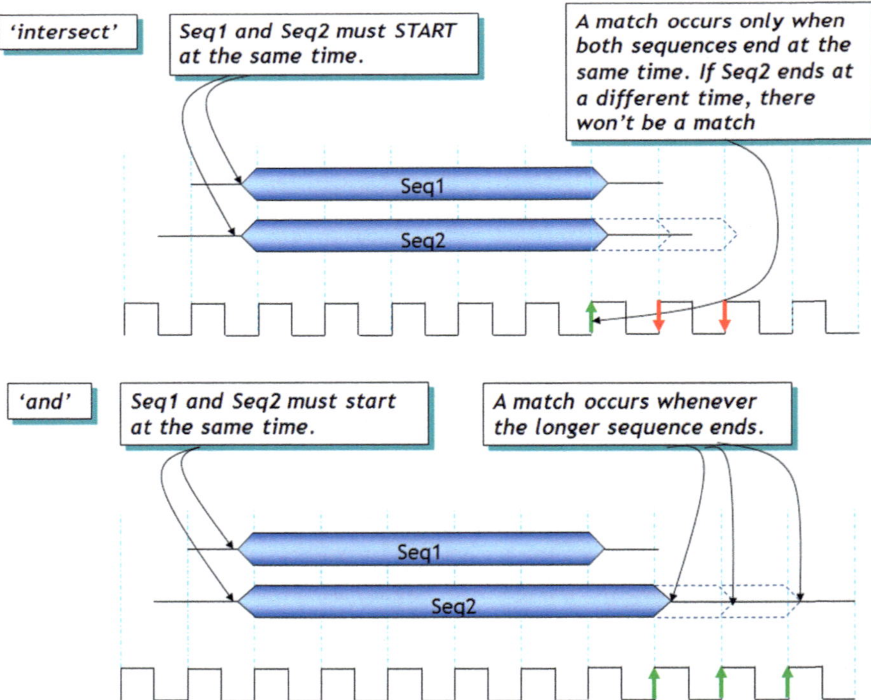

Fig. 8.48 *and* vs. *intersect*—what's the difference

8.28 first_match

first_match(Seq)

- **matches only the first of possibly multiple matches of the eval of Seq.**
- **useful for detecting the first occurrence in a *delay range*.**

8.29 Application: first_match

In Fig. 8.49, property "fms" says that on the first_match of "bcORef," "a" should rise. As you notice, the sequence "bcORef" has many matches because of the range operator and an "or." As soon as the *first* match of "bcORef" is noticed, the property looks for $rose(a). In the top log, that is the case and the property passes. ***Note that rest of the matches are now ignored***. In the bottom log, $rose(a) does not occur and the property fails—even though (and as noted in the log, of Fig. 8.49), (b &&c) is indeed true 3 clocks after *d* == 1 and even though the fact that this is an "or," the first_match looks for the very first match of either of the sequences in "bcORef" and looks for $rose(a) right after that. So, as soon as (e &&f) is true, the property looks for $rose(a) – which does not occur, and the property fails.

Figure 8.50 further explains "first_match." Annotations explain what's going on.

Figure 8.51 application clarifies $first_match further. This is the classic PCI bus protocol application. As the figure shows, the first-time frame_ && irdy_ are high (de-asserted) that the bus goes into IDLE state. Note that once frame_ and irdy_ are de-asserted the bus would remain in IDLE state for a long time. But we want the very first time that the bus transaction ends (indicated by frame_ && irdy_ high) that the bus goes into IDLE state. We do not want to evaluate any further busIdle conditions.

So, what would happen if you removed "first_match" from the above property? The property will continue to look for state==busidle every clock that frame_ && irdy_ is high. Those will be totally redundant checks.

Note that in all the examples above, we have used first_match () in the antecedent. Why? Because *the consequent (RHS) of a property behaves exactly like first_match by definition*. The consequent is not evaluated once its first match is found (without the use of first_match). But the antecedent will keep firing every time there is a match of its expression.

Hence, it makes sense to use first_match as part of antecedent sequence.

```
sequence bcORef;
  ( (##[2:5] (b && c)) or
    (##[2:5] (e && f))
  );
endsequence

property fms;
  first_match (bcORef) |=> $rose(a);
endproperty

baseP: assert property (@(posedge clk) d |-> fms) else gotoFail;
coverP: cover property (@(posedge clk) d |-> fms) gotoPass;
```

```
# run -all
# 5  CLK # 1 :: clk=1 d=0   b=0 c=0   e=0 f=0  a=0
#15  CLK # 2 :: clk=1 d=1   b=0 c=0   e=0 f=0  a=0
#25  CLK # 3 :: clk=1 d=0   b=0 c=0   e=0 f=0  a=0
#35  CLK # 4 :: clk=1 d=0   b=1 c=1   e=0 f=0  a=0
#45  CLK # 5 :: clk=1 d=0   b=1 c=1   e=0 f=0  a=1
#        45   property fms PASS
```

On the first match of (b && c), the property looks for $rose(a) the next clock; finds it and PASSes

```
#       55  CLK # 6 :: clk=1 d=1   b=0 c=0   e=0 f=0  a=0
#       65  CLK # 7 :: clk=1 d=0   b=0 c=0   e=1 f=1  a=0
#       75  CLK # 8 :: clk=1 d=0   b=1 c=1   e=0 f=0  a=0
#       85  CLK # 9 :: clk=1 d=0   b=0 c=0   e=0 f=0  a=1
#       85   property fms PASS
```

After d==1; (e && f) is found to be true but not in the required [2:5] clock range; the property next finds (b && c) to be true and $rose(a) the next clock; so it PASSes

```
#       95  CLK # 10 :: clk=1 d=1   b=0 c=0   e=0 f=0  a=0
#      105  CLK # 11 :: clk=1 d=0   b=0 c=0   e=0 f=0  a=0
#      115  CLK # 12 :: clk=1 d=0   b=0 c=0   e=1 f=1  a=0
#      125  CLK # 13 :: clk=1 d=0   b=1 c=1   e=0 f=0  a=0
#      125   property fms FAIL
#      135  CLK # 14 :: clk=1 d=0   b=0 c=0   e=0 f=0  a=1
```

(e && f) is true 2 clocks after d==1; but $rose(a) is not true 1 clock later. So the property FAILs. Note that (b && c) is true 3 clocks after d==1 and $rose(a) true 1 clock later. But since (e && f) was true FIRST, that $rose(a) had to be true the next clock.

Fig. 8.49 first_match—application

```
sequence bc;
  ( ##[0:$] (b && c), $display($stime,,,"FIRST MATCH b&&c") );
endsequence

property fms;
  first_match (bc) |=> $rose(a);
endproperty

baseP: assert property (@(posedge clk)  d |-> fms) else gotoFail;
coverP: cover property (@(posedge clk) d |-> fms) gotoPass;
```

Helpful Hint::

You can attach a subroutine with an expression in a 'sequence'.

More on this later....

On the first match of (b && c), the property looks for $rose(a) the next clock; finds it and PASSes

```
# run -all
#        5  CLK # 1 :: clk=1 d=0 b=0 c=0 a=0
#       15  CLK # 2 :: clk=1 d=1 b=0 c=0 a=1
#       25  CLK # 3 :: clk=1 d=0 b=0 c=1 a=0
#       35  CLK # 4 :: clk=1 d=0 b=1 c=0 a=0
#       45  CLK # 5 :: clk=1 d=0 b=0 c=1 a=1
#       55  CLK # 6 :: clk=1 d=0 b=1 c=0 a=0
#       65  CLK # 7 :: clk=1 d=0  b=1 c=1  a=0
#       65  FIRST MATCH b&&c
#       75  CLK # 8 :: clk=1 d=0 b=1 c=1 a=1
#       75    property ab PASS
```

(b && c) is true the same clock as d==1; and since the implication is overlapping in the assert statement, first match occurs the same clock. Property FAILs because $rose(a) is not true the next clock.

```
#       185  CLK # 19 :: clk=1 d=1 b=1 c=1 a=1
#       185  FIRST MATCH b&&c
#       195  CLK # 20 :: clk=1 d=0 b=1 c=1 a=0
#       195    property ab FAIL
```

Fig. 8.50 *first_match* Application

application

The first time PCI bus goes IDLE, the state machine should transition to busIdle state.

```
sequence busIdleCheck;
  (##[2:$] (frame_ && irdy_));
endsequence

property fms;
    @(posedge clk) first_match (busIdleCheck) |-> (state ==
busIdle);
endproperty
baseP: assert property (fms) ;
```

Fig. 8.51 first_match Application

For example, the following two assertions are equivalent and the use of first_match in them is redundant.

```
property ff;
    a |=> ##[1:5] b;// No need for first_match in the consequent here.
endproperty
property ff;
    a |=> first_match (##[1:5] b);// first_match is redundant here
endproperty
```

Note the following "cover" property. That further explains use of first_match.
Problem Statement:

```
abcProp: cover property (@ (posedge clk) a ##[1:4] b ##1 c);
```

Let us say, "a" is true at time 10 and "b" is true at time 20, thus meeting its requirement and then "c" is true at time 30. The entire sequence matches and will be considered "covered" (exercised). But if "b" remains true at time 20,30,40 and "c" remains true at 30,40,50, the coverage report will show multiple "coverage" of this sequence. Something we do not really want. The following will solve the problem.
Solution:

```
abcProp: cover property (@ (posedge clk) first_match (a ##[1:4]
##1 b ##1 c));
```

In this case, as soon as "a" is true and "b" is true the *first time* in ##[1:4] that "c" is evaluated to be true the next clock, the property is considered covered. No further evaluation of this sequence will take place until "a" is found asserted again.

In short, there can be many matches of a sequence, but you want the evaluation to stop on the very first match that you use "first_match."

8.30 *not* < property expr>

The "not" operator seems very benign. However, it could be easily misinterpreted because we are all wired to think positively—correct?

Figure 8.52 shows a use of "not." Whenever "cde" is true the property will fail because of "not" and pass if "cde" is not true. Please refer to an example in Fig. 8.53 and then an application thereafter.

8.31 Application: not Operator

First, please refer to "vacuous pass" in the Sect. 17.18 to understand the following application.

not (property_expr);

- If the property_expr evaluates to True, then the not (property_expr) evaluates to False.
- If the property_expr evaluates to False, then the not (property_expr) evaluates to True.

```
sequence cde;
  c ##1 d ##1 e;
endsequence

property nots;
  @(posedge clk) a |-> (not(cde));
endproperty

baseP: assert property (nots) else
gotoFail;
coverP: cover property (nots)
gotoPass;
```

If 'cde' matches, the property fails. If 'cde' does not match, the property passes

sequence 'cde' fails; so property 'nots' PASSes

```
#15  CLK # 2 :: clk=1 a=1 b=0 c=1 d=0 e=0
#25  CLK # 3 :: clk=1 a=0 b=0 c=0 d=0 e=0
#       25    property nots PASS
```

sequence 'cde' passes; so property 'nots' FAILs

```
115  CLK # 12 :: clk=1 a=1 b=0 c=1 d=1 e=1
125  CLK # 13 :: clk=1 a=0 b=0 c=0 d=1 e=0
135  CLK # 14 :: clk=1 a=0 b=1 c=1 d=0 e=1
135    property ab not cde FAIL
```

Fig. 8.52 not operator—basics

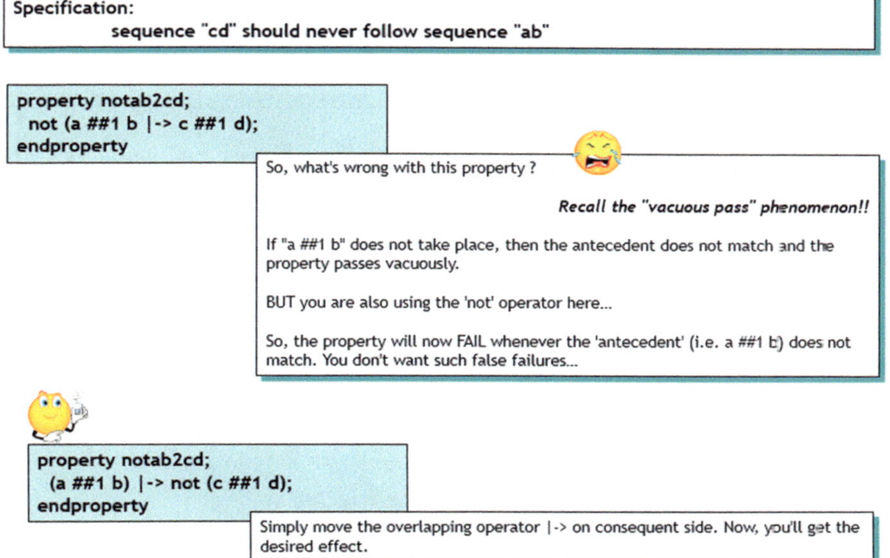

Fig. 8.53 not operator—application

Figure 8.53 shows a classic mistake engineers make when using "not" operator. Without the "not" in this example if the antecedent "a ##1 b" does not match, the property vacuously passes (*vacuous pass is discussed in Sect 17.18*) but nothing really happens. The property simply waits for the antecedent to be true so that the consequent can start its execution. However, since the antecedent has a "not" in front of it, as soon as the property sees that the antecedent does not match it will fail (*not* of vacuous pass). That is indeed detrimental to your design results where many false failures would pop up.

As shown at the bottom of the figure, simply move the implication operator "|->" on the consequent side and apply "not" on the consequent which has the same desired effect. You won't have to deal with vacuous pass.

Application in Fig. 8.54 is a very useful application. The specification says that once req is asserted (active high) that we must get an ack *before* getting another request. Such a situation occurs in many designs.

Let us examine the assertion. Property strictlyOneAck says that when "req" is asserted (active high) that!ack[*0:$] remains low until $rose(req). If this matches than the property fails (because of the "not").

In other words, we are checking to see that ack remains low until the next req, meaning if ack does go high before req arrives that the sequence (!ack[*0:$] ##1 $rose(req)) will fail and the "not" of it will make it pass. That is the correct behavior since we *do* want an ack before the next req.

Or looking at it conversely (and as shown in the log), if "ack" does remain low until the next "req" arrives that the sequence (!ack[*0:$] ##1 $rose(req)) will pass

Specification:
Once 'req' is asserted that you must get an 'ack' -*before*- the next request.

```
property strictlyOneAck;
  @(posedge clk) $rose(req) |=> (not (!ack[*0:$] ##1 $rose(req) ) );
endproperty
strictlyOneAckP: assert property (strictlyOneAck)
          else $display($stime,,,"\t Error: strictlyOneAck FAIL");
```

```
KERNEL:          0  clk=1 req=0 ack=0
KERNEL:      10000  clk=1 req=1 ack=0
KERNEL:      20000  clk=1 req=0 ack=0
KERNEL:      30000  clk=1 req=0 ack=0
KERNEL:      40000  clk=1 req=1 ack=0
KERNEL:      40000      Error: strictlyOneAck FAIL
KERNEL:      50000  clk=1 req=0 ack=1
```

Fig. 8.54 *not* operator—application

and the "not" of it will fail. This is correct also, because we do not want ack to remain low until next req arrives. We want "ack" to arrive before the next "req" arrives.

May seem a bit strange and this property can be written many different ways, but this will give you a good understanding of how negative logic can be useful

Exercise: What is a simpler way to write this property?

8.32 *if* (expression) property_expr1 *else* property_expr2

"if" "else" constructs are similar to their counterpart in procedural languages and obviously very useful. As the Fig. 8.55 annotates, we are making a decision in consequent based on what happens in the antecedent. The property "if" states that on "a" being true, either "b" or "c" should occur at least once, *any time* one clock after "a." If this antecedent is true, the consequent executes. Consequent expects "d" to be true if "b" is true and "e" to be true if "b" is false or "c" is true.

The simulation log in the bottom left of Fig. 8.55 shows that at time 15, "*a* == 1" and 1 clock later "b" is true as required. Since "b" is true, "d" is true 1 clock later at time 45. Everything works as required and the property passes. In the bottom right simulation log, "*a* == 1" at time 55 and "c" goes true at 85. This would require "e"

```
if (expression) property_expr1;

OR

if (expression) property_expr1 else property_expr2;
```

```
property ife;
  @(posedge clk) a ##1 (b || c) [->1] |->
  if (b)
    (##1 d)
  else
    (##1 e);
endproperty

baseP: assert property (ife) else gotoFail;
coverP: cover property (ife) gotoPass;
```

The property reads as follows::

@(posedge clk) 'a' followed by at least one 'b' OR 'c';

implies (|->)

that if 'b' is true than 1 clock later 'd' is true else 1 clock later 'e' is true.

```
5  CLK # 1 :: clk=1 a=0 b=0 c=0 d=0 e=0
15 CLK # 2 :: clk=1 a=1 b=0 c=0 d=0 e=0
25 CLK # 3 :: clk=1 a=0 b=0 c=0 d=0 e=0
35 CLK # 4 :: clk=1 a=0 b=1 c=0 d=0 e=0
45 CLK # 5 :: clk=1 a=0 b=0 c=0 d=1 e=0
45   property PASS
```

```
55 CLK # 6 :: clk=1 a=1 b=0 c=0 d=0 e=0
65 CLK # 7 :: clk=1 a=0 b=0 c=0 d=0 e=0
75 CLK # 8 :: clk=1 a=0 b=0 c=0 d=0 e=0
85 CLK # 9 :: clk=1 a=0 b=0 c=1 d=0 e=0
95 CLK # 10 :: clk=1 a=0 b=0 c=0 d=0 e=0
95   property FAIL
```

Fig. 8.55 *if ... else*

```
Specification :

On a TagCompare,

if there is a TagHit, start mesiCompare
else start an allocRead
```

```
application
property tagCheck;
  @(posedge clk) (State == TagCompare) ##1 (TagHit || TagMiss) |->

  if (TagHit)
    ##1 (State == mesiCompare)

  else
    ##1 (State == allocRead);

endproperty

baseP: assert property (tagCheck) else gotoFail;
coverP: cover property (tagCheck) gotoPass;
```

Fig. 8.56 if ... else—application

to be true 1 clock later, but it's not and the property fails. This is just but one way to use if-else and tie in antecedent with consequent.

Based on the analogy of the Fig. 8.55, a practical application is given in Fig. 8.56.

8.33 Application: if .. else

This property is self-explaining. On a TagCompare, if it's a hit, start MESI compare, else start a Read Allocation cycle.

8.34 "iff" and "implies"

p **iff** q is an equivalence operator. This property is true *iff* (if and only if) properties "p" and "q" are both true. When "p" and "q" are Boolean expressions "e1" and "e2," then e1 **iff** e2 is equivalent to e1 $<->$ e2.

p **implies** q is an implication operator. So, what's the difference between "implies" and the implication operator "|->"? In case of p |- > q, the evaluation of "q" starts at the match of "p." "p" is a sequence and "q" is a property. In case of p

implies q, *both* are properties and *both* "p" and "q" start evaluating at the *same* time and the truth results are computed using the logical operator "implies." There is no notion of a match of antecedent to trigger the consequent.

For example,

```
x ##2 y |-> a ##2 b;
```

vs.

```
x ##2 y implies a ##2 b;
```

In the case of implication operator "**|->**", evaluation of "a ##2 b" starts at the match of "x ##2 y." In the case of "**implies**," evaluation of both "x ##2 y" and "a ##2 b" start at the *same* clock tick and the property fails if either of the "x ##2 y" or "a ##2 b" fail.

Chapter 9
System Functions and Tasks

Introduction: This chapter discusses in detail the System Functions and Tasks such as $onehot, $onehot0, $isunknown, and $countones and Abort System Functions and Tasks such as $assertoff, $asserton, and $assertkill. Note that the 2009/2012 LRM introduces quite a few new abort functions such as $assertpasson, $assert-failon, $assertvacuousoff, and $assertcontrol). These are described in their entirety in Sect. 20.17.

A. B. Mehta, *SystemVerilog Assertions and Functional Coverage*,
https://doi.org/10.1007/978-3-030-24737-9_9

9.1 $onehot, $onehot0

$onehot and $onehot0 are quite self-explaining as shown in Fig. 9.1. *Note that if the expression is "Z" or "X" that $onehot or $onehot0 will fail. But will not fail if there are "x"s and "z"s on the bus but—at least—one "1." Read on.*

A simple application is described in Fig. 9.1. For any acknowledge of a bus grant there can only be one bus grant. This is very easily accomplished by $**onehot** as shown.

Note the results at time 25–35. There is an "*x*" and a "*z*" on the bus but since there is only one bit "1," it meets $**onehot** requirement and the property passes. What if that's not the result you want. You don't want an "*x*" or a "*z*" when searching for that one "1."

The next section covers $**isunknown**, but here's the solution to above dilemma. Write the property as follows and it will guarantee that the bus is indeed in a known state on—all—bits of the bus and that there is only one "1."

```
property bgcheck;
        (@posedge clk) bgack |-> ! ($isunknown(busgnt)) &&
        $onehot(busgnt);
endproperty
```

```
$onehot (<expression>)
```

Returns True if only one bit of the expression is a '1' (high).

```
property bgcheck;
  @(posedge clk) bgack |->
$onehot(busgnt);
endproperty
```

```
# run -all
#        5 clk=1 bgack=1 busgnt=xxxxxxxx
#        5   property bgcheck FAIL
#
#       15 clk=1 bgack=1 busgnt=00000001
#       15   property bgcheck PASS
#
#       25 clk=1 bgack=1 busgnt=x0000001
#       25   property bgcheck PASS
#
#       35 clk=1 bgack=1 busgnt=z0000001
#       35   property bgcheck PASS
#
#       45 clk=1 bgack=1 busgnt=11111111
#       45   property bgcheck FAIL
#
#       55 clk=1 bgack=1 busgnt=00000000
#       55   property bgcheck FAIL
```

```
$onehot0 (<expression>)
```

Returns True if all bits of the expression are '0' OR only one bit of the expression is a '1'.

Fig. 9.1 $onehot and $onehot0

9.2 $isunknown

$isunknown passes if the expression is unknown ("X" or "Z"). In other words, if the expression is not unknown, then the property will fail! *Hence, if you do want a failure on detection of an unknown ("X" or "Z"), then you have to negate the result of $isunknown.* Simple but easy to miss.

Simulation log in Fig. 9.2 clarifies the concept. Property "ucheck" states that if "bgack" is true that the "busgnt" is unknown. What? This is simply to show what happens if you use $isunknown without a "not."

9.2.1 Application $isunknown

Figure 9.3 shows two simple applications. First is identical to the one we just discussed, but with a "not."

```
@ (posedge clk) bgack |-> not ($isunknown(busgnt));
```

It says, if bgack is true (high) that busgnt must not be in unknown state Again, note that $isunknown returns a true on detection of an unknown. Therefore, if you

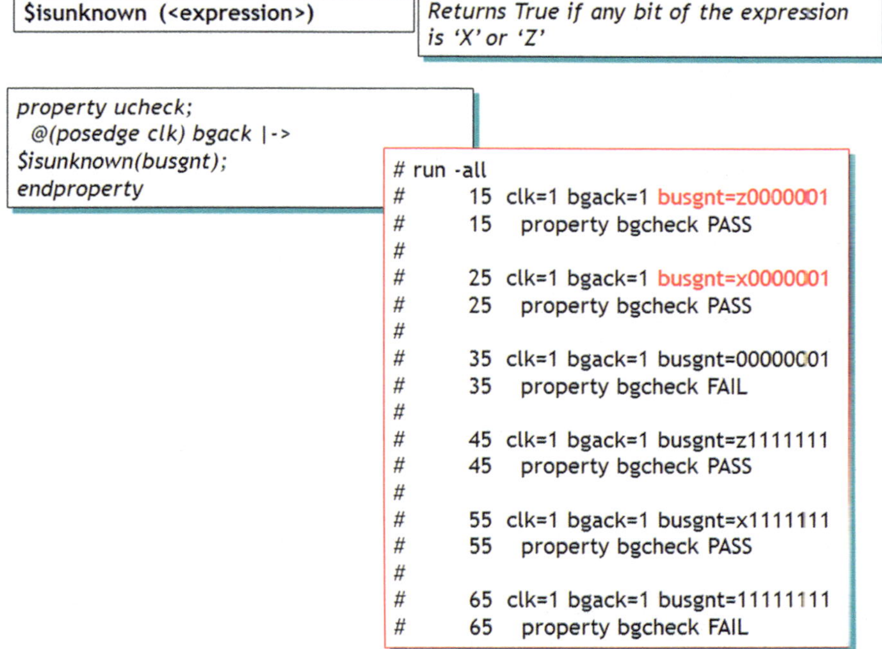

Fig. 9.2 $isunknown

Practical Note: Since $isunknown returns a true on detection of 'z' or 'x' in an expression, you may want to negate the results if you want a FAILures on 'x' or 'z' detection.

```
property ucheck;
  @(posedge clk) bgack |-> not
($isunknown(busgnt));
endproperty
```

```
#        15  clk=1 bgack=1 busgnt=z0000001
#        15    property bgcheck FAIL
#
#        25  clk=1 bgack=1 busgnt=x0000001
#        25    property bgcheck FAIL
#
#        35  clk=1 bgack=1 busgnt=00000001
#        35    property bgcheck PASS
#
#        45  clk=1 bgack=1 busgnt=z1111111
#        45    property bgcheck FAIL
#
#        55  clk=1 bgack=1 busgnt=x1111111
#        55    property bgcheck FAIL
#
#        65  clk=1 bgack=1 busgnt=11111111
#        65    property bgcheck PASS
```

Application

Specification :: Once a Cycle Starts, the control signals should not go Unknown.

```
property validControl;
  @(posedge clk) disable iff (busIdle || rst) adrStrobe |->
                               not ($isunknown( {cBE, cWrtAdr, cWrtData} ));
endproperty
```

Fig. 9.3 $isunknown Application

want a failure, you have to negate the result. This is shown in the simulation log above. The second application (bottom of Fig. 9.3) says that if "adrStrobe" is High that the control signals cannot be unknown.

9.3 $countones

$countones is very simple but powerful feature. Note that the system function can be used in a procedural block as well as in a concurrent property/assertion.

Figure 9.4 shows an application which states that if there is a bus grant acknowledge (bgack) that there can be only 1 bus grant (busgnt) active on the bus. Note that we are using $countones in a procedural block in this example. Note also that if the entire "busgnt" is unknown ("X") or tristate ("Z"), the assertion will fail. Figure 9.5 shows a very simple way to check for Gray Code compliancy.

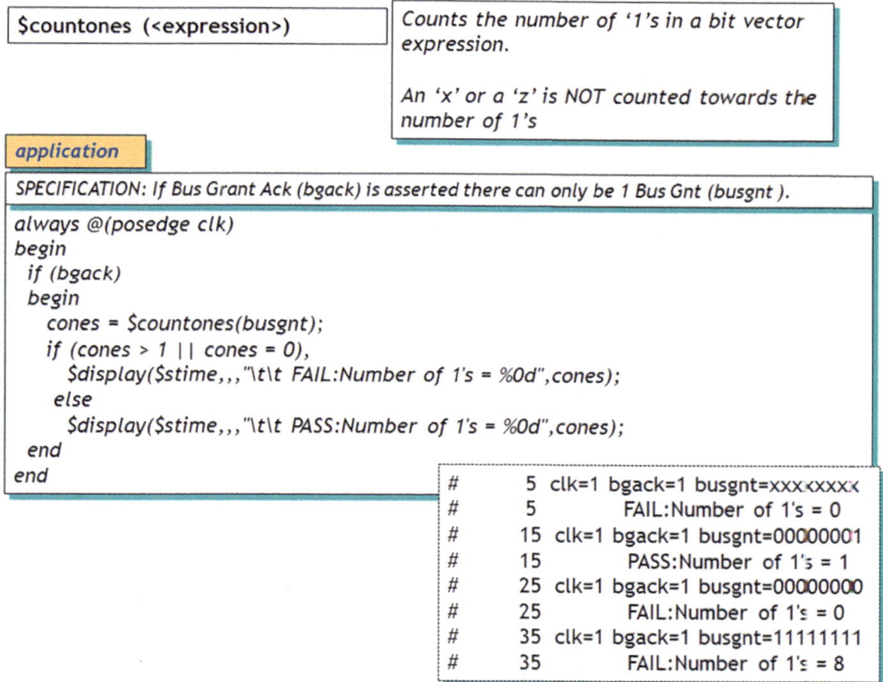

Fig. 9.4 $countones—basics and application

application

SPECIFICATION: Check that a bus conforms to Gray Code Transition

```
property CheckGrayCode (mySig);
   @(posedge clk) ($countones ($past (mySig) ^ mySig) <= 1);
endproperty

CGrayProp: assert CheckGrayCode (PipePointer);
```

Fig. 9.5 Application $countones

$countones returns the # of 1's in an expression. It can also be used to determine 'true'ness of the expressions.

In other words, it can be used for pass/fail indication.

If there is at least One '1', it's a pass, else it's a fail

```
property bgcheck;
  @(posedge clk) bgack |-> $countones(busgnt);
endproperty
```

```
#        15  clk=1 bgack=1 busgnt=00000001
#        15    property bgcheck PASS
#
#        25  clk=1 bgack=1 busgnt=00000000
#        25    property bgcheck FAIL
#
#        35  clk=1 bgack=1 busgnt=000000x1
#        35    property bgcheck PASS
#
#        45  clk=1 bgack=1 busgnt=000000z1
#        45    property bgcheck PASS
```

Fig. 9.6 $countones as Boolean

9.4 $countones (as Boolean) (Fig. 9.6)

9.5 $countbits

$countbits(e, list_of_control_bits) has the following arguments:

"**e**" = a bit vector, meaning a packed or an unpacked integral expression.

"list_of_control_bits" = $countbits returns the number of bits of its argument bit vector having the value of one of the control bits. These control bits can only have one of four values, namely, 1'b0, 1'b1, 1'bz, and 1'bx. You must specify at least one control bit and repeated control bits are ignored. You can have more than 1 control bit in the "list_of_control_bits."

For example,

If you want to make sure that none of the bits of a bus are in floating state (1'bz) when the bus is not driven, you can write the property as follows:

```
property checkFloat;
@(posedge clk) BusUndriven |-> ($countbits (bus, 1'bz) == 0);
endproperty
```

This property says that when "BusUndriven" signal is high, that none of the bits in "bus" are 1'bz. The way to read the property is: The $countbits(bus) returns the number of bits in "bus" that are in 1'bz state and sees that that count is zero, thus verifying that none of the bits are in 1'bz state.

Another example,

Make sure that none of the bits in a given vector are unknown. The following property can also be written using the system function $isunknown. The following accomplishing the same with $countbits. The example is to highlight that you can have more than 1 control bit.

```
property checkUnknown;
@(posedge clk) memread |-> ($countbits (data, 1'bx, 1'bz) == 0);
endproperty
```

$countbits(data) returns a count of bits that are in 1'bx and 1'bz state. If this count is zero, then there are no unknown bits.

9.6 $assertoff, $asserton, $assertkill

There are many situations when you want to have a global control over assertions both at module and instance level. Recall that "disable iff" provides you a local control directly at the source of the assertion.

As noted in Fig. 9.7, **$assertoff** temporarily turns off execution of all assertions. Note that if an assertion is under way when $assertoff is executed, the assertion *won't* be killed. You restart assertion execution on a subsequent invocation of **$asserton**. **$assertkill** will kill *all* assertions in your design *including* already executing assertion. And it won't automatically restart when the next assertion starts executing. It will restart executing only on the subsequent **$asserton. $asserton** is the default. It is required to restart assertions after a $assertoff or $assertkill.

Following is a typical application deployed by projects to suppress execution of assertions during reset or during an exception, if so required (Fig. 9.8).

$assertoff (level,[list of module, instance or assertion_identifier]);

$assertoff stops the checking of all specified assertions until a subsequent $asserton.

Note: If an assertion is already executing, it won't be affected.

$assertkill (level,[module/module instance or assertion_identifier]);

$assertkill shall abort execution of any currently executing specified assertions and then stop the checking of all specified assertions until a subsequent $asserton

$asserton (level,[module/module instance or assertion_identifier]);

$asserton shall re-enable execution of all specified assertions.

By default assertions are ON with an 'assert' statement

'level' = 0 turns on/off assertions at ALL levels under the given module/instance
= m (m>0) turns on/off assertions only at 'm' levels of hierarchy below the specified module / instance level.

assertion_identifier :: Name of the property or the label used with 'assert'

module, instance name :: Can be relative or full hierarchical

Fig. 9.7 $assertoff, $asserton, $assertkill—Basics

Example shows that assertions are turned OFF during 'reset_'; are turned ON after reset_ and are Killed OFF when a machine check exception is detected.

```
module assertion_control(input reset_,machinecheck_exception,
machinecheck_ISR_return );

always @(reset_) begin
  if (reset_ == 1'b0) $assertkill(0,top.pcim,top.axim);
  else $asserton(0,top.pcim,top.axim);
  // $assertkill(0,top);
end

always @(machinecheck_exception)
   $assertoff(0,top.datamodule.array);

always @(machinecheck_ISR_return)
   $asserton(0,top.datamodule.array);

endmodule
```


Fig. 9.8 Application Assertion Control

Chapter 10
Multiple Clocks

Introduction: This chapter describes multiply clocked sequences and properties and clock flow semantics (how does clock flow from one clock domain to another). It also discusses all the operators that work on multiply clocked properties such as "and," "or," "not," etc. It also describes nuances of Legal and Illegal conditions of such properties and sequences.

A. B. Mehta, *SystemVerilog Assertions and Functional Coverage*,
https://doi.org/10.1007/978-3-030-24737-9_10

10.1 Multiply Clocked Sequences and Properties

There are hardly any designs anymore that work only on a single clock domain. So far we have seen properties that work off of a single clock. But what if you need to check for a temporal domain condition that crosses clock boundaries. The so-called CDC (clock domain crossing) issues can be addressed by multiple clock assertions.

We'll thoroughly examine how a property/sequence crosses clock boundary. What's the relationship between these 2 (or more) clocks? How are sampling edges evaluated once you cross the clock domain. Note that in a singly clocked system, the sampling edge is always one—that is mostly the clock posedge or negedge. Since there are two (or more) clocks in multiply clocked system, we need to understand how the sampling edges cross boundary from one clock to another. I think it is best to fully understand the fundamentals before jumping into applications.

Note that there are differences in the way a "sequence" behaves for multiple clocks and the way a property behaves. Read on...

Figure 10.1 shows a simple multiply clocked sequence. It says at (posedge clk0) A is true and the *very next nearest strictly subsequent* (posedge clk1) B is true. Note the emphasis on "*very next*." That is because when you join two subsequences each of which runs on a different clock, you can transition only from one clock domain's sampling edge to the next clock domain's *very next* available sampling edge. This will be clearer when we dive a bit more into detail.

Multiply-clocked sequences are built by concatenating singly-clocked subsequences using the delay concatenation operators ##1 and ##0.

```
sequence mclocks;
  @(posedge clk0) A ##1 @(posedge clk1) B;
  //                ^^^^^^^^^^^^^^^^
endsequence
```

"##1 @(posedge clk1)" here does -not- necessarily mean a delay of one clock.

It means on a match of 'A' @(posedge clk0), *the ##1 moves the time to the nearest strictly subsequent posedge clk1* and the sequence ends at that point with a match of B.

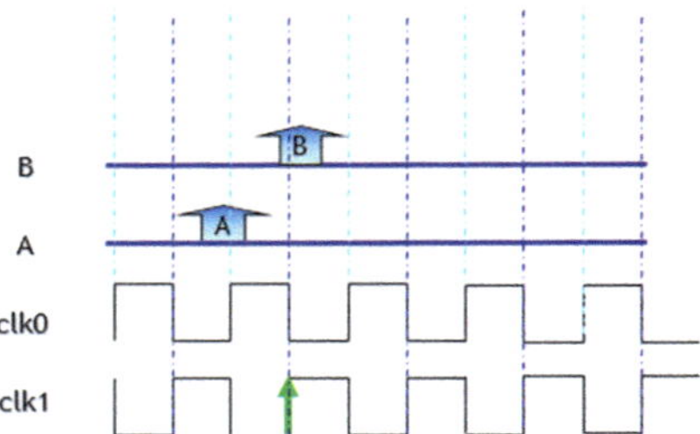

Fig. 10.1 Multiply clocked sequences—basics

Jumping ahead a bit, this "*very next nearest strictly subsequent edge*" semantic is why we use ##1 to cross clock boundaries. So, can you use ##2 when crossing clock boundaries? Nope. Read on.

10.2 Multiply Clocked Sequences

The timing diagram in Fig. 10.1 shows that at (posedge clk0), "A" is true. The clocks are out of phase, so the very next clock edge of clk1 is half a clock delayed from posedge clk0. At the posedge of clk1, "B" is sampled true and the sequence "mclocks" passes. The point here is that "##1 @ (posedge clk1)" waited only for ½ clk1 and not a full clk1 because the *very next nearest strictly subsequent posedge clk1* arrived within ½ clock period. The next clock can come in any time after clock0 and that will be the "very next" edge taken as the sampling edge for that subsequence.

Important Note:

The LRM 2005 requirement of ##1 between two subsequences have been removed from 1800 to 2009/2012. In the 2009/2012 standard you can have both ##1 and ##0 between two subsequences with different clocks. More on this coming up.

So, what happens if clk0 and clk1 are in phase? See Fig. 10.2 below. The explanation is in the figure itself. As self-evident, the sequence will wait for one full clk

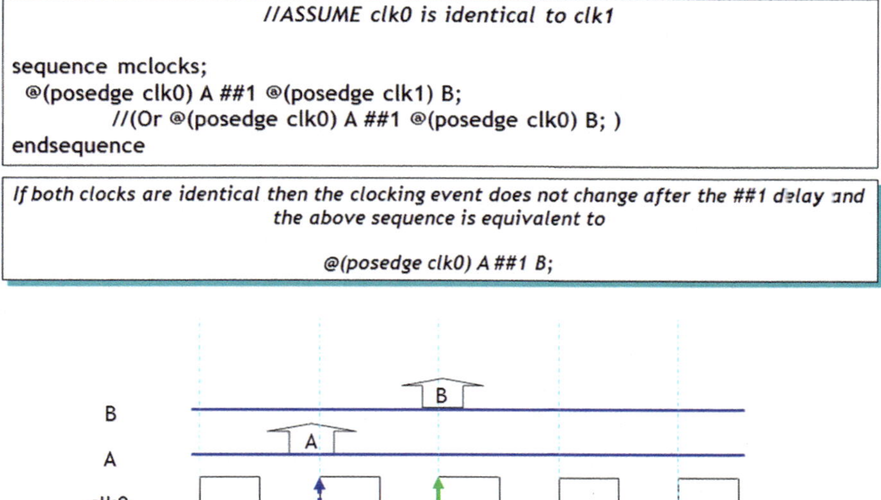

Fig. 10.2 Multiply clocked sequences—identical clocks

before sampling "B." But more importantly note that, *if* the clocks on the clock crossing boundary are identical (in phase and same period), then the following is true.

```
@ (posedge clk0) A ##1 @ (posedge clk1) B; is identical to
@ (posedge clk0) A ##1 @ (posedge clk0) B; is identical to
@ (posedge clk0) A ##1 B;
```

10.3 Multiply Clocked Sequences—Legal and Illegal Sequences

Before we move onto multiply clocked properties and based on our observations, let us quickly examine the legal and illegal cases of multiply clocked sequences. ***Again, these cases apply only to sequences and not to properties***.

As LRM puts it, Multi-clocked sequences are built by concatenating singly clocked subsequences using the single-delay concatenation operator ##1 or the zero-delay concatenation operator ##0. The single delay indicated by ##1 is understood to be from the end point of the first sequence, which occurs at a tick of the first clock, to the nearest strictly subsequent tick of the second clock, where the second sequence begins. The zero delay indicated by ##0 is understood to be from the end point of the first sequence, which occurs at a tick of the first clock, to the nearest possibly *overlapping* tick of the second clock, where the second sequence begins.

Bottom line is that you can only have ##1 or ##0 between two subsequences with different clocks. If the clocks are the same on both sides, then there is no such restriction. Figure 10.3 makes it clear.

10.4 Multiply Clocked Properties—"and" Operator

Note again that here we are discussing multiply clocked "and" in a *property* and not in a *sequence*. As mentioned above, such an "and" in a sequence is illegal.

The concept of "and" of two singly clocked properties have been discussed before. But what if the clocks in the properties are different? The important thing to note here is the concept of the very next strictly subsequent edge. In Fig. 10.4, at the posedge of clk0, "a" is sampled high. That triggers the consequent that is an "and" of "b" and "c." Note that "b" is expected to be true at the very next edge of clk1 (after the posedge of clk0). *In other words, even though there is a nonoverlapping operator in the property, we don't quite wait for 1 clock.* We simply wait for the very next posedge of clk1 to check for "b" to be true. The same story applies to "c." When both "b" and "c" occur as shown in Fig. 10.4, the property will pass. As with the singly clocked "and," the assertion passes at the match of the longest sequence "c."

Figure 10.5 shows another scenario to solidify the concept of "and" for multiply clocked assertions. The "and" is between two subsequences which use the same clock. The behavior is obvious but interesting. This is because "@ (posedge clk1) b

```
//ILLEGAL
sequence mclocks;
  @(posedge clk0) A ##2 @(posedge clk1) B;
  //          ^^^   ^^^                 ^^^
endsequence
```

This is **illegal** because you can't have any other clock delay except ##1 or ##0 between the two subsequences if the clocks are different on each side.

```
//LEGAL
sequence mclocks;
  @(posedge clk0) A ##2 @(posedge clk0) B;
  //          ^^^   ^^^                 ^^^
endsequence
```

This is **legal** because the clocks are SAME on both sides and this is equivalent to

```
sequence mclocks;
  @(posedge clk0) A ##2 B;
endsequence
```

```
//ILLEGAL
@(posedge clk0) A ##2 @(posedge clk1) B;
@(posedge clk0) A intersect @(posedge clk1) B;
@(posedge clk0) A and @(posedge clk1) B;
@(posedge clk0) A or @(posedge clk1) B;
@(posedge clk0) A not (@(posedge clk1) B);
```

ALL ILLEGAL

In short, for a "sequence", the only operator allowed between two subsequences with different clocks is ##1 or ##0

Fig. 10.3 Multiply clocked sequences—illegal conditions

```
property mclocks;
  @(posedge clk1) b and @(posedge clk2) c;
endproperty

baseP: assert property (@(posedge clk0) a |=> mclocks) else gotoFail;
coverP: cover property (@(posedge clk0) a |=> mclocks) gotoPass;
```

'B' and 'C' must be true at immediate next posedge of clk1 and clk2 respectively after the posedge of clk0

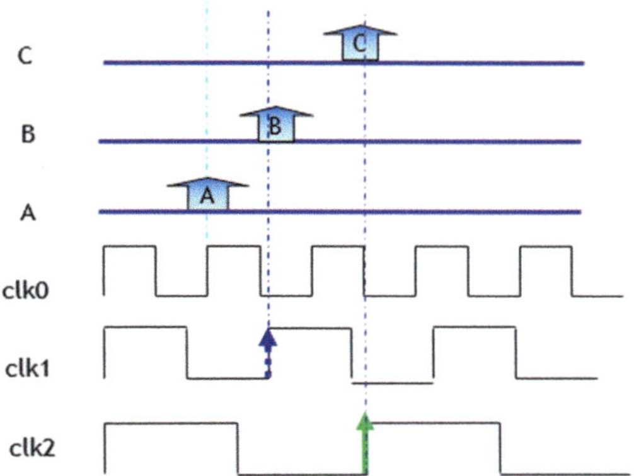

Fig. 10.4 Multiply clocked properties—"and" operator between two different clocks

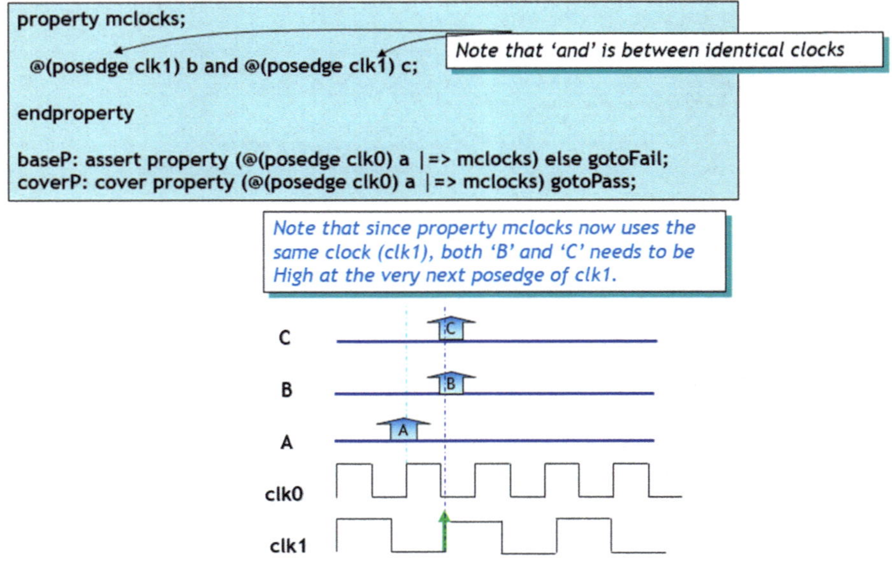

Fig. 10.5 Multiply clocked properties—"and" operator between same clocks

and @ (posedge clk1) c" acts essentially like "@ (posedge clk1) b **and** c." Hence, both the "b" and the "c" must now occur at the very next posedge of clk1.

In short, as we saw before, "@ (posedge clk1) b **and** @ (posedge clk1) c" is identical to "@ (posedge clk1) b **and** c."

10.5 Multiply Clocked Properties—"or" Operator

All the rules of "and" apply to "or"—except as in singly clocked properties—when either of the sequence (i.e., either the LHS or RHS of the operator) passes that the assertion will pass. The concept of "the very next strictly subsequent clock edge" is the same as with "and."

Please refer to Fig. 10.6 for better understanding of "or" of multiply clocked properties. The property passes when either @ (posedge clk1) "b" **or** @ (posedge clk2) "c" occurs. In other words, if @ (posedge clk2) "c" occurs before @ (posedge clk1) "b," the property will pass at @ (posedge clk2) "c."

10.6 Multiply Clocked Properties—"not"-Operator

"not" is an interesting operator when it comes to multiply clocked assertions.

The assertion in Fig. 10.7 below works as follows. At posedge clk0, "a" is true which triggers the consequent mclocks. The property mclocks specifies that @

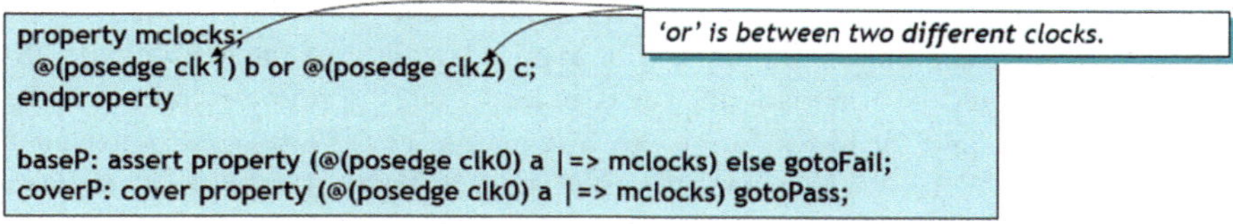

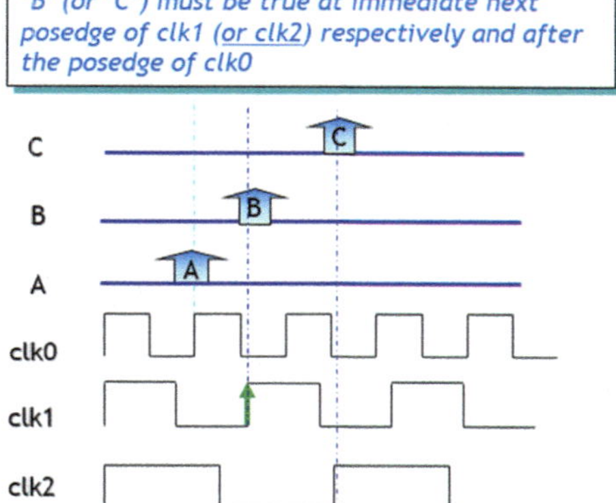

Fig. 10.6 Multiply clocked properties—“or” operator

```
property mclocks;
  @(posedge clk1) b and (not (@(posedge clk2) c) );
endproperty

baseP: assert property (@(posedge clk0) a |=> mclocks) else gotoFail;
coverP: cover property (@(posedge clk0) a |=> mclocks) gotoPass;
```

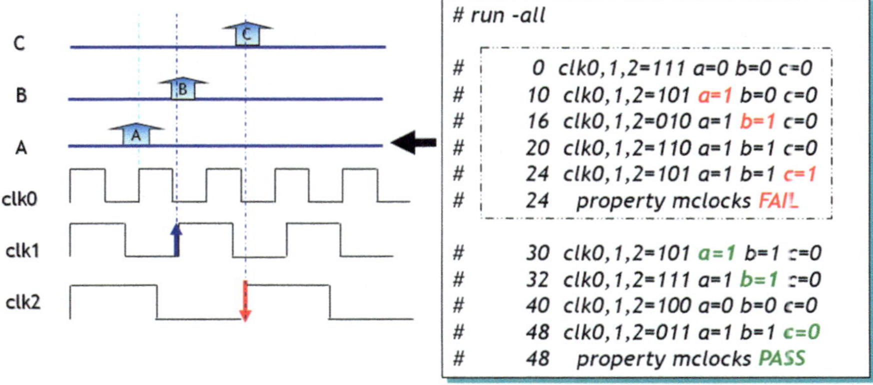

Fig. 10.7 Multiply clocked properties—“not” operator

posedge clk1 "b" needs to be true and "c" should—not—be true @ posedge clk2. The timing diagram shows that "a" is true at posedge clk0. At the subsequent edge of clk1 "b" should be true and since it is indeed true, the property moves along. Because of an "and" it looks for "c" to be *not* true at the very next subsequent posedge clk2 . Well, "c" is indeed true but since we have a "not" in front of @ (posedge clk2), the property will fail. The concept of "not" is the same as that of singly clocked properties except for the edge of the clock when it is evaluated. The simulation log clarifies the concept.

10.7 Multiply Clocked Properties—Clock Resolution

These rules are important to follow when you are dealing with multiply clocked properties (Fig. 10.8).

Figures 10.9 and 10.10 illustrate other important concepts. How do clocks apply (or flow) from one part of property to another? The description in the figure explains how this works.

1. In Fig. 10.9, property "mclocks," (posedge clk0) applies to "A" as well as "B" since "B" does not have an explicit clock. So far so good.
2. Then (posedge clk1) applies to "C." That also makes sense.
3. But what if clock is applied to "D" in the consequent since "D" does not have its own explicit clock?

```
property mclocks;
  @(posedge clk0) A |->
          if (D) @(posedge clk0) B;
endproperty
```

This is equivalent to
@(posedge clk0) A |-> if (D) B;

```
property mclocks;
  @(posedge clk0) A |->
          if (D) @(posedge clk0) B
          else @(posedge clk0) (~B);
endproperty
```

This is equivalent to
@(posedge clk0) A |-> if (D) B else (~B) ;

```
property mclocks;
  @(posedge clk0) A |->
          if (D) @(posedge clk0) B
          ##1 @(posedge clk1) Z
          else @(posedge clk0) (~B);
endproperty
```

This is equivalent to
@(posedge clk0) A |-> if (D) B ##1 @(posedge clk1) Z else (~B) ;

```
property mclocks;
  @(posedge clk0) A |->
          if (D) @(posedge clk0) B
          else @(posedge clk1) (~B);
endproperty
```

This used to be illegal in LRM 2005. But in LRM 2012 it is legal. You can indeed have an overlapping operator with two different clocks on LHS and RHS of implication operator.

Fig. 10.8 Multiply clocked properties—clock resolution

```
property mclocks;

  @(posedge clk0) A ##1 (b ##1 @(posedge clk1) C) |=> D;

endproperty
```

Here clk0 flows through 'A' then in the parenthesis to 'B' but not through 'C'; But once out of the parenthesis, it then flows through 'D'

```
//LRM: System Verilog 3.1a, Page 244:
//"The scope of a clocking event flows into parenthesized sub expressions and, if
//the sub expression is a sequence, also flows left-to-right across the parenthesized
//sub expression. However, the scope of a clocking event does not flow out of
//enclosing parenthesis".
```

```
sequence s1;
 @(posedge clk0) b ##1 c;
endsequence

sequence s2;
 @(posedge clk1) d ##1 e;
endsequence

sequence s;

 @(posedge clk) a ##1 s1 ##1 s2 ##1 f;

endsequence
```

Fig. 10.9 Multiply clocked properties—clock resolution—II

```
property mclocks;

  @(posedge clk1) @(posedge clk0) a |-> @(posedge clk0) b;

endproperty
```

This is equivalent to

```
property mclocks;
  @(posedge clk0) a |-> @(posedge clk0) b;
endproperty
```

Because, @(posedge clk1) is overridden immediately by @(posedge clk0)

Fig. 10.10 Multiply clocked properties—clock resolution—III

4. According to the 1800–2005 LRM, "D" will inherit (posedge clk0) and *not* the (posedge clk1). This is not quite intuitive. But LRM makes it very clear that "*the scope of a clocking event does not flow out of enclosing parenthesis.*"

5. In our case, ("B" ##1 @ (posedge clk1) C) is in parenthesis. So, once we are out of that parenthesis, (posedge clk1) does not flow forward but (posedge clk0) moves forward to the consequent "D."

Similarly, the bottom example in Fig. 10.9 shows how the clock would "flow" when we have multiple subsequences each with its own clock. Note that there are three different clocks in this sequence. (posedge clk) flows through "a." Then "s1" and "s2" use their own clocks as sampling edges (clocks) for their sequences. But once out of "s2," (posedge clk) is applied to "f"—not—the (posedge clk1) of "s2."

Figure 10.10 shows another interesting property. What will happen if you need to transition from one clock to another before checking for an expression/sequence? Well, you cannot quite do that. In Fig. 10.10, we show that (posedge clk1) is immediately followed by (posedge clk0). This does *not* mean that the property will wait first for (posedge clk1) then for (posedge clk0) and then apply (posedge clk0) to "a." It will simply override (posedge clk1) with (posedge clk0) and directly apply (posedge clk0) to "a." This is shown in the bottom of the figure in the equivalent property.

10.8 Multiply Clocked Properties—Legal and Illegal Conditions

Figures 10.11 and 10.12 are a recap as an easy reference to legal and illegal semantics of multiply clocked properties.

The topmost example shows that it's ok to have different clocks between the antecedent and the consequent, as long as the implication operator is nonoverlapping.

In 2012 LRM, the multi-clocked overlapping implication |-> has the following meaning: at the end of the antecedent the nearest tick of the consequent clock is awaited. If the consequent clock happens at the end of the antecedent, the consequent starts checking immediately. Otherwise, the meaning of the multi-clocked overlapping implication is the same as the meaning of the multi-clock nonoverlapping implication.

Also as shown in the third example, overlapping operator is perfectly legal if the clocks on both sides of the overlapping operator are the same.

And the last example is quite intuitive in that "=>"is equivalent to "|-> ##1." Hence, you can have different clocks on each side of overlapping operator.

Note also that as in the overlapping operator, the semantics of multi-clocked if/if-else operators is similar to the semantics of the overlapping implication.

For example,

```
@(posedge clk0) if (b) @(posedge clk1) c else @(posedge clk2) d
```

has the following meaning: the condition "b" is checked at posedge clk0. If "b" is true, then "c" is checked at the nearest, possibly overlapping posedge clk1, else "d" is checked at the nearest possibly overlapping posedge clk2.

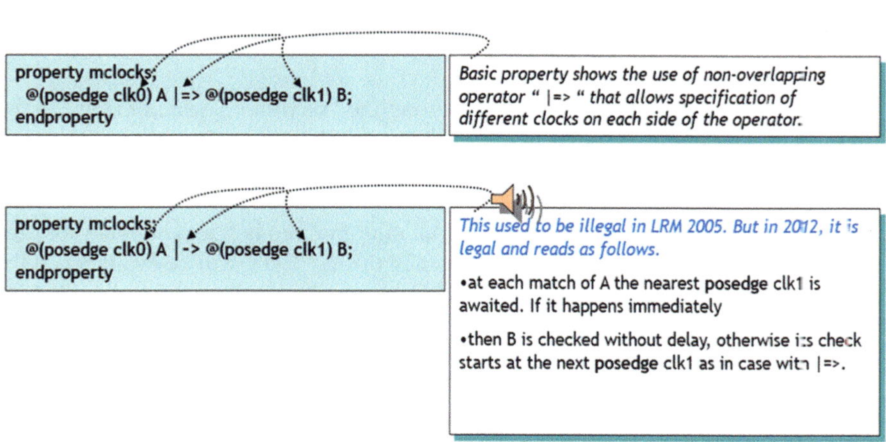

Fig. 10.11 Multiply clocked Properties—Legal and Illegal conditions

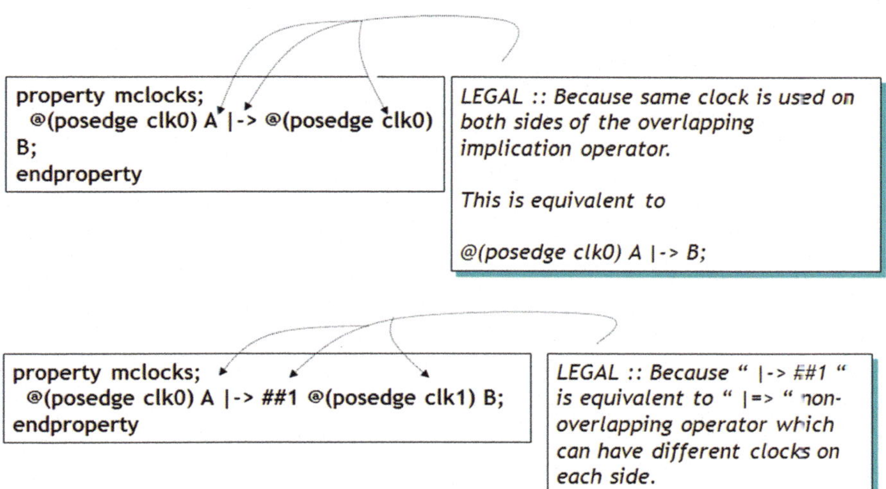

Fig. 10.12 Multiply clocked Properties—Legal and Illegal conditions

To summarize: clock flow provides that in a multi-clocked sequence or property, the scope of a clocking event flows left to right across linear operators (e.g., repetition, concatenation, negation, implication, followed-by, and the nexttime, always, eventually operators) and distributes to the operands of branching operators (e.g., conjunction, disjunction, intersection, if–else, and the until operators) until it is replaced by a new clocking event.

Here's an interesting example of how clock flows and how that makes a significant difference. What's the difference between the following two properties?

```
A1: assert property (
   @(posedge clk1) Frame_ |-> nexttime @(posedge clk2) IRDY);

A2: assert property (
   @(posedge clk1) Frame_ |-> ##1 @(posedge clk2) IRDY;
```

In case of A1, (posedge clk1) flows through to "nexttime," which means "nexttime" causes advance of (posedge clk1) to the strictly nearest (posedge clk1), after which it looks for a subsequent nearest (posedge clk2) to evaluate IRDY. So, the clock flow is (posedge clk1) to (posedge clk1) to (posedge clk2).

In case of A2, ##1 is the synchronizer, so after Frame_ is found high at (posedge clk1), the clock flows to (posedge clk2) which means IRDY will be evaluated at the very next strictly nearest (posedge clk2). Here the clock flow is (posedge clk1) to (posedge clk2).

Exercise: Can you figure out clock flow in the following property?

```
A3: assert property (
   @(posedge clk1) Frame_ |-> @(posedge clk2) nexttime IRDY;
```

Here is one more example to nail down the concepts of "clock flow" in multiply clocked properties.

```
a1: assert property
     (@(posedge clk) en && $rose(req) |=> gnt);
a2: assert property
     (@(posedge clk) en && $rose(req, @(posedge sysclk)) |=>
     gnt);
```

Both assertions a1 and a2 read: "whenever en is high and "req" rises, at the next cycle "gnt" must be asserted." In both assertions, the rise of "req" occurs if and only if the sampled value of "req" at the current posedge of clk is 1'b1 and the sampled value of "req" at a prior point is different from 1'b1. So where do the assertions differ? The assertions differ in the specification of the *prior* point. In a1 the prior point is the preceding posedge of clk, while in a2 the prior point is the *most recent* prior posedge of sysclk.

As another example,

```
always_ff @(posedge clk1)
Dreg <= $rose(Xreg, @(posedge sysclk));
```

Here, Dreg is updated in each time step in which posedge clk1 occurs, using the value returned from the $rose sampled value function in that time step. $rose compares the sampled value of the LSB of Xreg from the current time step (one in which posedge clk1 occurs) with the sampled value of the LSB of Xreg in the strictly prior time step in which posedge sysclk occurs. This example exemplifies how to detect a rising, falling, or past sampled value of a signal between two clock domains.

Chapter 11
Local Variables

Introduction: This chapter is entirely devoted to the dynamic Local Variables. Without the dynamic multi-threaded semantics and features of Local Variables, many of the assertions would be impossible to write. The chapter also lays out how operators such as "or" and "and" affect the workings of parallel threads forked off by Local Variables based assertions. There are plenty of examples and applications to help you weed through the semantics.

A. B. Mehta, *SystemVerilog Assertions and Functional Coverage*,
https://doi.org/10.1007/978-3-030-24737-9_11

Local variable is a feature you are likely to use very often. They can be used both in a sequence and in a property. They are called *local* because they are indeed local to a sequence and are not visible or available to other sequences or properties. Of course, there is a solution to this restriction, which we will study further in the section that follows.

Figure 11.1 points out key elements of a local var. The most important and useful aspect of a local variable is that it *allows multi-threaded application and creates a new copy of the local variable with every instance of the sequence in which it is used.* User does not need to worry about creating copies of local variables with each invocation of the sequence. Above application says that whenever "RdWr" is sampled high at a posedge clk, that "rData" is compared with "wData + 'hff" 5 clocks later. The example shows how to accomplish this specification. Local variable "int local_data" stores the "rData" at posedge of clk and then compares it with wData 5 clocks

Local variables are dynamic variables.

They are dynamically created when needed within an instance of a sequence and removed when the end of the sequence is reached.

'local vars' is one of the most powerful features of SVA language because it allows checking of complex pipelined behavior of the design.

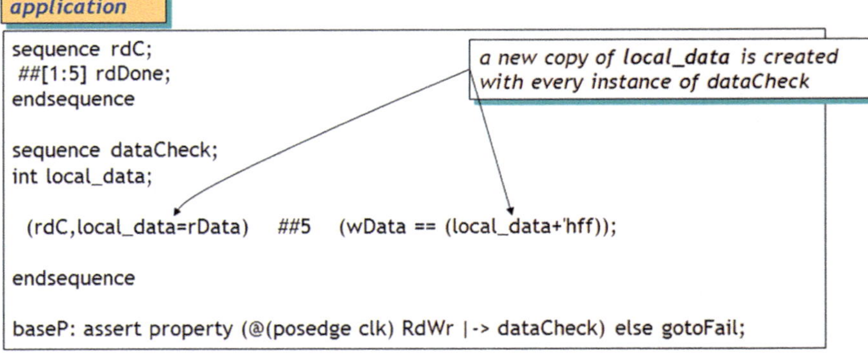

sequence datraCheck reads as ::

on matching 'rdC', store rData in the local var called local_data and ##5 clocks later wData must match local_data+'hff

Note that dataCheck is triggered when 'RdWr' is true. 'RdWr' can be true every clock and dataCheck would be triggered every clock. For every trigger of dataCheck, a new copy of local_data is created which will store rData and check for wData 5 clocks later.

Fig. 11.1 Local variables—basics

later. Note that "RdWr" can be sampled true at every posedge clk. Sequence "data_check" will enter every clock; create a new copy of local_data and create a new pipelined thread that will check for local_data+'hff with "wData" 5 clocks later.

Important Note*: The sampled value of a local variable is defined as the current value (and not the value in preponed region).*

Moving along, Fig. 11.2 shows other semantics of local variables. *Pay close attention to the rule that local variable must be "attached" to an expression while comparison cannot be attached to an expression*!!

As shown in Fig. 11.2, *a local variable must be attached to an expression when you store a value into it. But when you compare the value stored in a local variable, it must not be attached to an expression.*

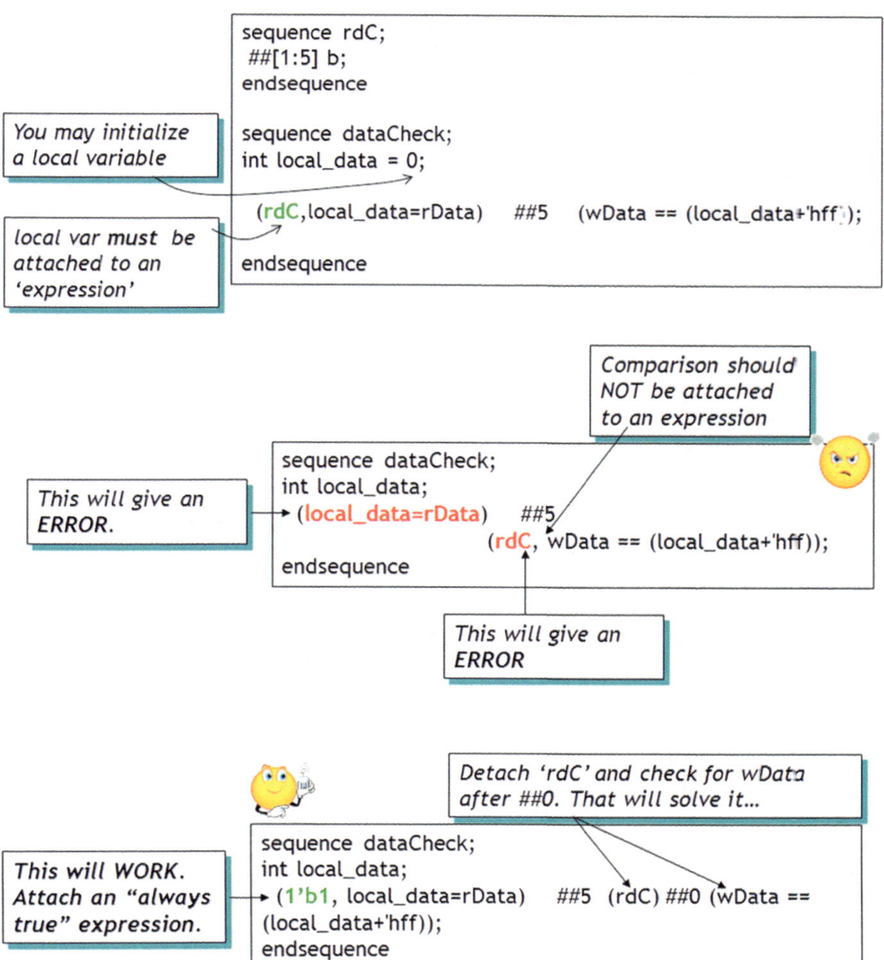

Fig. 11.2 Local variables—do's and don'ts

In the topmost example, "local_data=rData" is attached to the sequence "rdC." In other words, assignment "local_data = rData" will take place only on completion of sequence "rdC." Continuing with this story of storing a value into a local variable, what if you don't have anything to attach to the local variable when you are storing a value? Use 1'b1 (always true) as an expression. That will mean whenever you enter a sequence, that the expression is always true, and you should store the value in the local variable. Simple!

Note that local variables do *not* have default initial values. A local variable without an initialization assignment will be unassigned at the beginning of the evaluation attempt. An initialization assignment to a local variable uses the sampled value of its expression in the time slot in which the evaluation attempt begins. The expression of an initialization assignment to a given local variable may refer to a previously declared local variable. In this case the previously declared local variable must itself have an initialization assignment, and the initial value assigned to the previously declared local variable will be used in the evaluation of the expression assigned to the given local variable. More on this later.

Ok, so what if you want to compare a value on an expression being true? As shown in Fig. 11.2, you can indeed accomplish this by "detaching" the expression as shown. The resulting sequence (the last sequence in Fig. 11.2) will read as "on entering dataCheck, store rData into local_data, wait for 5 clocks and then if "b" (of sequence rdC) is true within 5 clocks, compare wData with stored local_data + 'hff".

Figure 11.3 points out a couple of other important features. First, there is no restriction in using a local variable in either a sequence or a property. In addition,

local variables can be used in 'sequence' or 'property'

```
sequence dataCheck;
int local_data;
  (rdC,local_data=rData)    ##1    ( wData == (local_data+'hff));
endsequence

property dataCheck;
int local_data;
  (rdC,local_data=rData)    |=>    ( wData == (local_data+'hff));
endproperty
```

```
sequence L_seq(Ldata);
int Ldata;
         (rdC, Ldata=rData);
endsequence
```

ERROR:: local var Ldata cannot be declared here because it is used as a formal argument.

Fig. 11.3 Local variables—and formal argument

you cannot declare a local variable as a formal and pass as an actual from another sequence/property. That makes sense, else why would it be called "local"?

In Fig. 11.4, we see that a local variable in a sequence is not visible to the sequence that instantiates it. The solution is quite straightforward. Instead of poking at the local variable directly, simply pass an argument to the sequence that contains the local variable. When the sequence L_seq updates the argument locally, it will be visible to the calling sequence (H_seq). Note that Ldata is not declared as a local variable in sequence L_seq (else that would be an error as we discussed). L_seq simply updates a formal and passes it to the calling sequence, where the actual is declared as a local variable. This is shown in the bottom of Fig. 11.4.

Figures 11.5, 11.6, 11.7, 11.8, 11.9, 11.10, 11.11, and 11.12 show finer rules. Keep them as reference when you embark upon complex assertions. Annotation in the figure explains the situation(s).

Figure 11.6 describes the semantics governing local variables when they are used in the OR of two sequences. The local variable must be assigned in both the sequences of an OR. However, what if you cannot really do that? There are a couple of solutions presented in Fig. 11.7.

Figure 11.9 describes semantics that govern an "and" of two sequences. In contrast to an "or" of two sequences, a local variable must *not* be attached to both sequences involved in an "and." The first solution is identical to that for an "or." Assign the local variable outside of the "and" of the two sequences as shown in the

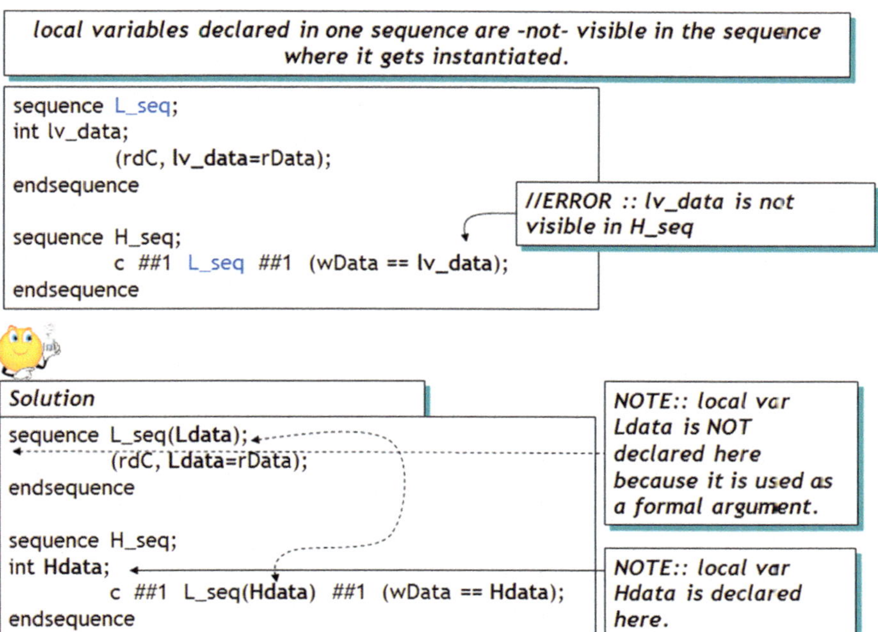

Fig. 11.4 Local variables—visibility

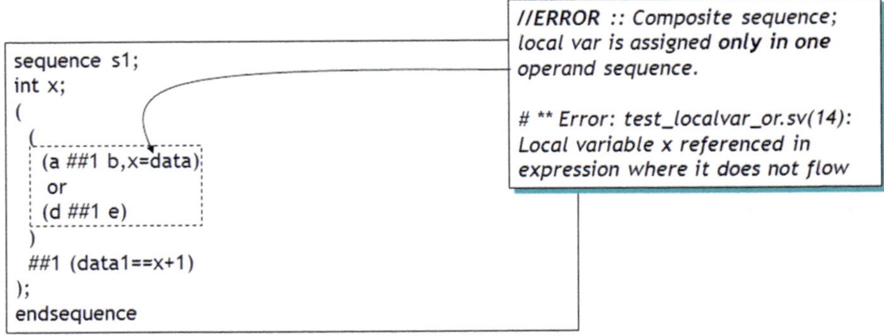

Fig. 11.5 Local Variable composite sequence with an "OR"

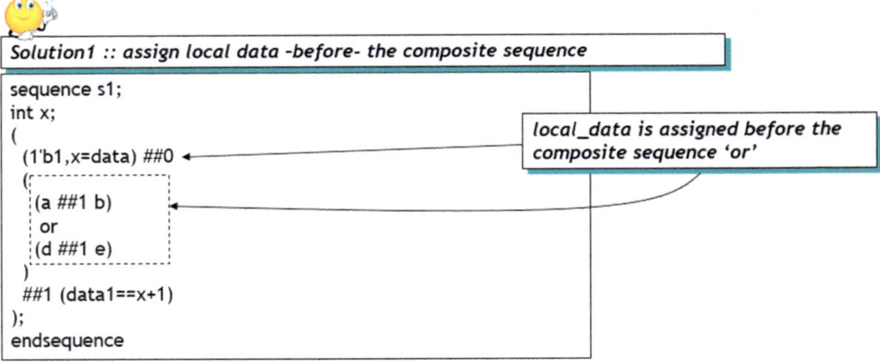

Fig. 11.6 Local Variables—for an "OR" assign local data—before—the composite sequence

figure. Alternatively, simply assign to the local variable in only one of two sequences, which is an obvious solution. Figure 11.9 shows solution #2 in addition to the solution #1 in Fig. 11.8.

Figure 11.10 describes further rules governing local variables. First, you can assign to multiple local variables, attached to a single expression. Second, you can also manipulate the assigned local data in the same sequence (as is the case for ldata2). But as before, there are differences in assigning to (storing to) local variables and comparing their stored value. *You cannot compare multiple local variable*

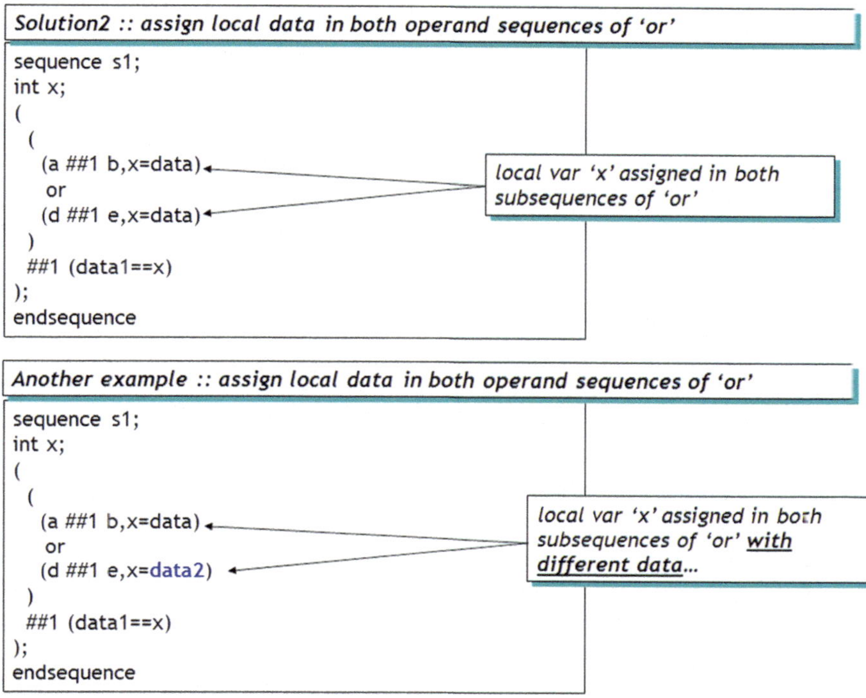

Solution2 :: assign local data in both operand sequences of 'or'

```
sequence s1;
int x;
(
  (
   (a ##1 b,x=data)
    or
   (d ##1 e,x=data)
  )
  ##1 (data1==x)
);
endsequence
```

local var 'x' assigned in both subsequences of 'or'

Another example :: assign local data in both operand sequences of 'or'

```
sequence s1;
int x;
(
  (
   (a ##1 b,x=data)
    or
   (d ##1 e,x=data2)
  )
  ##1 (data1==x)
);
endsequence
```

local var 'x' assigned in both subsequences of 'or' ***with different data...***

Fig. 11.7 Local Variables—assign local data in both operand sequences of "OR"

```
sequence s1;
int x;
(
  (
   (a ##1 b,x=data)
    and
   (d ##1 e,x=data)
  )
  ##1 (data1==x+1)
);
endsequence
```

//ERROR :: Composite sequence; local var is assigned in BOTH operand sequence.

*# ** Error: test_localvar_or.sv(14): Local variable x referenced in expression where it does not flow*

Solution1 :: assign local data -before- the composite sequence

```
sequence s1;
int x;
((1'b1,x=data) ##0
  (
   (a ##1 b)
    and
   (d ##1 e)
  )
  ##1 (data1==x+1));
endsequence
```

local_data is assigned before the composite sequence 'and'

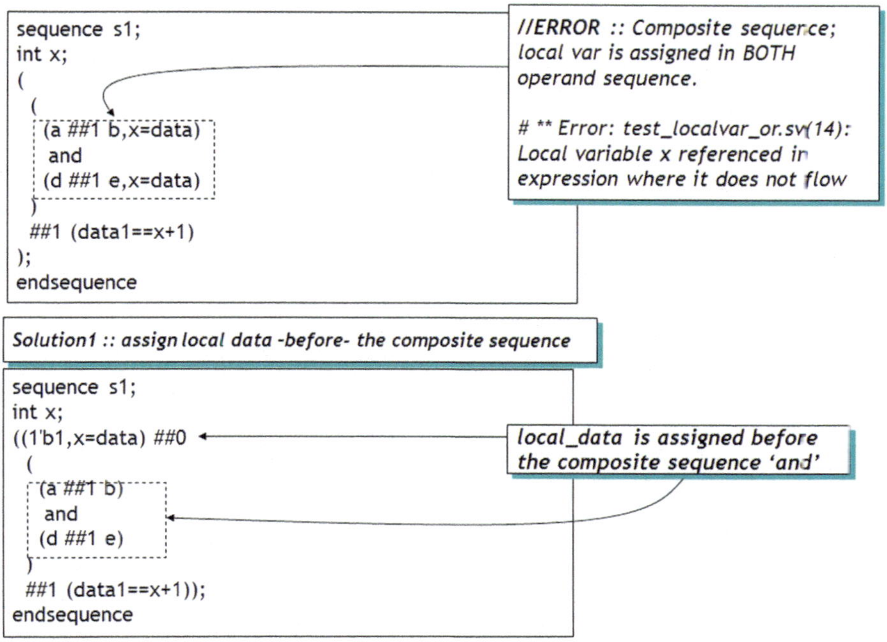

Fig. 11.8 Local Variables—"and" of composite sequences

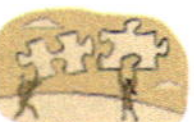

Solution2 :: assign local data in ONLY ONE operand sequences of 'and'

```
sequence s1;
int x;
(
  (
   (a ##1 b,x=data)
    and
   (d ##1 e)
  )
  ##1 (data1==x+1)
);
endsequence
```


Fig. 11.9 Local Variables—further nuances

```
sequence rdC;
 ##[1:5] b;
endsequence

sequence dSeq;
 ##2 d ##2 e;
endsequence

sequence dataCheck;
int ldata1, ldata2;
          (rdC, ldata1=rData, ldata2=retryData) ##5

          (dSeq, ldata2 = (ldata2+'hff)) ##0

          //(wData == ldata1, wretryData==ldata2);

          (wData == ldata1) ##0
          (wretryData == ldata2);

endsequence

baseP: assert property (@(posedge clk) a |=>
dataCheck) else gotoFail;
```


Fig. 11.10 Local Variables—further nuances

values in a single expression in a sequence as is the case in the line "// (wData == ldata1, wretryData ==! ldata2)." This is illegal. Of course, there is always a solution as shown in the figure. Simply separate comparison of multiple values in two subsequences with no delay between the two. The "Solution" annotation in the figure makes this clear.

```
property checkDelay;
int lv;
  (readReq, lv=dataDelay) |=> ##[0:lv] readData;
endproperty
```

ILLEGAL : cannot use a local variable in a delay range.
In general, a delay range cannot have variable delays. They must be constants (or formals that get assigned with a 'constant' actual).
Delay range operators need to be known at elaboration time. Hence they cannot be variables.

Fig. 11.11 Local variable cannot be used in delay range

ILLEGAL :: Cannot use a 'formal' to size a local variable in a property.
Size can only be a constant (or parameter) because it needs to be known at elaboration time.

```
property pr1 (int dSize, csig, enb=1'b1, logic pa, logic pb);

logic [dSize:0] Ldata;

  @(csig, Ldata=data) enb |-> pa ##2 pb;

endproperty

reqGnt: assert property ( pr1( 'd31, posedge clk, cStart, req, gnt) );
```

Fig. 11.12 Local Variables—cannot use a "formal" to size a local variable

Figure 11.11 shows that you cannot use a local variable in the range operator. But, it's not the local variables fault. It's the fact that we cannot have variable delay in either #m or #[m:n] delay operators. From software point of view, the delay range operators need to be known at elaboration time. Hence they cannot be "variables." From hardware point of view, this is a bummer!

Figure 11.12 shows that you cannot use a "formal" to size a local variable. Again, "size" of a vector (bus) declaration can only be a constant. Again, there is a software reason and a hardware reason.

Following points out further cases of legal/illegal declarations of local variables.

```
property illegal_legal_declarations;
     data;     //ILLEGAL. 'data' needs an explicit data type.
     logic data = 1'b0; //LEGAL. Note that unlike SystemVerilog
                           variables, local variables have no
                        //default initial value. Also, the assignment
                        can be any expression and
                         // need not be a constant value
     byte data [ ];     //ILLEGAL - dynamic array type not
                        allowed.
endproperty
```

Also, you can have multiple local data variable declarations as noted above. And a second data variable can have dependency on the first data variable. *But the first data variable must have an initial value assigned.* Here's an example.

```
property legal_data_dependency;
    logic data = data_in, data_add = data + 16'h FF; //LEGAL
endproperty

property illegal_data_dependency;
    logic data, data_add = data + 16'FF;
//ILLEGAL. 'data' used in expression assignment of 'data_add'
is not initialized.
endproperty

sequence illegal_declarations (
output logic a, // ILLEGAL: 'local' keyword is not specified
with direction.

local inout logic b, c = 1'b0, // ILLEGAL: default actual
argument illegal for inout

local d = expr, // ILLEGAL: type must be specified explicitly

local event e, // ILLEGAL: 'event' type is not allowed

local logic f = g // ILLEGAL: 'g' cannot refer to the local
variable declared below. It must be resolved upward from this
declaration

);
```

Note one more example of how you can declare a local variable used for initialization directly as an input. First, the sequence with the traditional way of initializing BSize.

ONE:

```
sequence burst (logic FRAME_, BurstSize = 4)
logic abc = 1'b0, BSize = BurstSize;
@(posedge clk)
    FRAME_ |=> ....
endsequence
```

Since the BurstSize is solely used for sizing the local variable BSize, you can parameterize it as follows, where now the actual will determine the BSize.

TWO:

```
sequence burst (logic FRAME_, local input logic BSize)
logic abc = 1'b0;
@(posedge clk)
    FRAME_ |=> ....
endsequence
```

The keyword "local" specifies that BSize is an "argument local variable" while the direction "input" specifies that BSize will receive its initial value from the actual argument expression.

Note that in the first sequence the BSize initialization takes place @(posedge clk). Here the declaration assignments are performed when the evaluation reaches alignment with @(posedge clk) and at that point the value in the formal argument BurstSize is assigned to BSize as its initial value. In the second sequence, the initialization of BSize takes place when the actual changes its value assigns to formal BSize. Similarly, an "output" "argument local variable" outputs its value to the actual argument whenever the sequence matches. For "inout," it obviously acts both as "input" and "output." So, the rules specified for "input" applies when it acts as an "input" and the rules for "output" apply when it acts as an "output."

Note that "argument local variables" precede the "body local variables." If a sequence or property has both "input" "argument local variables" and the "body local variables" with declaration assignments, the initialization assignment of the "input" "argument local variables" are performed first.

On the similar line of thought, the following rules also apply. See the "sequence" below.

```
sequence local_IO (
    local       byte a;
local inout     byte b;
local input     logic c;
local output    byte d;
);
endsequence
```

Following rules apply to the local variables of sequence local_IO.

1. If a direction is specified for an argument, then the keyword "local" must also be specified.
2. If the keyword "local" is specified, then the data type must also be explicitly specified.
3. If the keyword "local" is specified without a direction, then a default direction of "input" is understood.
4. An "input" argument local variable may be declared with an optional default actual argument which can be any expression.
5. An "output" or "inout" argument local variable *cannot* be declared with a default actual argument because the actual argument must specify the local variable that will receive the "output" value.
6. An "argument local variable" can also be declared as "output" or "inout," but *only* in a sequence declaration. An "argument local variable" of a property must be of direction "input."
7. An "output" "argument local variable" outputs its value to the actual argument whenever the sequence matches.
8. An "input" receives its initial value from the actual argument.
9. For "inout," it obviously acts both as "input" and "output" and the rules for "input" apply when it is of direction "input" and the rules for "output" apply when it is of direction "output."
10. As stated above, it is important to understand that the "sampled" values are used for all terms that are *not* "local" while the "current" values are used for terms that are "local."
11. The actual argument bound to an "argument local variable" of direction "output" or "inout" must itself be a local variable.

A special note on the use of method ".triggered" with a local variable. A local variable passed into an instance of a named sequence to which sequence method (.triggered) is applied, is not allowed. For example, the following is illegal.

```
sequence check_trdy (cycle_begin);
        cycle_begin ##2 irdy ##2 trdy;
endsequence

property illegal_use_of_local_with_triggerd;
      bit local_var;
      (1'b1, local_var = CB) |-> check_trdy(local_var).
      triggered;  //ILLEGAL
endproperty
```

Some more rules on referencing local variables.

A local variable can be referenced in expressions such as:

- Array indices
- Arguments of task and function calls
- Arguments of sequence and property instances
- Expressions assigned to local variables
- Boolean expressions
- Bit-select and part-select expressions

However, a local variable *cannot* be referenced in the following:

- Clocking event expressions (even though LRM is ambiguous on this)
- The reset expression of a "disable iff"
- The abort condition of a reset operator (accept_on, sync_accept_on and the reject forms of these expressions. See Sect. 20.16)
- Expressions that are compile time constants, e.g., [*n], [-> n], ##n, and [=n], and the constant expressions of ranged forms of these operators
- An argument expression to a sampled value function ($rose, $fell, $past. etc.)

11.1 Application: Local Variables

The application in Fig. 11.13 is broken down as follows.

```
($rose(read),localID=readID
    On $rose(read), the readID is stored in the localID.
not (($rose(read) && readID==localID) [*1:$])
```

Once a 'read' has been issued, another 'read' for the same readID cannot be re-issued until a readAck with the same ID has returned.

```
property checkRead;
  int localID;
  ($rose(read),localID = readID) |=>
                             not (($rose(read) && readID==localID) [*1:$]) ##0
                             ($rose(readAck) && readAckID == localID);
endproperty

baseP: assert property (checkRead) else
                             $display($stime,,,"\tproperty FAIL");
```

Fig. 11.13 Local variables—application

Then we check to see if another read ($rose(read)) occurs and it's readID is the same as the one we stored for the previous Read in localID. We continue to check this consecutively until *##0 ($rose(readAck) && readAckID == localID)* occurs.

If the consecutive check does result in a match, that would mean that we did get another $rose(read) with the same readID with which the previous read was issued. That's a violation of the specs. This is why we take a "not" of this expression to see that it turns false on a match and the property would fail and end.

If the consecutive check does not result in a match until *##0 ($rose(readAck) && readAckID == localID)* arrives, then we may get a readAck with the same readAckID with which the original read was issued. The property will then pass (or fail if readAckID is not equal to localID).

In short we have proven that once a "read" has been issued that another "read" from the same readID cannot be re-issued until a "readAck" with the same ID has returned.

Example: This example shows a simple way to track time. Here, on falling edge of Frame_, rising edge of IRDY cannot arrive for at least MinTime.

Solution:

```
property FrametoIRDY (integer minTime);
integer localBaseTime;
     @(posedge clk) ($fall(Frame_), localBaseTime = $time)
     |=>
     $rose(IRDY) && $time >= localBaseTime + minTIme);
endproperty
measureTime:   assert   property   (FrametoIRDY   (.minTIme
(MINIMUM_TIME)));
```

Local variable examples are scattered throughout the book. Some are found in following sections.

Section Clock delay range operator: ##[m:n]: multiple threads 8.5.

Section Asynchronous FIFO TEST-BENCH AND ASSERTIONS 18.2.

Chapter 12
Recursive Property

Introduction: This chapter discusses Recursive property, its semantics and practical applications.

A. B. Mehta, *SystemVerilog Assertions and Functional Coverage*,
https://doi.org/10.1007/978-3-030-24737-9_12

Recursive property simply states that a condition holds. The property calls itself with a correct nonoverlapping implication operator and correct antecedent and consequent relation. As shown in Fig. 12.1 (top "property rc1"), if "ra" is true *and* at next clock rc1(ra) is true that "rc1" should recur on itself. Note that the antecedent in "rc1" is 1'b1, meaning the antecedent is always true. This allows for an easy and correct "and" of an expression with an antecedent/consequent implication. Note also that you must use a nonoverlapping operator in a recursive property.

The topmost example in Fig. 12.1 (baseP: assert property (@ (posedge clk) $fell(rst_) |- > rc1(bStrap)) else gotoFail;) specifies that when $fell(rst_) is true, the consequent rc1(bStrap) is invoked. "rc1" property takes "bStrap" as the input and does an "and" of the input with (1'b1 |=> rc1(ra)). This means that the implication 1'b1 is always true and that rc1(ra) will be called every posedge of clk to recur on itself. The property will continue to go into loop on itself until bStrap is sampled 0.

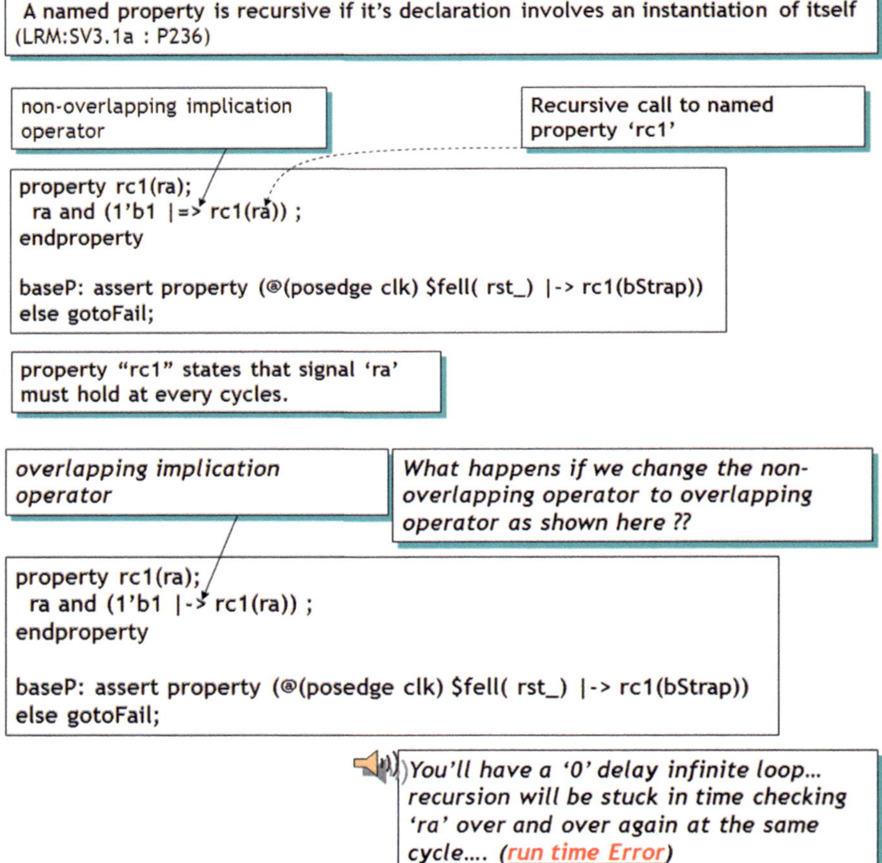

Fig. 12.1 Recursive property—basics

In that case the "and" will fail and so would the property. In other words, we have checked to see that after $fell(rst_), the "bStrap" (bootstrap signal) does not get de-asserted (i.e., go Low).

Note: If you use an overlapping operator, the property will recur on itself in zero time, essentially trapping the simulator in a zero-delay loop causing simulation to hang. This will result in a Run Time Error.

But what good does the property in Fig. 12.1 really do. It will check for a signal to hold forever. How do you apply such a model to real world? What you really need to check is that a condition holds until another condition remains true. If that happens, the property passes, else it fails. Now that is more practical. Let us understand this with the following example.

12.1 Application: Recursive Property

As shown in Fig. 12.2, with an "or" condition, the recursive property becomes useful. The specification says that we need to make sure that "intr" is held true until "iack" is asserted. Let's see how that works. Property rc1 says that either iack is true or intr is true that the property calls rc1 recursively. Now, if the "intr" falls (goes low) before an iack, the **(intr and (`true |=> rc1(intr, iack)))** will fail because this is an AND with "intr." On the other hand, if iack arrives first, the "**iack or ...**" condition will pass because this is an OR. In other words, we are recursive on "intr" to see that it holds true until iack arrives.

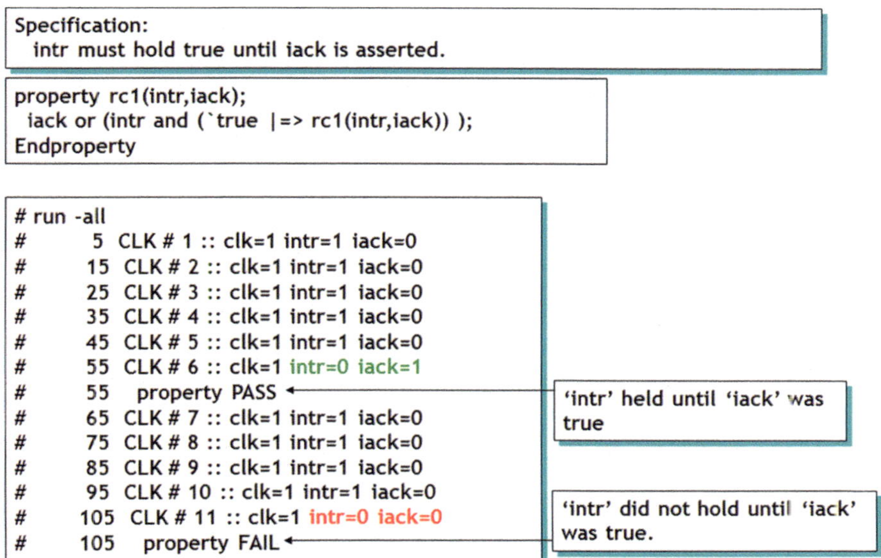

Fig. 12.2 Recursive property—application

```
Specification:
        For a Dcache write miss (missDCache), start miss allocation (allocStart)
the next clock and issue a readMem the clock after.

        Check to see that above matches and on a match make sure that Write
Data is Held (wDataH asserted) until mem read completes (readC)
```

```
sequence missAlloc;
        missDCache ##1 allocStart ##1 readMem;
endsequence

sequence rc1(ra,rb);
        rb or (ra and (`true |=> rc1(ra,rb)) );
endsequence

property s_rc1(genericSeq,sra,srb);
        (genericSeq,tdisp) |=> rc1(sra,srb);
endproperty

baseP: assert property (@(posedge clk) s_rc1(missAlloc,wdataH,readC))
                                    else gotoFail;

task tdisp;
$display($stime,,,"Dcache Miss to Alloc to Read Mem sequence matches");
endtask
```

"sequence missAlloc" is passed as an actual parameter to "property s_rc1". Only on a match of the missAlloc the check for Write Data Hold until Read complete is triggered.

Fig. 12.3 Recursive property—application

At time 55 in the simulation log, iack = 1 and intr = 0. Since intr was equal to "1" the previous clock, it held itself until iack arrived. Hence, the property passes. On the other hand, at time 105, intr goes "0" *before* iack goes 1. In other words, intr did not hold itself until iack arrived and the property fails. Figure 12.3 shows another interesting application.

12.2 Application: Recursive Property

The specification of this property reads as "on detection of missAlloc, see that wdataH is held until readC is true." In Fig. 12.3, property "s_rc1" checks to see that "misAlloc" is true, upon which it triggers "rc1." Property rc1 in turn holds true (i.e., recursive) until readC is true. If "wdataH" goes low before readC goes high, the property (and hence the entire assertion) will fail (because wdataH is the LHS of an "and" condition in sequence rc1). If readC arrives before wdataH, the "or" condition in the sequence rc1 passes and hence the assertion will pass.

This way we have made sure that wdataH is held until readC completes. If this is not quite apparent at first, please refer to Fig. 12.2 to understand how "property rc1" works.

Note that we are passing the entire sequence "missAlloc" as an actual to the formal "genericSeq" of property s_rc1. This is a very useful way to use a sequence as an actual to sequence formal.

Another point I'd like to make here (and as I've maintained throughout the book) is that you need to break down a complex specification into smaller sequences and then tie them all together in a property. So, the first thing we did in this example was to create a sequence called "missalloc" that ties in the relationship among missDCache, allocStart, and readMem. This relationship was the requirement in the first part of the specification. Then we made sure that we establish a relationship required by the specification that write data is held until readC completes. We did this using the recursive sequence "rc1." Once these two relationships were modeled, all we need to do was to tie these two through a property. We determined that the antecedent in property "s_rc1" is the sequence "missAlloc" and consequent is the sequence "rc1." So, we tied the two sequences in the property "s_rc1." Thus, a complex specification was modularly modeled into an assertion.

Further nuances are described in Fig. 12.4 with annotations.

Figure 12.5 shows that "disable iff" is not allowed in a recursive property. Well, there is a simple solution to that problem, which is shown in the bottom of the figure. Separate the requirement of "disable iff" and the recursive nature of the property in two properties. The recursive property does not contain "disable iff" and the "property rillegal" disables rLegal with the "disable iff" condition.

Operator 'not' cannot be applied to recursive property instances.

```
property rIlligal;
  c |-> a and (`true |=> not (rIlligal) );
                          //  ^^^
endproperty
```

*** Error:
test_recursive_restrictions.sv(18):
Operator "not" can not be applied
to recursive properties.*

If p is a recursive property, then, in the declaration of p, every instance of p must occur after a positive advancement in time

```
property riLegal;
  c |-> a and (`true |-> riLegal);
                   //  ^^^
endproperty
```

***Error: (vsim-8312)
test_recursive_restrictions.sv(61):
Use of recursion in property rLegal
without positive advance in time is
illegal.*

Fig. 12.4 Recursive property—further nuances I

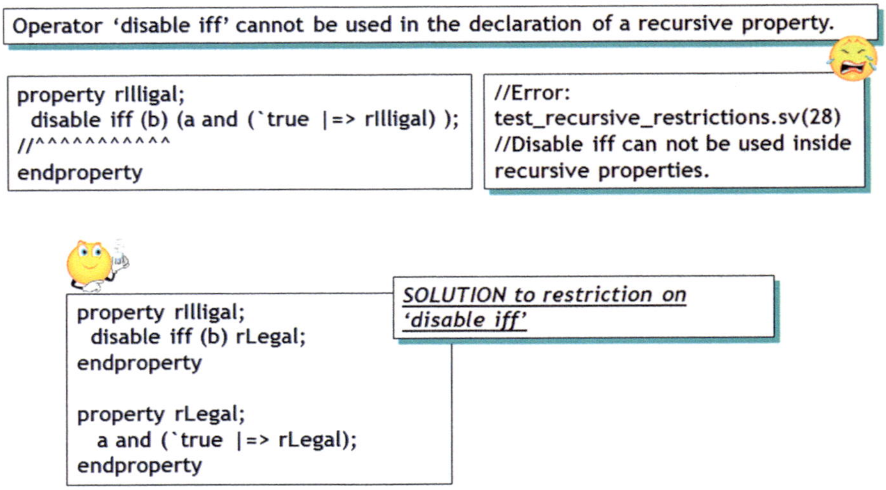

Operator 'disable iff' cannot be used in the declaration of a recursive property.

```
property rIlligal;
  disable iff (b) (a and (`true |=> rIlligal) );
//^^^^^^^^^^^
endproperty
```

```
//Error:
test_recursive_restrictions.sv(28)
//Disable iff can not be used inside
recursive properties.
```

SOLUTION to restriction on 'disable iff'

```
property rIlligal;
  disable iff (b) rLegal;
endproperty

property rLegal;
   a and (`true |=> rLegal);
endproperty
```

Fig. 12.5 Recursive property—further nuances II

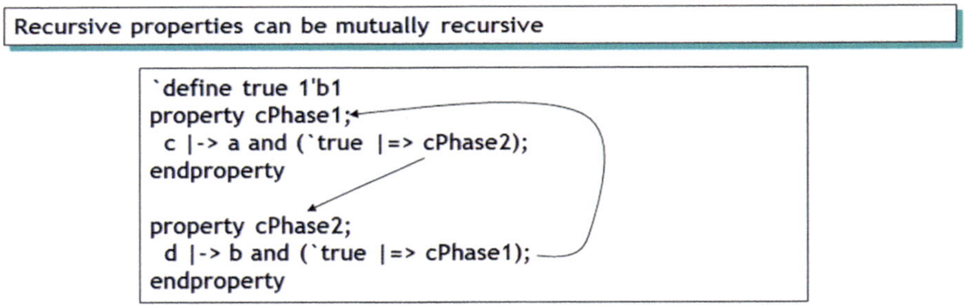

Recursive properties can be mutually recursive

```
`define true 1'b1
property cPhase1;
  c |-> a and (`true |=> cPhase2);
endproperty

property cPhase2;
  d |-> b and (`true |=> cPhase1);
endproperty
```

Fig. 12.6 Recursive property—mutually recursive

Figure 12.6 shows that two recursive properties can indeed be mutually recursive. "cPhase2" calls "cPhase1" and "cPhase1" in turn calls "cPhase2." Why would this not end up in a zero-delay loop? First, there is the 1 clock wait because of the nonoverlapping operator in both properties. Second, each property has an antecedent and the property will execute only if the antecedent is true. So, if "cPhase2" calls "cPhase1," then "cPhase1" will first wait for "c" to be true and then trigger the recursive part of the property. The same happens when "cPhase1" calls "cPhase2."

Also note that the operators accept_on, reject_on, sync_accept_on, and sync_reject_on (Sect. 20.16) have not been covered yet, but they may be used inside a recursive property. But, strong operators s_nexttime, s_eventually, s_always, s_until, and s_until_with *cannot* be applied to any property expression that instantiates a recursive property.

Chapter 13
Endpoint of a Sequence (.triggered and .matched)

Introduction: This chapter describes the endpoint sequence detection methods such as .triggered (.ended in 2005 LRM) and .matched and their operation with overlapping and nonoverlapping operators. Legal, illegal conditions and plenty of examples and applications are presented.

A. B. Mehta, *SystemVerilog Assertions and Functional Coverage*,
https://doi.org/10.1007/978-3-030-24737-9_13

13.1 .triggered (Replaces .ended)

Before we learn how .triggered works, here's what has changed in the 1800-2009/2012 standard.

The 2009/2012 standard gets rid of .ended and in place supports .triggered. In other words, .triggered has the same exact meaning as .ended, only that .triggered can be used both where .ended gets used as well as where .triggered was allowed in previous versions. In other words, .triggered can be used in a sequence as well as in procedural block and in level sensitive 'wait' statement.

Following from the 2012 LRM:

IEEE Std 1800–2005 17.7.3 required using the .ended sequence method in sequence expressions and the .triggered sequence method in other contexts. Since these two constructs have the same meaning but mutually exclusive usage contexts, in this version of the standard, the .triggered method is allowed to be used in sequence expressions, and the usage of .ended is deprecated and does not appear in this version of the standard.

Note that the entire discussion devoted to .triggered in this chapter applies directly to .ended. In other words, you can replace .triggered with .ended in this chapter and you will get the same results.

If you are mainly interested in the *end* of a sequence regardless of when it started, .triggered is your friend. The main advantage of methods that detect the endpoint of a sequence is that you do *not* need to know the start of the sequence. All you care for is, when a sequence ends.

Figure 13.1 shows that behavior. Sequence 'branch' is a complete sequence for a branch to complete. It could have started any time. The property endCycle wants to make sure that the 'branch' sequence has indeed ended when endBranch flag goes high. In other words, whenever $rose(endBranch) is detected to be true that the next clock, branch must end. This is indeed very powerful and useful feature. This makes the assertion intuitive as well.

IMPORTANT NOTE: What if you simply write the assertion as "at the end of 'branch' see that endBranch goes high" as in "branch(a, b, c, d) |=> $rose(endBranch)". What's wrong with that? Well, what if $rose(endBranch) goes high when the 'branch' is still executing? That $rose(endBranch) would go unnoticed until the end of the sequence 'branch'. The .triggered operator would catch this. If endBranch goes high when sequence 'branch' is still executing, the property will fail. That's because the property endCycle expects 'branch' to have ended when endBranch goes high. Since endBranch could have risen prematurely, the property will see that at $rose(endBranch) the 'branch' sequence has not ended, and the property would fail. The forward-looking property would not catch this. This is a very important point. Please make a note of it.

Also, note that the clock in both the source and destination sequence must be the same. But what if you want the source and destination clocks to be different? That is what '.matched' does, soon to be discussed.

Figure 13.2 explains further nuances.

.triggered is a method on a sequence (that returns true or false).

Whenever the end point of a sequence is reached, .triggered will be true -regardless- of when the sequence started.

- *.triggered allows another way to create smaller subsequences leading to more complex ones.*

Any sequence that will have a method attached to it must have an explicit clock.

"Use of a method on an unclocked sequence is illegal".

application

```
sequence branch(a ,b ,c ,d);
 @(posedge clk) $fell(a) ##[1:5] $rose(b) ##1 c [=2] ##1 d;
endsequence

property endCycle;
 @(posedge clk) $rose( endBranch ) |=> branch(a, b ,c , d).triggered;
endproperty
```

The source and destination clocks -must- be the same.

Fig. 13.1 .triggered—end point of a sequence

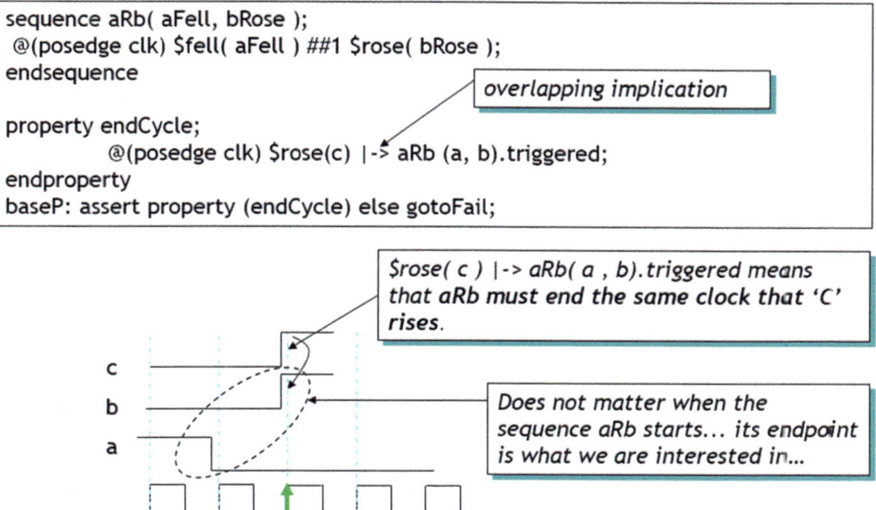

Fig. 13.2 .triggered with overlapping operator

In the example, $rose(c) implies that the sequence aRb must have ended. Key point to note here is that the implication operator is overlapping. This means that when $rose(c) occurs that at the same clock, sequence aRb should end. As shown in Fig. 13.2, when $rose(c) is true that at the same clock $rose(b) must occur and the previous clock $fell(a).

Figure 13.3 describes the same example but with the nonoverlapping operator.

You must understand this very carefully, as simple as it looks. Nonoverlapping states that when $rose(c) is true that *at the next clock* the sequence aRb must end. Hence, as shown is the figure, the property looks for aRb to end one clock *after* $rose(c). This maybe intuitive but easy to miss.

Let us revisit an example I had presented in Fig. 8.15. This figure is again depicted here for easy reference. Note that in this example, we see how a "consecutive repetition" operator allowed us to design a property for state transition checks. What if you want to model the same property using .triggered?

First, state transition checks using "consecutive repetition" operator—same as shown in Fig. 8.15.

Specification:

The state machine must follow specified state transitions.

```
`define readStart (read_enb ##1 readStartState)
`define readID (readStartState ##1 readIDState)
`define readData (readIDState ##1 readDataState)
`define readEnd (readDataState ##1 readEndState)

sequence checkReadStates;

  @(posedge clk)
  `readStart           ##1
  `readID      [*1:$]  ##1  //`readID && !`readData
  `readData    [*1:$]  ##1  //`readData && !`readEnd
  `readEnd     ;

endsequence
```

Next, the same property written using ".triggered" method.

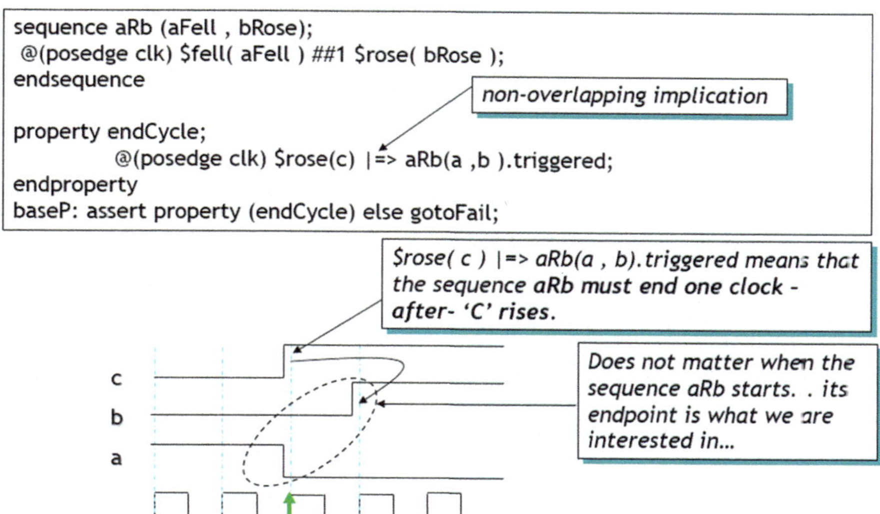

Fig. 13.3 .triggered with nonoverlapping operator

```
Specification:
Make sure that the state machine follows the specified transitions

sequence readStart; @(posedge clk) read_enb ##1 readStartState; endsequence
sequence readID; @(posedge clk) readStartState ##1 readIDState; endsequence
sequence readData; @(posedge clk) readIDState ##1 readDataState; endsequence
sequence readEnd; @(posedge clk) readDataState ##1 readEndState; endsequence

property checkReadStates;
 @(posedge clk)
        readStart.triggered  ##[1:$]
        readID.triggered     ##[1:$]
        readData.triggered   ##[1:$]
        readEnd
        ;
endproperty

sCheck: assert property (checkReadStates) else $display ($stime,,,"FAIL");
cCheck: cover property (checkReadStates) $display ($stime,,,"PASS");
```

Let us examine the property using .triggered. Instead of using `define, we have to explicitly declare different sequences for state transitions (because the .triggered method can only be attached to a named sequence). Also, note that in each sequence there is an explicit @(posedge clk). This is because .triggered method cannot be attached to a non-clocked sequence. The main property checkReadStates shows how .triggered is used in place of the consecutive repetition operator. We wait for each sequence to "end" and see that the next sequence then ends within 1 clock to whenever, since we don't know how long will the next state transition (e.g., from readStart to readID) take place. This is the reason for ##[1:$] after each .triggered sequence. And that each of these "ends" occurs in a specified order. This example also illustrates how multiple sequences are used in a property, and as I've said before, it is best to break down a complex property (as this) in multiple small sequences and then build the master property.

Here's the entire test-bench with the property and the simulation log. Please study the simulation log to solidify your concept of .triggered.

```
module state_transition;
logic readStartState, readIDState, readDataState, readEndState;
logic clk, read_enb;

sequence readStart; @(posedge clk) read_enb ##1 readStartState;
endsequence
sequence readID; @(posedge clk) readStartState ##1 readIDState;
endsequence
sequence readData; @(posedge clk) readIDState ##1 readDataState;
endsequence
sequence readEnd; @(posedge clk) readDataState ##1 readEndState;
endsequence

property checkReadStates;
  @(posedge clk)
    readStart.triggered     ##[1:$]
    readID.triggered        ##[1:$]
    readData.triggered      ##[1:$]
    readEnd
    ;
endproperty

sCheck: assert property (checkReadStates) else $display ($stime,,,
"FAIL");
cCheck: cover property (checkReadStates) $display ($stime,,,
"PASS");
```

```
initial
begin
    read_enb=1; clk=0;
    @(posedge clk) readStartState=1;
    @(posedge clk) readIDState=1;
    @(posedge clk) @(posedge clk); readDataState=1;
    @(posedge clk) @(posedge clk); readEndState=1;
end

initial $monitor($stime,,,"clk=",clk,,
        "read_enb=%0b",read_enb,,,
        "readStartState=%0b",readStartState,,
        "readIDState=%0b",readIDState,,
        "readDataState=%0b",readDataState,,
        "readEndState=%0b",readEndState);

always #10 clk=!clk;
endmodule
```

```
/*
#    10   clk=1   read_enb=1   readStartState=1   readIDState=0
     readDataState=0 readEndState=0
#    20   clk=0   read_enb=1   readStartState=1   readIDState=0
     readDataState=0 readEndState=0
#    30   clk=1   read_enb=1   readStartState=1   readIDState=1
     readDataState=0 readEndState=0
#    40   clk=0   read_enb=1   readStartState=1   readIDState=1
     readDataState=0 readEndState=0
#    50   clk=1   read_enb=1   readStartState=1   readIDState=1
     readDataState=0 readEndState=0
#    60   clk=0   read_enb=1   readStartState=1   readIDState=1
     readDataState=0 readEndState=0
#    70   clk=1   read_enb=1   readStartState=1   readIDState=1
     readDataState=1 readEndState=0
#    80   clk=0   read_enb=1   readStartState=1   readIDState=1
     readDataState=1 readEndState=0
#    90   clk=1   read_enb=1   readStartState=1   readIDState=1
     readDataState=1 readEndState=0
#    100   clk=0   read_enb=1   readStartState=1   readIDState=1
     readDataState=1 readEndState=0
#    110   clk=1   read_enb=1   readStartState=1   readIDState=1
     readDataState=1 readEndState=1
#    120   clk=0   read_enb=1   readStartState=1   readIDState=1
     readDataState=1 readEndState=1
#    130   PASS
```

```
#     130   clk=1   read_enb=1   readStartState=1   readIDState=1
      readDataState=1 readEndState=1
#     140   clk=0   read_enb=1   readStartState=1   readIDState=1
      readDataState=1 readEndState=1
#     150 PASS
#     150   clk=1   read_enb=1   readStartState=1   readIDState=1
      readDataState=1 readEndState=1
#     160   clk=0   read_enb=1   readStartState=1   readIDState=1
      readDataState=1 readEndState=1
#     170 PASS
*/
```

The .triggered method may be applied to a named sequence instance, with or without arguments, an untyped formal argument, or a formal argument of type sequence, as follows:

```
sequence_instance.triggered
or
formal_argument_sequence.triggered
```

When method .triggered is evaluated in an expression, it tests whether its operand sequence has reached its end point at that particular point in time. The result of .triggered does not depend upon the starting point of the match of its operand sequence.

Here's an example of .triggered usage in a procedural assignment using level sensitive control.

```
sequence  busGnt;
  @ (posedge clk) req ##[1:5] gnt;
endsequence
initial begin
   wait (busGnt.triggered) $display($stime,,, "Bus Grant given");
end
```

Note that .matched cannot be used in the above "wait" statement. That is illegal. ".matched" is discussed in the next section.

Here's an example that shows the use of .triggered in sequences.

```
sequence abc;
  @ (posedge gclk) a ##[1:5] ##1 b [*5] ##1 c;
endsequence

sequence myseq;
  @ (posedge gclk) d ##1 abc.triggered ##1 !d;
endsequence
```

Another example:
Upon detection of IRDY End, Tabort must remain de-asserted until Frame_ End.
Solution:

```
sequence IRDY;
        IRDY_start ##2 IRDY_end;
endsequence

sequence Frame_;
        Frame_Start ##[1:16] Frame_end;
endsequence

busyPr: assert property (@(posedge clk)
        IRDY |-> !Tabort until Frame_.triggered;
```

Finally, following is *ILLEGAL*.

```
logic x,y,z;
//....
Xillegal: assert property (@(posedge clk)
                    Z |-> (x ## 3).triggered  //ILLEGAL
```

Why is this illegal? Recall that .triggered can only be applied to a ‘named sequence instance” or a formal argument. (x ## 3) is neither of the two.

Following is *ILLEGAL* as well because the clocks on two sides of the implication differ. They need to be the same.

```
sequence clk1Seq;
            @(posedge clk1) x ##2 y;
endsequence
Zillegal:assertproperty(@(posedgeclk)z|->clk1Seq.triggered);
//ILLEGAL
```

Also, note that .triggered can be used as a subexpression. Here’s an example:

```
default clocking @(posedge clk); endclocking
sequence ab;
    a ##2 b;
endsequence

sequence cd;
    c ##2 d;
endsequence
```

```
sequence cdtr;
    z ##[2:5] cd.triggered;
endsequence

ftr: assert property (ab |-> nogo until cdtr.triggered);
```

The order of execution for the consequent is cd, cdtr, ftr. Note that the order of sequence evaluation is statically determined at compile time. A proper order of evaluation must always be found.

13.2 .matched

The main difference between .triggered and .matched is that .triggered requires both the source and destination sequences have the same clock. .matched allows you to have different clocks in the source and destination sequences.

Since the clocks can be different, understanding of .matched gets a bit complicated. But it follows the same rules as that for multiply clocked properties. As shown in Fig. 13.4, sequence "e1" uses "clk" as its clock while sequence "e2" uses "sysclk"

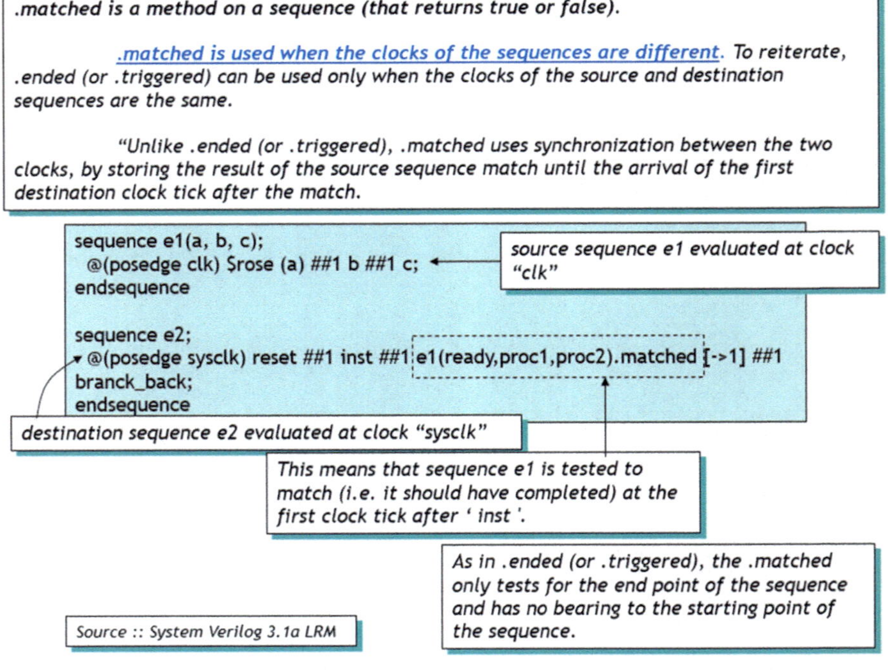

Fig. 13.4 .matched—Basics

as its clock. Sequence “e2” says that after “reset”; 1 clock later; “inst” must be true and 1 “clk” later sequence “e1” must match (i.e., end) at least once and 1 “sysclk” later branch_back must be true. This is a very interesting way of “inserting” a .matched (or .triggered for that matter) within a sequence. Sequence “e1” is running on its own and in parallel to the calling sequence. What we really care for is that it matches (ends) when we expect it to.

Let us look at some examples that will make the concept clearer.

Figure 13.5 shows that on the rising edge of “c” we want to check that “aRb” sequence matches. Let us look at the timing diagram. When, $rose(c) is true at posedge of clk1 (clk1 = 3), it looks for the end of the sequence “aRb.” Sequence “aRb” started during clk = 3 and it ends at clk = 4. Note that the end of “aRb” (i.e., match of $rose(bRose)) is exactly at the very *next subsequent* “clk” edge after “clk1” edge. That’s what the property asked for “@ posedge clk1) $rose (c) |= > @ (posedge clk) aRb(a, b).matched;”

The key point to note here (as in multiply clocked properties) is the clock crossing boundary. From clk1 to clk with 1 clock delay in-between (as implied by |= > nonoverlapping implication) does not mean 1 full clock. It simply means the *very next subsequent edge* of “posedge clk” after “posedge clk1.” Please refer to multiply clocked properties (Sect. 10.1) if this concept is not clear.

Figure 13.6 uses overlapped implication and identical (in-phase) clocks (the clocks must be the same because we are using overlapping implication). Hence

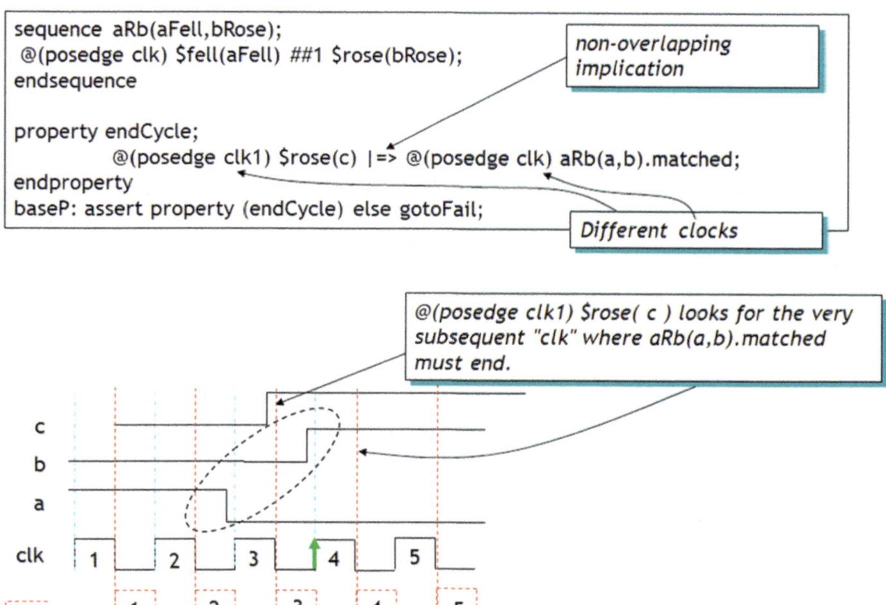

Fig. 13.5 .matched with nonoverlapping operator

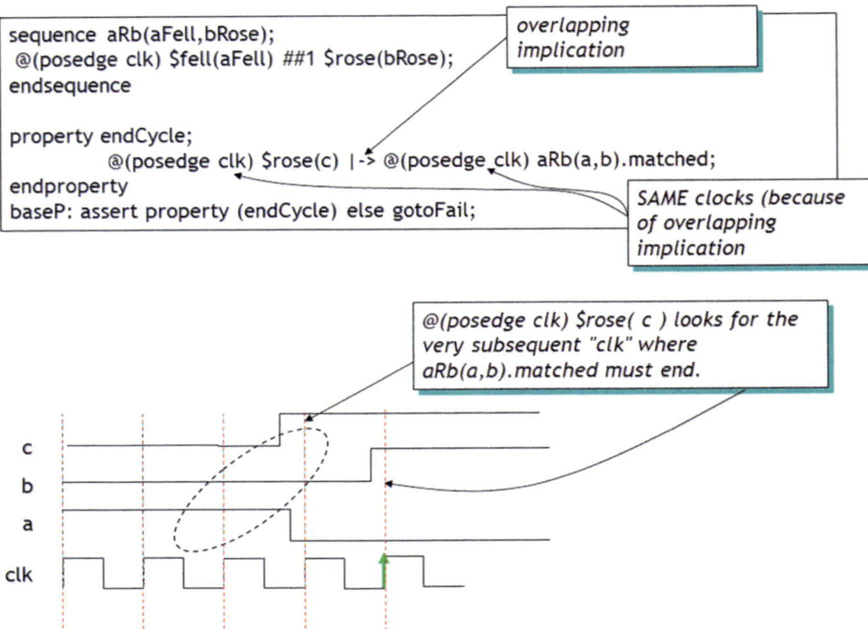

Fig. 13.6 .matched—overlapped operator

when $rose(c) is true, it looks for the very *next subsequent* (posedge clk) overlapping with its (posedge clk). But the very next (posedge clk) is (obviously) 1 clock later. So, 1 clock after $rose(c), the sequence sees that aRb has matched (i.e., ended).

13.3 Application: .matched

In Fig. 13.7, sequence "RdS" uses Busclk while the property checkP uses sysclk. "@ (negedge sysclk) RdS.matched" means that at the negedge of sysclk, the sequence RdS (which is running off of Busclk) must end. "RdS" could have completed slightly (i.e., when the very preceding posedge of Busclk would have arrived) earlier than the negedge sysclk. That is ok because we are transitioning from Busclk to sysclk (as long as the sequence RdS completes at the *immediately preceding* posedge Busclk).

Let us now look at how clock "flows" or gets inferred in sequences and properties that use .triggered and .matched methods.

```
sequence RdS;
  @(posedge Busclk) $fell (as_) ##1 rd ##[1:5] oe_;
endsequence

property checkP;
  @(negedge sysclk) RdS.matched |=> rdDataLatch ##0 ($isunknown
(data) == 0);
endproperty
```

The matched value of sequence RdS is sampled at the negedge of sysclk and if it is true (i.e. the sequence has completed) it implies that rdDataLatch is asserted the next negedge sysclk and that the data is not unknown at that clock

Fig. 13.7 .matched—Application

```
module clock_inference;
logic a, b, c, d;

default clocking cb @(posedge clk_d); endclocking


sequence e4;
  $rose(b) ##1 c;
endsequence

a1: assert property (@(posedge clk_a) a |=> e4.triggered);
//Since the leading edge in 'a1' is posedge clk_a, 'e4' infers
posedge 'clk_a' as per clock flow rules

sequence e5;
  @(posedge clk_e1) a ##[1:3] e4.triggered ##1 c;
endsequence
```

/∗ Note that the leading clock edge here is "clk_e1." So, "e4" will infer posedge clk_e1 as per clock flow rules wherever e5 is instantiated (with/without a method) ∗/

```
a2: assert property (@(posedge clk_a) a |=> e5.matched);
```

/* Here, "a" triggers "e5.matched" with the leading clock as "posedge clk_a." "e5" in turn triggers "e4.triggered" with the leading clock as "posedge clk_e1." Hence, by clock flow rules, "e4" infers "posedge clk_e1" from sequence "e5."
*/

```
    sequence e6(f);
        @(posedge clk_e2) f;
    endsequence

a3: assert property (@(posedge clk_a) a |=> e6(e4.triggered));
```

/*Similar to argument above, the clock flows from "posedge clk_a" to "posedge clk_e2" and sequence "e4" infers
"posedge clk_e2"
*/

```
    always @(posedge clk_a) begin
        @(e4); d = a;
    end
```

/* This is very interesting and important to understand. Sequence "e4" infers "**default clocking** cb @(posedge clk_d); **endclocking**" and not posedge clk_a as there is more than one event control in this procedure.
*/

```
    endmodule
```

Chapter 14
"expect"

Introduction: This chapter describes the procedurally *blocking* statement "expect" and its differences with the "assert" statement.

A. B. Mehta, *SystemVerilog Assertions and Functional Coverage*,
https://doi.org/10.1007/978-3-030-24737-9_14

"expect" takes on the same syntax (not semantics) as "assert property" in a procedural block. *Note that "expect" must be used only in a procedural block*. It cannot be used outside of a procedural block as in "assert" or property/sequence (recall that "assert property" can be used both in the procedural block and outside of it in a sequence or a property). So, what's the difference between "assert property" and "expect"?

"expect" is a blocking statement while "assert property" is a non-blocking statement. Blocking means, the procedural block will wait until "expect" sequence completes (pass or fail). For "assert property" non-blocking means that the procedural block will trigger the "assert property" statement and continue with the next statement in the block. "assert property" condition will continue to execute in parallel to the procedural code. Note that "assert property" behavior is the same whether it's outside or inside a procedural block. It always executes in parallel on its own thread with the rest of the logic.

Please refer to the simulation log in Fig. 14.1. You notice that the procedural code *waits* for "expect" to complete (i.e., blocks execution of procedural code) and only

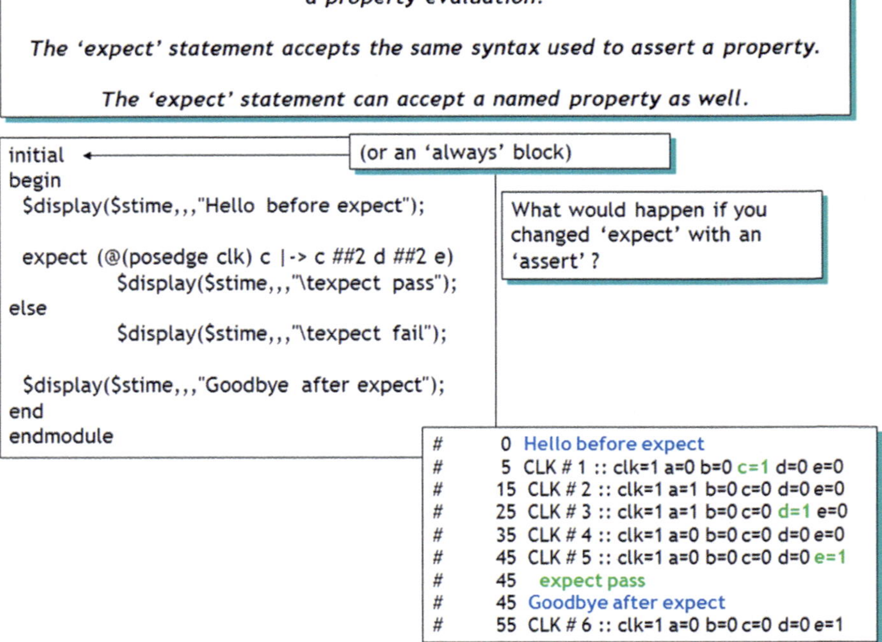

Fig. 14.1 "expect"—Basics

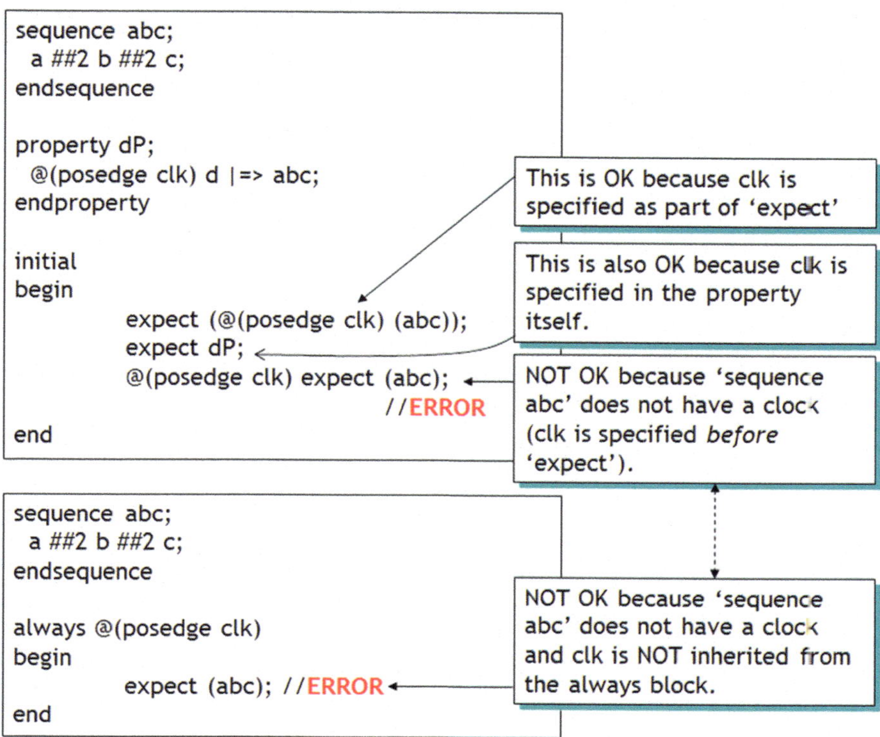

Fig. 14.2 "expect"—Error conditions

on completion of "expect" that it executes the subsequent $display. Figure 14.2 highlights further nuances of "expect" semantics. Annotations in the figure are self-explanatory. *Key point is that "expect" does not inherit a clock from its preceding procedural clock.* It needs an explicit clock in the sequence (or property) it "expects" or with its own declaration.

Finally,

The expect statement can appear anywhere a wait statement can appear. Because it is a blocking statement, the property can refer to automatic variables as well as static variables.

For example, the task below waits between 1 and 10 clock ticks for the variable data to equal a particular value, which is specified by the automatic argument value.

The second argument, "success," is used to return the result of the expect statement: 1 for success and 0 for failure.

```
integer data;
...
task automatic wait_for (integer value, output bit success);
   expect (@(posedge clk) ##[1:10] data == value) success = 1;
    else success = 0;
endtask

initial begin
    bit ok;
    wait_for (23, ok);
...
end
```

Chapter 15
"assume" and "restrict" for Simulation and Formal (Static Functional) Verification

Introduction: This chapter describes "assume" and "restrict" statements and their usage for "static formal" (or "static functional") and "constrained random" methodologies.

A. B. Mehta, *SystemVerilog Assertions and Functional Coverage*,
https://doi.org/10.1007/978-3-030-24737-9_15

15.1 "assume" Statement

This is an interesting operator. "assume" specifies the property as an assumption for the environment (stimulus, response, etc.). They may be used by simulators to constrain the random generation of free checker variable values or by formal tools to constrain the formal computation. The most useful environment for "assume" is that of static formal verification. Static formal is a method whereby the formal algorithm exercises all possible combinational and temporal possibilities of inputs to exercise all possible "logic cones" of a given logic block and checks to see that the assertion holds. During such verification if you do not specify any constraints (i.e., for a five input (a, b, c, d, e) block, if you don't specify any constraints such as "assume" $a = 0$ and $b = 1$), then the static formal will try to explore all possible combinations of the five input in both combinatorial and temporal (if required) domain. Without any constraints provided via "assume," the static formal tool may experience something called "state space explosion" problem. As the description suggests, the tool may give up if too many inputs are unconstrained. This is where the "assume" statement comes into picture.

So, how does it behave in simulation? The examples in Fig. 15.1 simply show that "assume" without any other property reliant on assumed property will act like

'assume' is useful mainly for 'static formal' and 'constrained random' dynamic simulation where you need to constraint the environment using the conditions specified in a property.

- *'assume' statement allows properties to be considered as assumptions.*
- *When a property is 'assume'd the tools constrain the env. so that the property holds.*
- *For 'formal' analysis, there is no obligation to verify that the assumed property holds. The statement is simply assumed true and the scope of formal is constrained according to the assumed property.*
- *For simulation, the 'assume'd property must be checked (as in 'assert') and reported if it fails to hold.*

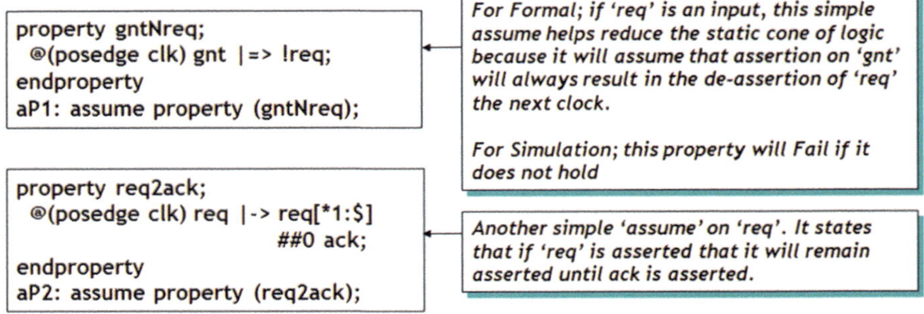

Fig. 15.1 "assume" and formal verification

"assert." If the property associated with the assume statement is found to be false, the property fails.

In the following example, we are defining two properties that will be used in "assume" with weighted distribution (further on distribution after the example). Then we define two properties that will be asserted. *The properties that are 'assume'd must hold for the asserted properties*. As we discussed above, if assumed properties fail, the results of "assert" are not valid. Simulator will show FAIL message whenever 'assume'd property fails. That would be an indication that the "assert" results are not valid.

```
property pr1;
     @(posedge clk) !reset_n |-> !req;
endproperty

property pr2;
     @(posedge clk) ack |=> !req;
endproperty

a1:assume property (@(posedge clk) req dist {0:=40, 1:=60} );
assume_req1 : assume property (pr1);
assume_req2 : assume property (pr2);

property pa1;
     @(posedge clk) !reset_n || !req |-> !ack;
endproperty

property pa2;
     @(posedge clk) ack |=> !ack;
endproperty

assert_ack1 : assert property (pa1);
assert_ack2 : assert property (pa2);
```

15.2 "restrict" Statement

As we discussed above, the "assume" statement will fail if it does not hold its assumed property. Restrictions are treated as assumptions in Formal Verification, but they are completely ignored in simulation. They are meant for limiting formal proofs to particular cases. In formal verification, for the tool to converge on a proof of a property or to initialize the design to a specific state, it is often necessary to constrain the state space.

Other than the keyword "restrict," the restriction syntax is the same as the syntax of assertions and assumption. Unlike assertions and assumptions, restrictions have only concurrent form (cannot be used in procedural blocks), and they cannot have action blocks.

LRM −1800 2017 added this feature. Its semantic is identical to "assume" but in contract *"restrict" property won't fail if it does not hold* because the "restrict" statement is simply ignored in simulation.

```
restrict property ( property_spec ) ;
```

For example, for Formal, you want to restrict the logic evaluation to only PCI_read cycles, you can do the following:

```
restrict property (@(posedge clk) PCI_Cycle_Type == PCI_read);
```

Simulation will simply ignore this property, but Formal will restrict the state space logic cone evaluation of PCI block to only PCI_read cycles.

Note that *neither* the "assume" nor the "restrict" can have an action block.

15.3 "dist" (Distribution Operator) and "inside" Operator

A short introduction to the "dist" operator. This is a SystemVerilog operator and a detailed discussion is beyond the scope of this book.

"dist" operator specifies weighted distribution of the item (an integral SystemVerilog expression) to which the weighted distribution is applied. For example,

```
data dist {100 := 1; 500 := 3, 700 := 5}}
```

"data" has a weight of "1" for it to be 100, a weight of "3" for it to be "500", and a weight of "5" for it to be 700. In other words, "data" value 100 will occur less often than "data" value 500 or 700. The same for "data" value 500 compared to 700. Again, "data" is equal to 100, 500, or 700 with a weighted ratio of 1-3-5. "dist" in this sense constraints design inputs from your test-bench.

However, when it comes to assertions, the "dist" operator works a bit different. When a property with "dist" is used within an "assert" or "cover" or "assume" or "restrict" assertion statement, the "dist" operator is equivalent to "inside" operator, and the weight specification is *ignored*. Here's a simple example:

```
a1_assume: assume property (probe_assume) ( {req dist {0: 40,
1:60} )
property probe_assume;
     @(posedge clk) req |=> ack;
endproperty
```

As noted above, "assume" will use the weighted distribution. But "assume" will ignore this weighted distribution. So, assertions have provided another operator called "inside." So, the above property (with an "assume") is the same as the one below (with an "assert").

```
a1_assert: assert property (probe_assert) ( req inside {0, 1] )
property probe_assert;
        @(posedge clk) req |=> ack;
endproperty
```

In the preceding example, signal "req" is specified with distribution in assumption a1_assume and is converted to an equivalent assertion a1_assertion.

It should be noted that the properties that are assumed must hold in the same way with or without biasing. When using an assume statement for random simulation, the biasing simply provides a means to select values of free variables, when there is a choice of selection at a particular time.

Chapter 16
Clock Domain Crossing (CDC) Verification Using Assertions

Introduction: This chapter explores one of the most critical aspects of multi-clocked designs, that being the clock domain crossing (CDC) logic. How do you use assertions to model CDC behavior and catch bugs thereof?

A. B. Mehta, *SystemVerilog Assertions and Functional Coverage*,
https://doi.org/10.1007/978-3-030-24737-9_16

SystemVerilog Assertions (SVA) are a great way to check for sequential domain conditions at clock boundaries. The CDC signals crossing from one clock domain to another are perfect candidates to check for using SVA. SVA fully supports multi-clock domain assertions as well as multi-threaded local variables to make full proof checkers to see that your CDC synchronizers (whatever the design style) work as promised. Note that the assertions presented here can be used for both Simulation Based Checking and Formal based checking (static functional).

Here's a description of how a two-flop synchronizer works (Fig. 16.1). For a fast TxClk to a slow RxClk, if the CDC signal (TxData) is only pulsed for one fast-clock cycle, the CDC signal could go high and low between the rising edges of a slower clock and not be captured into the slower clock domain. This is shown in Fig. 16.2. In the figure, TxData goes high and then goes low (1 High Pulse) *in-between the RxClk period.* In other words, this High pulse will not be captured by the RxClk. That results into the Rx1Data remaining at the previously captured state of "0" and so does RxData. The High pulse on TxData is dropped by the receive logic which will result in incorrect behavior in the Receive Logic.

Hence, a two-flop synchronizer won't work when the Transmit Clock is faster than the Receive Clock.

One potential solution to this problem is to assert the TxData signal (i.e., the CDC signal) for a period that exceeds the cycle time of the Receive clock. This is shown in Fig. 16.3. The general rule of thumb is that the minimum pulse width of the Transmit signal be 1.5× the period of the Receive clock frequency. The assumption is that the CDC signal will be sampled at least once by the receive clock. The issue with this solution will arise if an engineer mistakes this solution to be a general-purpose solution and miss the Transmit (CDC) signal period requirement. This is where SystemVerilog Assertions come into picture. Put an assertion on the CDC signal for its period check when crossing from high frequency to the low frequency domain.

For any synchronizer design, there will be assumptions on TxData stability. Should it remain stable for 2 clocks? 3 clocks? This is to make sure that the CDC signal Rx1Data has enough time to filter the metastability region and pass the

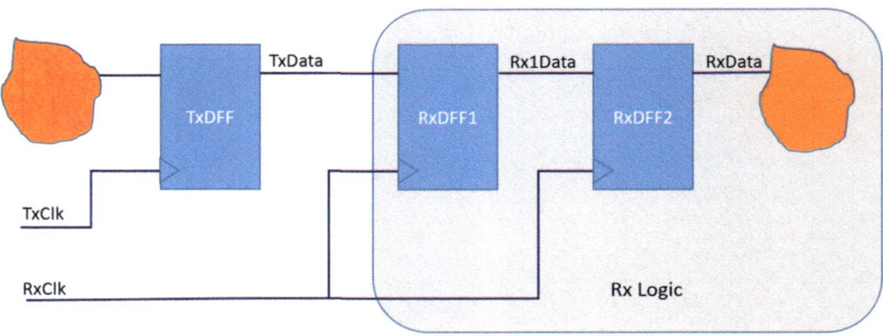

Fig. 16.1 Two-flop synchronizer

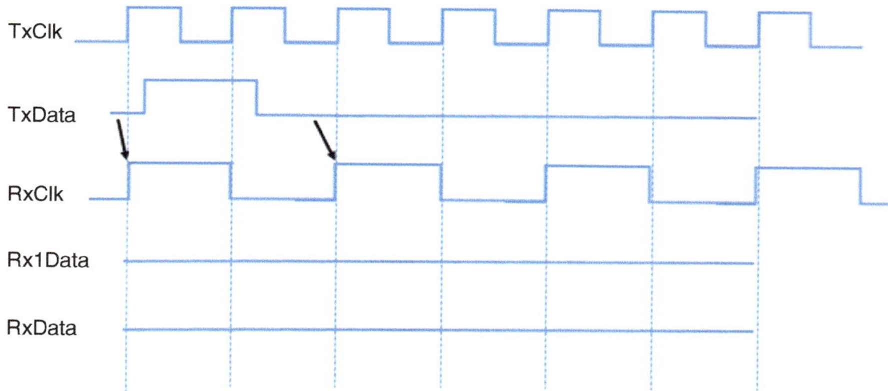

Fig. 16.2 Faster transmit clock to slower receive clock—two-flop synchronizer won't work

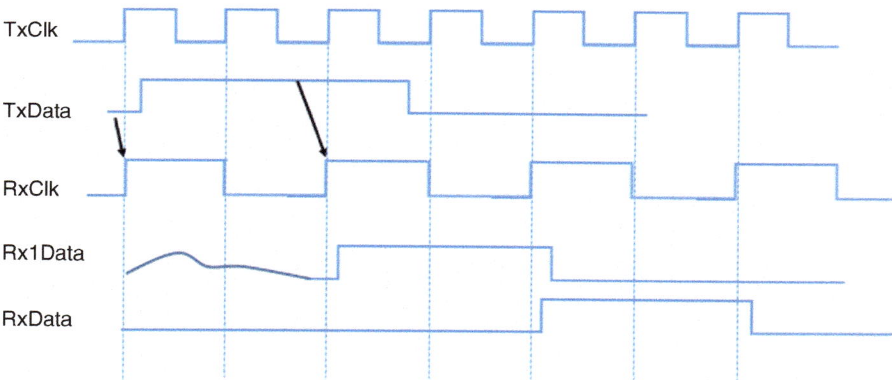

Fig. 16.3 Lengthened transmit pulse for correct capture in receive clock domain

correct value to RxData (the output). Let us go with the assumption that TxData should remain stable for 2 clocks every time it assumes a new value (i.e., it changes). This assumption is required since we assume that TxClk is faster than RxClk.

Here's a simple assertion to check for TxData stability.

```
property TxData_stable;
  @(posedge Txclk) $changed(TxData) |=> $stable(TxData) [*2];
endproperty

assert property (TxData_stable);
```

Let us now see how to make sure that this two-flop single bit synchronizer correctly transfers data so that RxData === TxData after metastability filter.

```
property Tx_to_Rx_CDC_DataCheck;
local Data;

@(posedge Txclk) ($changed(TxData)) |=>
   (1'b1, (Data = TxData)) ##1
   @(posedge RxClk) (Rx1Data === Data) ##1 (Rxdata === Data);
endproperty: Tx_to_Rx_CDC_DataCheck

assert property (Tx_to_Rx_CDC_DataCheck);
```

First the assertion checks that TxData has changed at posedge of TxClk. If it has, we first store the TxData into the dynamic local variable Data. 1'b1 is required because local data store must be attached to an expression. Since we don't have any condition, we simply say "always true" is the expression. "always true" means always store TxData into Data, whenever TxData changes. Then, we check at the CDC boundary clock RxClk that the data has indeed transferred to Rx1Data by comparing Rx1Data with the stored TxData (in Data). One clock later the RxData must match the TxData that was transmitted on TxClk. This guarantees that the CDC 1-bit two-flop synchronization works as intended. Again, note that the assumption of TxClk faster than RxClk must be adhered to.

Exercise: Write a simple assertion to check for asynchronous Glitch on TxData. The above solution assumes no Glitch on TxData.

Ok, now let us write a comprehensive assertion for a multi-bit gray code counter-based data transfer across CDC region. The write data are written to fifo_in on wclk (write Clock); and read from fifo_out on rclk (read clk). The assertion must make sure that whatever data were written into FIFO at the write pointer that the same data is read out from FIFO when read pointer is equal to the write pointer.

```
sequence rd_detect(ptr);
##[0:$] (read_en && !empty && (aff1.rd_ptr == ptr));
endsequence

property data_check(wrptr);
integer ptr, data;
@ (posedge wclk) disable iff (!wclk_reset_n || !rclk_reset_n)
    (write_en && !full, ptr=wrptr, data=fifo_in,
  $display($stime,"\t Assertion Disp wr_ptr=%h data=%h", aff1.
  wr_ptr, fifo_in))

|=>

  @ (negedge rclk) first_match(rd_detect(ptr),
    $display($stime,,," Assertion Disp FIRST_MATCH ptr=%h
    Compare data=%h fifo_out=%h", ptr, data, fifo_out))
        ##0 (fifo_out === data);
endproperty
```

```
dcheck : assert property (data_check(aff1.wr_ptr)) else
$display($stime,,,"FAIL: DATA CHECK");
dcheckc : cover property (data_check(aff1.wr_ptr))
$display($stime,,,"PASS: DATA CHECK");
```

In this assertion, data_check property checks to see that FIFO is not full. If so, saves wr_ptr into the local variable "ptr" and the data from fifo into local variable "data" and display that so that we can easily see how the assertion is progressing during simulation.

If the antecedent is true, the consequent says that the first match of rd_ptr being the same as wr_ptr (note wr_ptr was stored in local variable "ptr") that the read data is the same as the write data (note write data were stored in local variable "data" in the antecedent).

Sequence rd_detect(ptr) is used as an expression to first_match. It says that wait from now until forever until you detect a read and its rd_ptr is equal to the wr_ptr (which is stored in the local variable "ptr" in the antecedent).

Many such assertions can be written to see that your synchronizer design works. As an exercise, try writing simple assertions for your synchronizer design.

16.1 Automated CDC Verification

Figure 16.4 shows a proposed methodology for automatic CDC verification which is being offered by many EDA vendors. Here's a brief description of the different components of such a methodology.

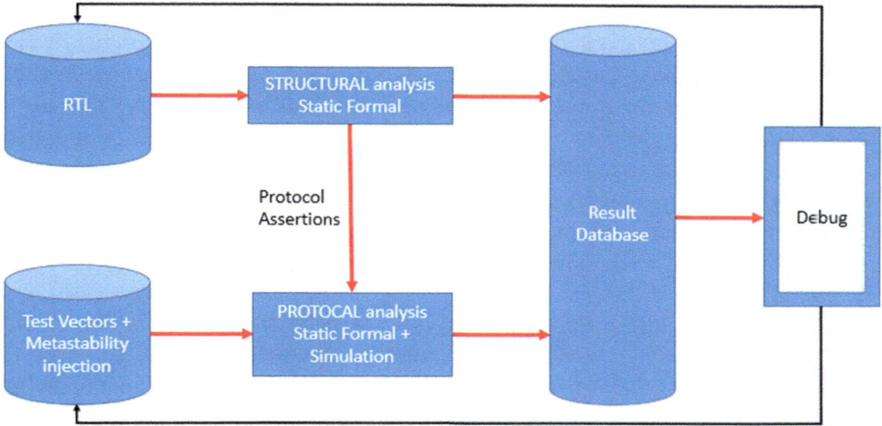

Fig. 16.4 Clock domain crossing—automated methodology

16.1.1 Step 1: Structural Verification

Identify RTL blocks (not the entire SoC RTL) that have CDC signals at play. Feed such RTL blocks to the static formal Structural analysis tool. This tool will identify CDC Synchronization “structures” within your logic and analyze to see if they meet the requirements. For example, a single bit CDC synchronization will work with a 2-flop synchronizer. But for a multi-bit synchronizer, the 2-flop solution won’t work. You may need an asynchronous FIFO based solution or a gray counter (where only 1 bit change at a time). The tool will analyze such situations and provide a structural analysis report. The results are also stored in a UCDB style database for further debug analysis. This step should find issues with missing and incorrectly implemented synchronizers and potential re-convergence problems.

More important in this step is to automate derivation of SystemVerilog Assertions. For example, for a 2-flop synchronizer, the input data should remain stable for at least 1.5× the Receive clock when you are crossing from faster clock to slower clock. The structural analysis tool will (should) automatically write such assertions for the next stage of protocol verification. There are many such constraints/checks that need to be provided to the protocol analyzer. The structural analysis tool “knows” what type of structure has been designed and thereby should be able to create assertions for protocols that the structure needs to adhere to.

You need to evaluate the structural analysis results in an EDA vendor provided debug tool and either accept the recommendation or reject them and implement the best structure that you envision. Don’t worry; the protocol analyzer will grab you if your structure does not meet synchronization protocol requirements.

16.1.2 Step 2: Protocol Verification

Once the structural analysis is complete, the assertions (either automatically created or manually) will be input to the protocol analyzer. The Static Formal method employed in the protocol analyzer will try all possible combinations of inputs (both in temporal and combinational domain) to the RTL block and see if any of the assertions FAIL. These assertions ensure that the CDC signal is stable when going from the TX to the RX domain; the multiple-bit CDC data is gray coded, or it is stable when it is sampled. The results will show failures which need to be analyzed to correct the synchronizer. Multiple iterations of this step will make sure that the logic will survive under all conditions of input and that the metastability has been addressed.

In addition to static formal, you may also want to simulate using the created assertions. For example, you feel comfortable with sweeping clocks to check for re-convergence logic. Or you want to deploy the so-called static + simulation hybrid methodology to check for the structural integrity against required protocol specification.

16.1.3 Step 3: Debug

Of course, debug is a big part of this strategy. The results from Structural and Protocol analysis are stored in an UCDB style database. The debug tool will associate the structure against the protocol and show the relationship. It will also help you debug failing assertions. EDA tools do support such debug capabilities.

Based on the debug results you will either change the RTL and/or change the input test vectors and metastability injection strategy.

This loop will continue until there are no more assertions that fire and the metastability issues are completely resolved.

Chapter 17
Important Topics

Introduction: This chapter addresses many important topics such as testing the testbench, triggering concurrent assertions from procedural blocks, calling subroutines, sequences as formal arguments, as antecedent and as triggering condition in a sensitivity list. It also describes further nuances such as how to design "variable delay" using a "counter," effects of blocking nature of an "action" (pass/fail) block, cyclic dependencies, vacuous pass of an assertion, empty sequences, etc.

A. B. Mehta, *SystemVerilog Assertions and Functional Coverage*,
https://doi.org/10.1007/978-3-030-24737-9_17

17.1 Test the Test-Bench

Following assertions are important to note. We are checking our own test-bench! Yes, that is important. This is a simple test-bench, but the idea is that as your test-bench develops complex code that there is a good chance you will make mistakes. So why not use assertions to catch those errors as well. Very simple assertions are presented for a synchronous FIFO design. In real life, there will be a lot more assertions in a complex test-bench. This example simply highlights writing assertions from test-bench point of view.

```
module (...)
/*
In the first assertion we check that if FIFO is full so that
we do not keep writing to it, since this particular FIFO does
not have a pushback signal.
Similarly, the second assertion checks to see that if the FIFO
is empty that we do not keep reading it!
*/

/*-----------------------------------------
  Checks for the Test-bench
------------------------------------------*/
property check_full_write_en;
   @ (posedge wclk) disable iff (!wclk_reset_n)
    full |-> !write_en;  //Stop the testbench from writing to
    a FIFO that is full.
endproperty

check_full_write_enA : assert property (check_full_write_en)
else $display($stime,,,"%m FAIL: check_full_write_en");
check_full_write_enC : cover property (check_full_write_en)
$display($stime,,,"%m PASS: check_full_write_en");

property check_empty_read_en;
   @ (posedge rclk) disable iff (!rclk_reset_n)
        empty |-> !read_en; //Stop the testbench from reading
        from an empty FIFO.
endproperty

check_empty_read_enA: assert property (check_full_write_en)
else $display($stime,,,"FAIL: %m check_full_write_en");
check_empty_read_enC: cover property (check_full_write_en)
$display($stime,,,"PASS: %m check_full_write_en");

//Here's where your testbench code goes.
endmodule
```

17.2 Embedding Concurrent Assertions in Procedural Block

Yep, you can indeed assert a concurrent property from a procedural block. Note that property and sequence itself are declared outside of the procedural block.

Off the bat, what is the difference than between embedding an immediate assertion vs. embedding a concurrent assertion in the procedural block? Well, immediate assertion is invoked with "assert" while embedded concurrent assertion is invoked with "assert property." A concurrent assertion that is embedded in a procedural block is the regular concurrent assertion "assert property" (i.e., it can be temporal). In other words, an immediate assertion embedded in the procedural code will complete in zero time, while the concurrent assertion may or may not finish in zero time. But is the concurrent assertion in the procedural code blocking or non-blocking? Hang on to this thought for a while.

Figure 17.1 points out that the *condition* under which you want to fire an assertion is already modeled behaviorally and do not need to be duplicated in an assertion (as an antecedent). The example shows both ways of asserting a property. `ifdef P shows the regular way of asserting the property that we have seen throughout this book. `else shows the same property being asserted from an always block. But note that in the procedural block the assertion is preceded by "if (bState == CycleStart)" condition. In other words, a condition that could be already in the behavioral code is used to condition an assertion. If the property was "asserted" outside of the procedural block, you would have to duplicate the condition that is in the procedural block, as an antecedent in the property.

Let us turn our attention back to embedded concurrent assertion being blocking or non-blocking. What happens when you fire a concurrent assertion from the

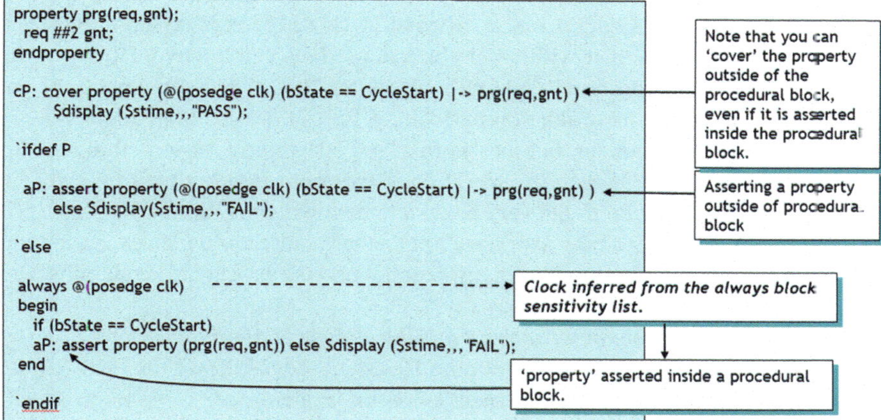

Fig. 17.1 Embedding concurrent assertions in procedural code

```
property pr1(req,gnt);
  cstart |=> req ##2 gnt;
endproperty

property pr2;
  $fell(aStrobe) |=> $fell(Wenb) || $fell(Renb);
endproperty

cP: cover property (@(posedge clk) pr1(req,gnt) )
          $display ($stime,,,"PASSpr1");
cA: cover property (@(posedge clk) pr2)
          $display ($stime,,,"PASSpr2");

always @(posedge clk)
begin
  if (CheckWrite) begin
  assert property (pr1(req,gnt)) else
$display($stime,"FAILpr1");
  assert property (pr2) else $display($stime,"FAILpr2");
  end
end
```

Procedural trigger of a concurrent assertion runs as a separate parallel thread to the procedural code (in other words, it is non-blocking).

These two assertions will fire in parallel.

Second assertion will NOT wait for the first one to complete

```
#       10 clk=1 cstart=0 req=0 gnt=0 aStrobe=1 Wenb=1 Renb=1
#       30 clk=1 cstart=1  req=0 gnt=0 aStrobe=0 Wenb=1 Renb=1
#       50 clk=1 cstart=0 req=1 gnt=0 aStrobe=1 Wenb=0 Renb=1
#       50 PASSpr2
#       70 clk=1 cstart=0 req=0 gnt=0 aStrobe=0 Wenb=1 Renb=1
#       90 clk=1 cstart=0 req=0 gnt=1 aStrobe=1 Wenb=1 Renb=0
#       90 PASSpr1
#       90 PASSpr2
```

Fig 17.2 Concurrent assertion embedded in procedural code is non-blocking

procedural code, and it does not finish in zero time? What happens to the procedural code that follows the concurrent assertion? Will it stall until the concurrent assertion finishes (blocking)? Or will the following code continue to execute in parallel to the fired concurrent assertion (non-blocking)? That's what Fig. 17.2 explains.

There are two properties "pr1" and "pr2." They both "consume" time, i.e., they advance time. The procedural block '**always** @ (posedge clk)' asserts both these properties one after another without any time lapse between the two. How will this code execute? The procedural code will encounter "assert property (pr1 …)" and fire it. "pr1" will start its evaluation by looking for cstart to be high and follow on through with the consequent. In other words, "pr1" is waiting for something to happen in time domain. *But the procedural code that fired it won't wait for "pr1" to complete*. It will move on to the very next statement which is "assert property (pr2) " and fire it as well. So, now you have "pr1" that is already under execution and "pr2" that is just fired both executing in parallel and the procedural code moves on to other code that sequentially follows.

In short, a concurrent assertion in procedural code is non-blocking.

As shown in the simulation log, at time 10, (@ posedge clk) we fire "pr1." At the *same* time (since the very next statement is "assert property (pr2 …)") we fire "pr2." At time 30, "cstart==1" and "aStrob==0." This means the antecedent condition of

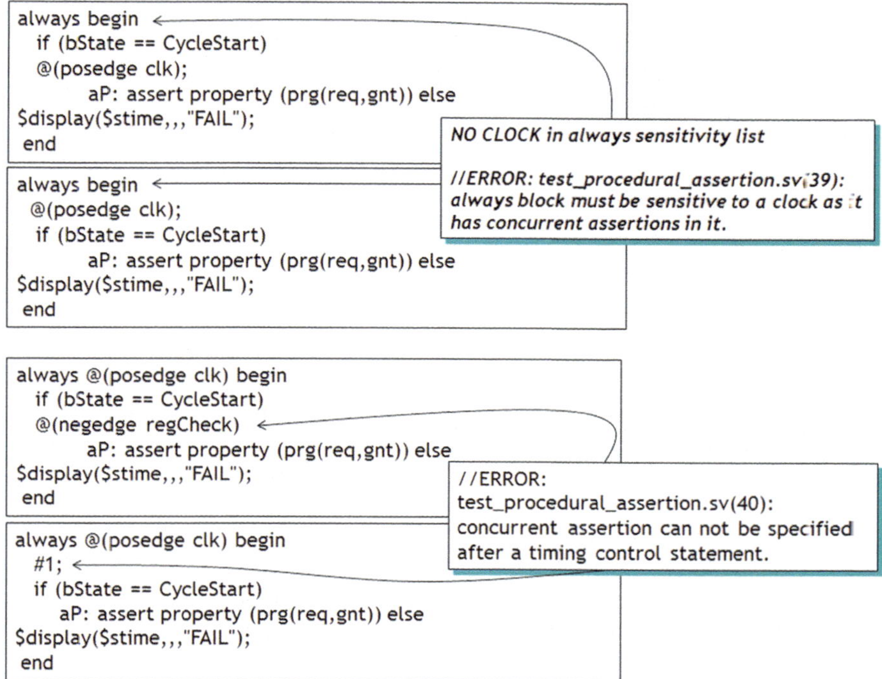

Fig. 17.3 Embedding concurrent assertions in procedural code—further nuances

both "pr1" and "pr2" have been met. At 50, "Wenb==0" which completes the property "pr2" and the property passes as shown at time 50 in simulation log. Therefore, the first thing you notice here is that even though "pr1" was fired first, "pr2" finished first. In other words, since both properties were non-blocking and executing on their own parallel threads, there is no temporal relationship between them or among "pr1," "pr2" and the procedural code. Following the same line of thought, see why both "pr1" and "pr2" pass at the same time at 90.

See the rules on inferring clock edge for the assertion in the procedural block. In addition, other nuances of semantic are noted in Fig. 17.3. In short, the clock inference for the embedded assertion comes from the "always" block edge sensitivity. It is not derived from any other temporal domain condition (edge or level) that is embedded in the procedural code.

Note that procedural code in context of placing concurrent assertion means only the "initial" and the "always" blocks.

One note on "automatic" variables in the procedural code and their use in an assertion (in contrast to "static" variables that we have used so far). An "automatic" variable is sampled *at the time* the assertion attempt is started. This is in contrast to "static" variable, which in the assertion expression are always *sampled* at the sampling edge in the preponed region as with any other concurrent assertion that we have seen so far. To reiterate, since this is a very important point, the value of an "automatic" variable is *not* sampled at the sampling edge rather captured at the time it arrives at the assertion attempt that value of the variable is then used throughout the assertion evaluation.

Note further Legal and Illegal conditions when it comes to embedding concurrent assertions in procedural code. Concurrent assertions can only be placed in an "initial" or an "always" block.

```
property q1;
     $rose(a) |-> ##[1:5] b;
endproperty

property q2;
     @(posedge clk) q1;
endproperty

property q5;
     @(negedge clk) b[*3] |=> !b;
endproperty

always @(negedge clk) begin
a1: assert property ($fell(c) |=> q1); // LEGAL: contextually
inferred leading clocking
                                     // event, @(negedge clk)
a3: assert property ($fell(c) |=> q2); // ILLEGAL: multi-
clocked property with contextually
                            // inferred leading clocking event
a4: assert property (q5); // LEGAL: contextually inferred
leading clocking event, @(negedge clk)
end
```

Following are all ILLEGAL. Their clocks cannot be inferred.

```
always @(clk) begin
       a = b + c;
       a1: assert property (z |=> d | e);  //ILLEGAL
       .....
end
```

In the above example, we don't have any edge operators (posedge or negedge). Hence, a leading clock cannot be inferred, i.e., "clk" will not be applied to property a1. *An event expression without an edge operator is not allowed.*

```
always @(posedge (dma_intr | intr) begin
    intrIn = intr;
    a1: assert property (z |=> d | e); //ILLEGAL
    ....
end
```

Here, "intr" is reused in the procedural code. Hence, @(posedge dma_intr | intr) cannot be considered the leading clock for the assertion. Clock is not inferred in this case.

```
always @(posedge dma_intr or posedge intr) begin
    ........
    a1: assert property (z |=> d | e); //ILLEGAL
    ........
end
```

Here there are two edges. Hence which one should be the leading clock? That cannot be determined and there won't be any clock inference. Only one valid event expression can be specified in the event control for the inferred clock.

Here's a quick note on how procedural concurrent assertions work in a "for loop."

```
logic [7:0] a;
logic [15:0] b;
always @(posedge clk) begin
    If (!reset) begin
                for (int i = 0; i < 4; i++) begin
                ....
                z1: assert property (a [i] ##[1:16] b[i]);
                end
end
end
```

In this example, four assertions, z1_1, z1_2, z1_3, and z1_4 are initiated when "for loop" starts executing. Clock inferred is @(posedge clk).

Let's look at another interesting example of firing concurrent assertions from a procedural block.

```
initial begin
        clk = 0;
        int i = 10;   //Note that 'i' is initialized to 10
        before the for loop starts.
        forever begin #10; clk = ~clk; end //meaning posedge
        clk won't occur until time 10
end

always @(posedge clk) begin
       for (i=0; i<10; i++) begin
              a1: assert property (req[i] && ack[i]);
    end
end
```

The concurrent assertion "a1" triggers at time 10, @(posedge clk). You would think that there will 10 assertion threads initiated one for each value of "i." *Nope*. It will only check (or fire the thread) for " req[10] && ack[10]."

What's going on?

During the first invocation of @(posedge clk), the value of "i" was initialized to 10. So, when the for loop starts, *the value of "i" will be sampled 10 in the preponed region*. So, "i" will remain 10 and will not change as the for loop progresses because every time the value of "i" will be 10 as sampled from the preponed region. After the first for loop execution, the last value of "i" will be 10. So, again when the next posedge clk arrives, the sampled value of "i" will be 10. The sampled value of "i" will always be 10, its final value from the previous execution of the for loop. The "assert property" will never see "i" values from 0 to 9. So, req[0] && ack[0], req[1] && ack[1], … req[9] && ack[9] threads will never materialize. *This is an important point to note when debugging*.

So, how would you solve this problem? See the code below:

```
initial begin
        clk = 0;
        int i = 10;   //Note that 'i' is initialized to 10
        before the for loop starts.
        forever begin #10; clk = ~clk; end //meaning posedge
        clk won't occur until time 10
end

always @(posedge clk) begin
        for (i=0; i<10; i++) begin
        a1: assert property (req[const'(i)] && ack[const'(i)]);
        //Note const' cast
end
end
```

Note that now we have cast "i" to a const'. *The sampled value of a const' cast expression is defined as the current value of its argument.* In other words, the value of "i" will *not* be the one from preponed region. It will be the actual value at the time of execution. So, "i" will indeed go through its values from 0 to 9.

With the const cast, now you will have 10 assertion threads fired. req[0] && ack[0], req[1] && ack[1], … req[9] && ack[9].

This rule of const' cast (or automatic variable) applies to the action block as well. The same rules that apply to procedural concurrent assertion arguments also apply to variables appearing in their action blocks. Thus, constant or automatic values may be used in action blocks as well as the assertion statements themselves, where they behave as inputs to the action block that shall not be modified as shown in the example below.

```
always @(posedge clk) begin
        for (i=0; i<10; i++) begin
                a1:assertproperty(req[const'(i)]&&ack[const'(i)])
                else
                $error ("Assertion a1 FAIL: Index i = %0d",
                const'(i)); //Action Block
      end
end
```

Lastly, if you want to clearly see the difference between the value of a variable cast as const' vs. its sampled value in preponed region, change the above $display as shown below.

```
$error ("Assertion a1 FAIL: Index Const i = %0d
:: Sampled i=%0d", const'(i), $sampled(i) );
```

To reiterate:

When a past or a future value of a const cast expression is referenced by a sampled value function, the current value of this expression is taken instead. This is true also for automatic variables. Automatic variables—also—act the same way as const' cast. In other words, the sampled value of an automatic variable is defined as the current value of its arguments, not the value from the preponed region.

17.3 Calling Subroutines on the Match of a Sequence

Attaching a subroutine to an expression is an excellent feature and a great boon to debugging effort and other applications. The subroutine calls, like local variable assignments, appear in the comma-separated list that follows the sequence. The subroutine calls are said to be *attached* to the sequence. It is an error to attach a subroutine call to a sequence that admits an empty match.

For example, if you'd like to know exactly when an expression is executed in a complex sequence (this is just but one example), you can "attach" a Verilog task to the expression and display the conditions you are interested in. Figure 17.4 explains this scenario.

As shown in Fig. 17.4, you can attach a subroutine to an expression, a sequence or to an expression or subsequence within a sequence. Note that the attached subroutine will execute on *successful* completion of either the expression or the sequence to which it is attached. For example, (not(cde,tdisp1)) in the topmost example of Fig. 17.4 means that "tdisp1" will execute when "cde" reaches its true conclusion. Else, it won't execute.

Figure 17.5 further explains when a subroutine is executed. First, at the top of the figure you notice that we "attach" a local variable as well as a subroutine ($display task). Since $rose(ptrue) is the sequence to which the subroutine is attached, it will execute only when $rose(ptrue) is true. Similarly, the next part of the figure shows (pout == (local_data+5), $display (...)), where (pout==(local_data+5)) is the expression to which the $display subroutine is attached. Again, the subroutine $display will execute only if (pout == (local_data+5)) is true.

You will also notice that $display in this figure (for most part) displays local variables. This is one of the important use of $display as a subroutine because *local variables cannot be accessed from an action block (pass or fail).*

In Fig. 17.6, the subroutine is a Verilog "task"—lvar_seq_trigger which in turn contains a $display. But *I don't want you to run away with the idea that the only subroutine you can attach is the one that $displays something!* For example, you can use

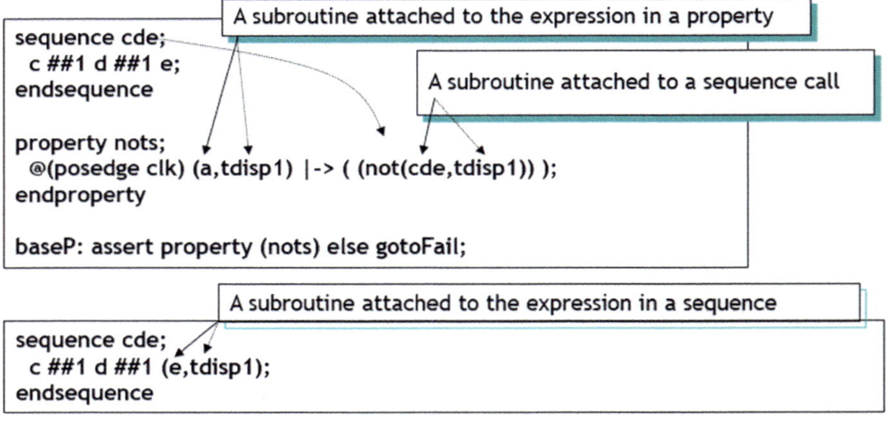


```
task tdisp1;
  $display($stime,,,"%m c=%b d=%b e=%b",c,d,e);
endtask
```

Fig. 17.4 Calling subroutines

```
sequence lvar_seq(pin,pout);
  int local_data;
  ($rose(ptrue),local_data = pin,$display($stime,,,"pin=%0d",pin) )
//                                ^^^^^^^^^^^^^^^^^^^^^^^^^^^^^^
//            This will be executed when $rose(ptrue) is detected..
  ##5
  (pout == (local_data+5),$display($stime,,,"pout=%0d",pout) );
//                          ^^^^^^^^^^^^^^^^^^^^^^^^^^^^^^^
//                          This will be executed ONLY IF
//                          the sequence MATCHES.
endsequence

property lvar;
  @(posedge clk) cStart |-> lvar_seq(pipe_in,pipe_out);
endproperty
```

Fig. 17.5 Calling subroutines—further nuances

```
sequence lvar_seq(pin,pout);
  int local_data;
  ($rose(ptrue),local_data = pin,lvar_seq_trigger(local_data))
  ##5
  (pout == (local_data+5),lvar_seq_match(pin,pout,local_data));
endsequence

property lvar;
  @(posedge clk) ptrue |-> lvar_seq(pipe_in,pipe_out);
endproperty

baseP: assert property (lvar) else gotoFail;
coverP: cover property (lvar) gotoPass;

task lvar_seq_trigger;
input ldata;
  $display($stime,,,"%m ldata=%0d",ldata);

  //$display($stime,,,"%m ldata=%0d",lvar_seq.pin);
  // ** Error: Hierarchical access to formal parameter pin' of 'lvar_seq' is illegal.
endtask

task lvar_seq_match;
input tpin,tpout,ldata;
  $display($stime,,,"%m pin=%0d pout=%0d ldata=%0d",tpin,tpout,ldata);
endtask
```

Fig. 17.6 Application: Calling subroutines and local variables

a subroutine to collect coverage information (using covergroups and coverpoints). We will see this with an example when we discuss Functional Coverage. *Since the attached subroutine can be task, you can think of many possible applications.*

The idea behind attaching a "task" to the expression is that you can do whatever that Verilog allows you to do in a "task" except that you cannot access the local variables of the sequence that invoked the "task." But you can indeed pass a local variable as an argument to the "task" as shown in Fig. 17.6.

"sequence lvar_seq" has a local variable called "local_data." This local variable is passed to "lvar_seq_trigger" as an actual argument. "task lvar_seq_trigger" in turn uses that as an input "ldata" and displays it. This is one way you can pass a local variable to the attached subroutine.

But note that you cannot access a variable (local or not) hierarchically from the attached subroutine (for example), task lvar_seq_trigger. This is shown with an ERROR in the figure. Here we tried to hierarchically access variable "pin" in "sequence lvar_seq" by using "lvar_seq.pin." That is a violation.

To recap:

All subroutine calls attached to a sequence are executed at every successful match of the sequence (i.e., at the end of the sequence).

For each successful match, the attached calls are executed in the order they appear in the list.

Assertion evaluation does *not* wait on or receive data back from any attached subroutine. The subroutines are scheduled in the Reactive region, like an action block.

Actual argument expressions that are passed by value use *sampled* values of the underlying variables and are consistent with the variable values used to evaluate the sequence match.

An automatic variable may be passed as a constant input for a subroutine call from an assertion statement in procedural code. An automatic variable cannot be passed by reference nor passed as a nonconstant input to a subroutine call from an assertion statement in procedural code.

Local variables can be passed into subroutine calls attached to a sequence. Any local variable that is assigned in the list following the sequence, but before the subroutine call, can be used in an actual argument expression for the call. If a local variable appears in an actual argument expression, then that argument must be passed by value.

Tasks, task methods, void functions, void function methods, and system tasks can be called at the end of a successful non-empty match of a sequence.

Finally, arguments passed to a subroutine must be by value or by reference ("ref" or "const ref").

17.4 Sequence as a Formal Argument

SystemVerilog assertions are indeed powerful as evident from this feature. *You can send an entire sequence as an actual argument to a property or another sequence.* One obvious advantage is that you may reuse a sequence in different properties as

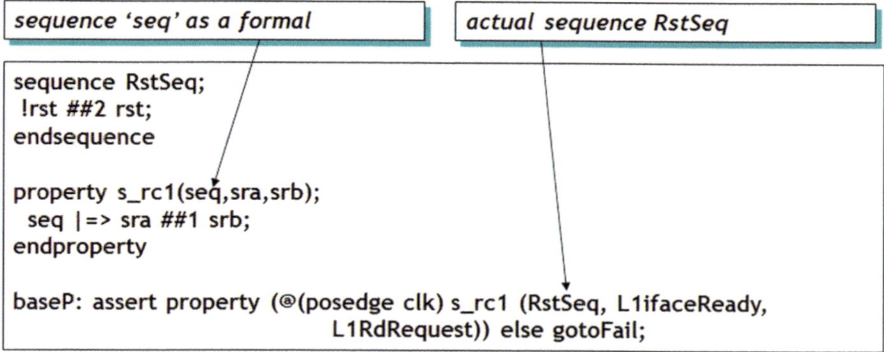

Fig. 17.7 Sequence as a formal argument

an actual to the property's formal argument. One example of this is the Reset Sequence as shown in Fig. 17.7. Reset sequence is often used in different properties as an antecedent. Write it once and pass it to different properties as an actual argument. That is reusability with observability and debuggability. The sequence that is passed to a property can (obviously) be used on both the antecedent and consequent side.

17.5 Sequence as an Antecedent

Since a sequence can be passed as an actual argument there are many advantages. We saw one in Fig. 17.7. Here's another as shown in Fig. 17.8. Here, we define a simple sequence "seq" and pass it to property "s_rc1." In this property, we use "seq" (i.e., c_seq in the property) as an antecedent.

As with any antecedent, the property will wait for antecedent to be true and then imply the consequent. Now, with many operators (e.g., "throughout") we have observed that the LHS and RHS of the operator is equally responsible for failure. If either side fails that the operator and hence the sequence/property fails. Important thing to note here is that in the cases of "throughout" the operator was used in the consequent and not in antecedent. Anything that fails in consequent causes the property to fail. Here, a sequence is used in the "antecedent" meaning even if the sequence in antecedent fails, the property will not fail. Instead, the property will simply wait for the sequence "c_seq" to be eventually true, after which it will execute the consequent. This makes sense because consequent fires only when antecedent is sampled to be true.

```
sequence seq;
 c ##1 d ##1 e;
endsequence

property rc1(ra,rb);
  rb or (ra and (`true |=> rc1(ra,rb)) );
endproperty

property s_rc1(c_seq,sra,srb);

        (c_seq,tdisp) |=> rc1(sra,srb);

endproperty

baseP: assert property (@(posedge clk) s_rc1(seq,a,b)) else gotoFail;
```


Important Point on sequence used as antecedent ::

Since c_seq is on the LHS of the implication, that the property 'WAITS" until it is true before implying rc1.

In other words, if c_seq fails, the property *won't* fail.

property s_rc1 simply waits until sequence 'seq' matches and then fires evaluation of the consequent. If 'seq' does not match, the property won't fail and won't start evaluation of the consequent.

Fig. 17.8 Sequence as an antecedent

Short end of the story is that no matter what you have in the antecedent, it will not cause a failure. Antecedent's job is to evaluate its expression/sequence and on sampling it to be true, imply the consequent. If the antecedent is not true, the property will pass vaccuously.

17.6 Sequence in Sensitivity List

Let us take the use of a "sequence" even further. Guess what? You can use a sequence for event control either in the sensitivity list or as an explicit edge sensitive control in a procedural block. You can use this feature very effectively because modeling a temporal domain condition in SVA is far easier than using behavioral Verilog. You can design certain condition as a sequence and then use it in your procedural behavioral Verilog code as shown in and.

Figure 17.9 shows that the "always" block waits for sequence "sr1" to complete after which it displays its PASS result.

Exercise: Can you figure out why there is no FAIL report? Was "**always @** (sr1)" triggered when "gnt" did not follow "req" after 2 clocks? Please experiment and see what happens.

The sequence in Fig. 17.10 says that @ (posedge clk) if $rose(read) is sampled high (edge) that there should be at least 1 readC (read complete). The "initial" block waits for this sequence to complete (using @ (ReadComplete) and then issues the next Read.

```
sequence sr1;
        @(posedge clk) req ##2 gnt;
endsequence

always @(sr1)
  $display($stime,,,"req ##2 gnt PASS");
```

Using sequence as an event trigger for always block.

Triggers only when the sequence matches

```
# run -all
#        5  clk=1 req=0 gnt=0
#       15  clk=1 req=1 gnt=0
#       25  clk=1 req=0 gnt=0
#       35  clk=1 req=0 gnt=1
#       35  req ##2 gnt PASS

#       45  clk=1 req=1 gnt=0
#       55  clk=1 req=0 gnt=0
#       65  clk=1 req=0 gnt=0

#       75  clk=1 req=1 gnt=0
#       85  clk=1 req=0 gnt=0
#       95  clk=1 req=0 gnt=1
#       95  req ##2 gnt PASS
```

Fig. 17.9 Sequence in procedural block sensitivity list

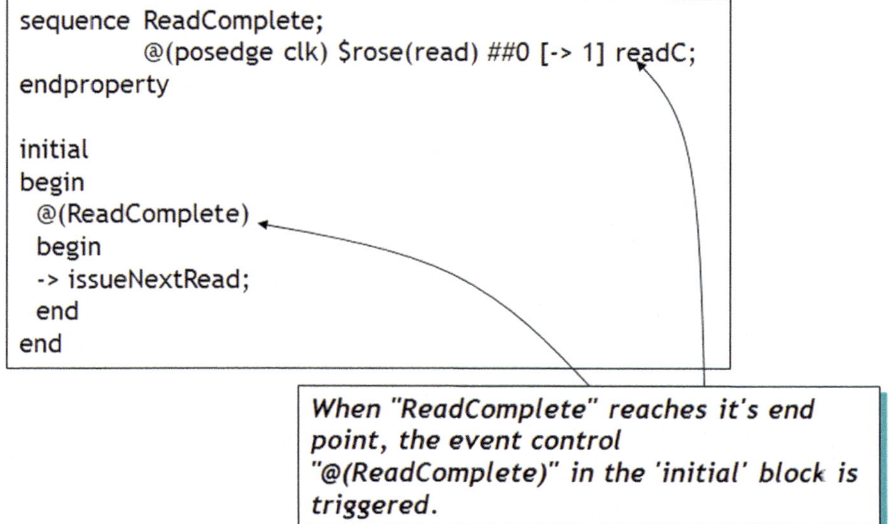

Fig 17.10 Sequence in "sensitivity" list

As you can see, this feature is extremely powerful in using the power of sequence in designing your SystemVerilog test-bench code.

17.7 Building a Counter

Ok, that is enough of sequences (for a while, at least). Let us see how we can effectively use local variables and the consecutive repetition operator to build something!

Let us build a counter. Why? There are many applications where you will be able to use this example. For example, you want to make sure that an incoming packet on a network generates an interrupt when its payload reaches a maximum threshold.

As shown in Fig. 17.11, the property checkCounter declares a local "int" called "LCount." It waits for a rising edge on "startCount" and on this rising edge, it stores "initCount" into "LCount" (initCount is defined elsewhere in your procedural code).

It then waits for 1 clock and increments LCount by 1 and continues to do so at every clock until LCount reaches maxCount. The consecutive repetition operator [*0:$] does the counting. In other words, "LCount=LCount+1" repeats at every posedge clk until you reach "LCount == maxCount." Once the maxCount is reached, the antecedent implies at the same clock (overlapping implication) that the intr be asserted.

Quick Note: In this book, as you have noticed, instead of giving large applications and then describe a limited set of SVA features, I have chosen to describe each operator with simple applications so that you clearly understand the workings of the operator and apply the operator features to design your assertions.

Following is an example of how to build a counter using the consecutive repetition operator [* m]

```
property checkCounter;
int LCount;
  @(posedge clk) disable iff (!rst_n)

          (
                    ($rose(startCount), LCount=initCount ) ##1

                    (1, LCount = LCount+1)[*0:$] ##1 (LCount == maxCount) |->

                    (Intr == 1'b1)
          );
endproperty
assert property (checkCounter);
```

Fig. 17.11 Application: Building a counter using Local Variables

17.8 Clock Delay: What If You Want *Variable* Clock Delay?

Figure 17.11 built a counter which can be used to create a variable delay model. Note that SVA allows only constant fixed delays with its delay operator. So, the example in Fig. 17.13 demonstrates a strategy to get around this limitation.

Figure 17.12 describes a typical specification. We need to check for variable latency based on the position of a "read" in the read queue. In other words, if the "read" is at the end of the queue, "read" will complete with maximum latency. On the contrary if it's at the beginning of the queue, it will complete with minimum latency. Or with a latency anywhere in the middle.

Since the delay (or range delay) operator does not allow variable delay, how would you model this with one generic assertion? You do not want to create a separate assertion for each of the fixed latency. That is the problem statement. Now let us see how we solve this. Consider this example as an idea generator.

The concept in Fig. 17.13 is identical to building a counter example. Here instead of incrementing the local variable we decrement it until it reaches zero. In this application readLatency is defined in your procedural code and it changes based on the position of Read in the read queue. That part of the code is not shown here

```
sequence Sab;
  a ##2 b;
endsequence

property ab;
          @(posedge clk) z  |-> Sab;
endproperty
```

Recall that in ##<delay>

<delay> *must* be a positive integer (a constant)

But what if you want to have variable clock delay in a property ?

Consider the following application

Check for read latency which varies depending on the position of *read* in a queue.

- If it's at the end of the queue, it carries maximum latency
- if it's at the beginning of the queue, it carries minimum latency
- and something in-between carries in-between latency.

How would you code it ?

One way would be to have a different property for every possible latency; each property using a fixed latency.

Another would be to simply check for 'max' latency which catches all cases. But this could be dangerous in critical mission applications where min. latency must also be met.

Wouldn't it be better to have a single property which can use a 'variable' for ##<variable> where the 'variable' can be assigned different values ?

Fig. 17.12 Variable Delay—Problem Statement

Here's pseudo-code of what you *want* to accomplish : ***NOTE this is just pseudo-code and WON'T work because SVA does not allow variable ## delay***

```
property read_latency_check;
  @(posedge clk) disable iff (!rst_n) ($fell(rd_)) |->
                         ##[readLatency] (read_data == expected_data);
endproperty

assert property (read_latency_check);
```

NOT ALLOWED: variable 'readLatency' as ## delay.

Possible Solution:

```
property read_latency_check;
int Ldelay;
  @(posedge clk) disable iff (!rst_n)

        (
                ($fell(rd_), Ldelay=readLatency ) ##1

                (1, Ldelay = Ldelay-1)[*0:$] ##1 (Ldelay==0) |->

                (read_data == expected_data)
        );
endproperty
assert property (read_latency_check);
```

Fig. 17.13 Variable Delay—Solution

When the property "read_latency_check" is asserted, it will assign the readLatency to Ldelay on assertion ($fell(rd_)) and decrement it at every posedge clk, until it reaches 0.

```
(1, Ldelay = Ldelay-1)[*0:$] ##1 (Ldelay==0)
```

Ldelay is decremented consecutively at every posedge clk until Ldelay==0. Need for "1" in (1, Ldelay = Ldelay-1)? Why? Recall that we can assign to a local variable only when that assignment is attached to an expression. Since we do not have any explicit expression, we simply use "always true" as an expression.

You can continue to change readLatency from procedural code based on the position of "read" in your read queue and use the same property to check for different latencies of read in the read queue. If the readLatency that you assigned in your test-bench is greater than or less than the required latency, the expected_data at the end of the latency would not match read_data and the assertion will fail.

This simple example has very powerful application capabilities.

17.9 What If the "Action Block" Is Blocking?

We have seen that the assertion of a property ("assert property") allows you two "action" blocks. One is triggered when the property passes and the other when it fails.

This action block can contain any procedural code that SystemVerilog supports. The procedural block can have temporal domain "delay" (e.g., @ (posedge clk) or "wait sig"). That is when you need to carefully weigh in the consequences. If there is no "delay" in the block, life is straightforward. The "assert property" triggers the block without any delay; the block executes in 0 time; returns and the property moves along with its execution. But if there are delays in the action block, here's what happens.

The property in Fig. 17.14 says "assert property (pr1) else failtask." If the property fails, call a task called "failtask." "failtask" in turn waits for 4 @ (posedge clk) and returns from the task. The 4-clock delay is just an example for temporal delay.

Now let us look at the simulation log in Fig. 17.15. At time 30, req=1, so the property pr1 moves along. At time 70, gnt=0 which is a failure condition because gnt should have been "1" at that time. Since there is a failure, the "failtask" is invoked at time 70 (the time of failure). The "failtask" waits for 4 posedge clks, which are displayed in the log file at time 90, 110, 130, 150. While the "failtask" was waiting for its clocks to complete, at time 110, req goes high again. So, the property starts executing and expects "gnt" to be high at 150. And again, gnt is '0,'

```
property pr1;
  @(posedge clk) req |-> ##2 gnt ;
endproperty

reqGnt: assert property (pr1) else failtask;

task failtask;
   $display($stime,,,"FROM failtask - 0");

   @(posedge clk) $display($stime,,,"FROM failtask - 1");
   @(posedge clk) $display($stime,,,"FROM failtask - 2");
   @(posedge clk) $display($stime,,,"FROM failtask - 3");
   @(posedge clk) $display($stime,,,"FROM failtask - 4");

endtask
```

Fig. 17.14 Blocking action block task

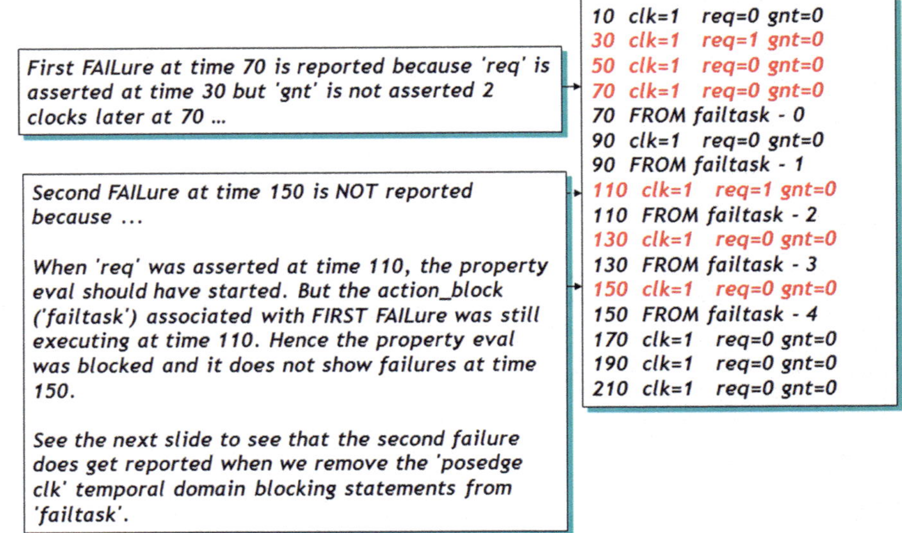

Fig. 17.15 Blocking action block simulation log

so the property should fail. But it does not!! Why? Because at time 150, the "failtask" was still completing its 4th clock wait. Since the 4th clock wasn't over by the time the "gnt" based failure came along at 150, the failure got suppressed because first invocation of "failtask" wasn't over.

The point is, if you call a procedural block that does not complete in "0" time, and if the next trigger of the property antecedent comes along causing another pass/fail, the next trigger of the action block associated with pass/fail won't happen. So, be careful in using time lapse in your action block(s).

Figure 17.16 simulation logs highlight the same point. There is one example with time lapse in the action block and another without.

As shown in, the properties on LHS and RHS are identical, except that LHS action block calls a "failtask" that elapses time (4 clocks). The RHS on the other hand calls "failtask" that does not elapse any time.

The "req" condition is also identical in both simulation logs. But the RHS shows two invocations of "FROM failtask - 0" while the LHS log shows only one invocation (as explained above) because the action block is "blocking" and does not allow reentry into an already executing action block.

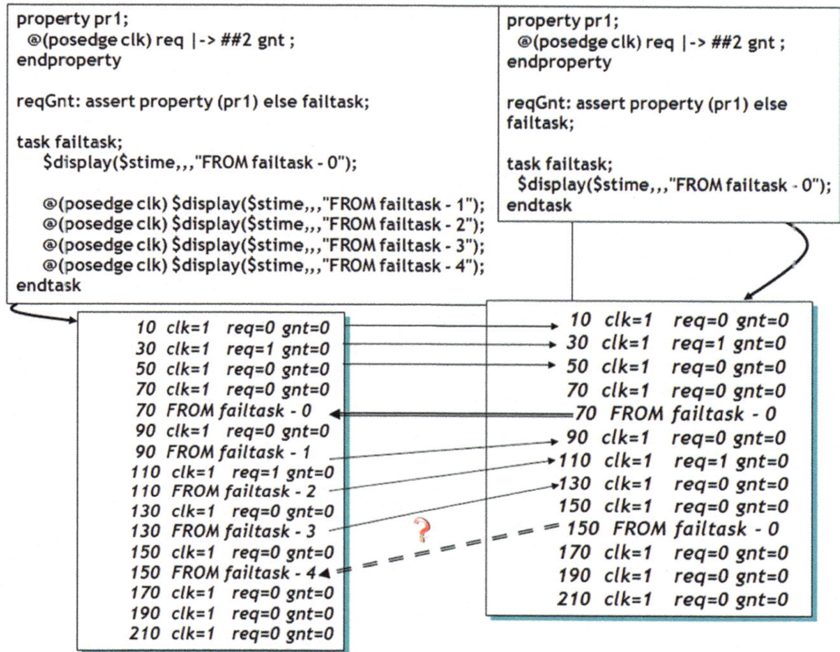

Fig. 17.16 Blocking vs. non-blocking action block

17.10 Nested Implications in a Property. Be Careful…

Can you have multiple nested implications in a property? Sure, you can. However, you need to very carefully understand the consequences of nested implications in a property. Let us look an example and understand how this works.

In Fig. 17.17, the property mclocks (at first glance) looks very benign. But play close attention and you will see two implications. @ (posedge clk) if "a" is true that implies "'bSeq' ##1 c," which implies "dSeq." One antecedent implies a consequent which acts as the antecedent for another consequent.

Now, let us look at the simulation log. At time 175, a=1, so the property starts evaluation and implies "bSeq ##1 c." At time 185 "bSeq" matches, so the property now looks for ##1 c. At time 195, c is—not—equal to "1" but the property does not fail. Wow! Reason? Note that "bSeq ##1 c" is now an antecedent for "dSeq" and as we know if there is no match on antecedent that the consequent won't be evaluated, and the property won't fail. Here that seems to apply even though "bSeq ## 1 c" is a consequent, it is also an antecedent. Language anomaly? Not really, but the behavior of such properties is not quite intuitive. Since "bSeq ##1 c" did not match, the entire property is discarded, and the property again waits for the next "a==1" to start all over again.

Confusing? Well, it is. Hence, please don't use such nested implication properties unless you are absolutely sure that that's what you want. I've seen engineers use it because the logic seems intuitive, but the behavior is not.

```
sequence bSeq;
 ##[1:5] b;
endsequence

sequence dSeq;
 ##2 d ##2 e;
endsequence

property mclocks;
  @(posedge clk) a |-> bSeq ##1 c |-> dSeq;
endproperty
```

```
#     165 CLK# 17 :: clk=1 a=0 b=0 c=0 d=0 e=1
#     175 CLK# 18 :: clk=1 a=1 b=0 c=1 d=0 e=0 //a=1 so start
#     185 CLK# 19 :: clk=1 a=0 b=1 c=0 d=0 e=0
                     //b=1 next clock; so 'bSeq' matches

#     195 CLK# 20 :: clk=1 a=0 b=0 c=0 d=0 e=0
                     //But c NE 1 the next clock
                     //and the property does NOT fail.

#     205 CLK# 21 :: clk=1 a=0 b=0 c=0 d=0 e=0
#     215 CLK# 22 :: clk=1 a=0 b=0 c=1 d=0 e=0

//So, when 'c' does go '1' a couple of clocks later, the 'dSeq' seq
//won't start eval at that time. The fact that 'c' did not go '1'
//the very next clock after 'bSeq' matches that the entire
//property will simply not get evaluated for pass or fail; until //'a' is asserted
again.

#     225 CLK# 23 :: clk=1 a=0 b=0 c=0 d=0 e=0
#     235 CLK# 24 :: clk=1 a=0 b=0 c=0 d=1 e=0
#     245 CLK# 25 :: clk=1 a=0 b=0 c=0 d=0 e=0
#     255 CLK# 26 :: clk=1 a=0 b=0 c=0 d=0 e=1
#     265 CLK# 27 :: clk=1 a=0 b=0 c=0 d=0 e=1
```

Fig. 17.17 Nested implications in a Property

My suggestion is to use only a single implication. It will keep your code unambiguous. For example,

Following two properties are equivalent.

```
P1: assert property (@posedge clk) req |-> ##2 gnt |-> ##2 gntAck;
P1: assert property (@posedge clk) req ##2 gnt |-> ##2 gntAck;
```

Following two are equivalent.

```
P2: assert property (@posedge clk) req |=> gnt |=> gntAck;
P2: assert property (@posedge clk) req ##1 gnt |=> gntAck;
```

And finally following two are equivalent

```
P3: assert property (@posedge clk) a |-> b |-> c;
P3: assert property (@posedge clk) a ##0 b |-> c;
```

So, as you can see, it is indeed possible to write properties with a single implication operator. Keep it simple.

17.11 Subsequence in a Sequence

A sequence can be embedded in another sequence. The embedded sequence can be called a subsequence. Figure 17.18 shows that sequence "abc" is embedded into sequence "abcRule." The embedded subsequence infers the clock from the parent sequence, if the subsequence does not have an explicit clock of its own.

Also, as shown in Fig. 17.19, a sequence can be used both as an antecedent and/or a consequent.

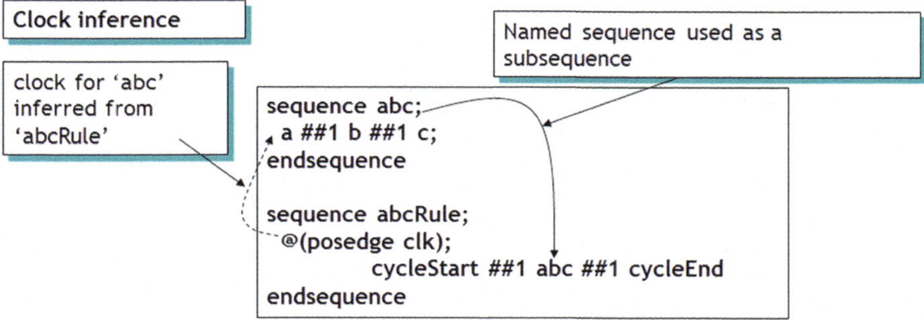

Fig. 17.18 Subsequence in a sequence—clock inference

Use as antecedent

```
sequence s1;
  @(posedge clk) a ##1 b ##1 c;
endsequence

sequence s2;
  @(posedge clk) d ##1 e;
endsequence

property s1tos2;
@(posedge clk) s1 |=> s2;
```

'sequence' can be used either as an antecedent or as a consequent or both.

Fig. 17.19 Subsequence in a Sequence

As shown, the antecedent is sequence "s1," which implies sequence "s2" as consequent. Here, each sequence has its own explicit clock. However, if that were not the case, the subsequences would inherit (infer) the clock from property s1tos2.

The reason for pointing out this usage is, again, to emphasize that it's best to break down a property into smaller sequences and then build the larger overall property. The smaller the sequence, the better it is for debuggability and controllability.

17.12 Cyclic Dependency—Mutually Recursive Property

As shown in Fig. 17.20, you can indeed have cyclic dependency between properties but *not* among sequences. But *note that the cyclic dependency between properties is only between consequent of the property, not the antecedent.*

What is the use of this feature? If you want to check for continuous toggle between two states of a state machine, you can use this property. The property as shown in this example will never complete until simulation ends.

Note also that you *cannot* do something like "c |=> d ##1 e ##1 p1" where p1 is a property. You cannot use another property as a subsequence for cyclic dependency. You will get the following Error.

∗∗ Error: massert.v(42): Illegal SVA property value in RHS of "##" expression.

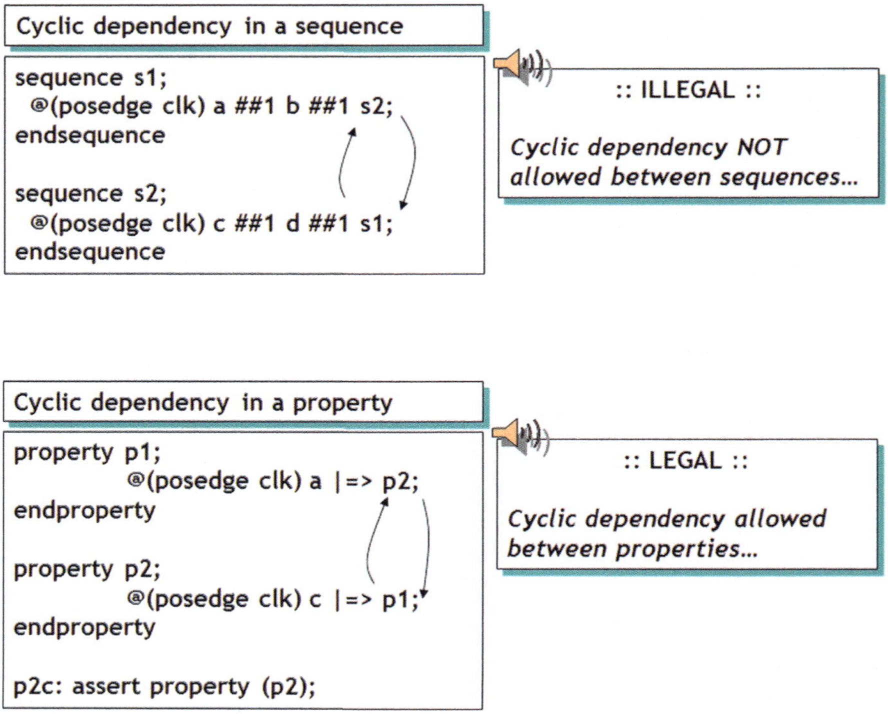

Cyclic dependency in a sequence

```
sequence s1;
  @(posedge clk) a ##1 b ##1 s2;
endsequence

sequence s2;
  @(posedge clk) c ##1 d ##1 s1;
endsequence
```

:: ILLEGAL ::

Cyclic dependency NOT allowed between sequences...

Cyclic dependency in a property

```
property p1;
        @(posedge clk) a |=> p2;
endproperty

property p2;
        @(posedge clk) c |=> p1;
endproperty

p2c: assert property (p2);
```

:: LEGAL ::

Cyclic dependency allowed between properties...

Fig. 17.20 Cyclic dependency

17.13 Refinement on a Theme…

Figure 17.21 shows very simple use of Boolean expressions to effectively model commonly required specification.

17.14 Simulation Performance Efficiency

In Fig. 17.22, the top property rdyProtocol says that if rdy is true then you must get a rdyAck. We have designed that using the constant delay range. Nothing wrong with that, but (as seen from simulation results), the infinite range-based design runs

sequence 'abc' states that 'a' is followed by 'b' followed by 'c' (all with a single clock delay). Simple enough ….

```
sequence abc;
        @(posedge clk) a ##1 b ##1 c;
endsequence
```

In many applications, however, it maybe required that 'a' remains asserted when b is asserted and then 'a and b' remain asserted when 'c' is asserted. The sequence here will do the trick …

```
a ##1 a & b ##1 a & b & c;
OR (with good use of parenthesis)
a ##1 (a & b) ##1 (a & b &c);
```

OR if it is required that 'a' and 'b' must get deasserted the very next clock after they are found asserted, then this sequence will do the trick.

```
a ##1 !a & b ##1 !a & ! b & c;
OR
a ##1 (!a & b) ##1 (!a & !b & c);
```

Fig. 17.21 Refinements on a theme

```
property rdyProtocol;
  @(posedge clk) rdy |-> ##[1:$] rdyAck;
endproperty
assert property(rdyProtocol);
```

Avoid long or infinite time ranges.

```
property rdyProtocol;
  @(posedge clk) rdy |-> rdyAck [-> 1];
endproperty
assert property(rdyProtocol);
```

A more simulation efficient way of expressing the 'infinite' range requirement…

Fig. 17.22 Simulation Performance Efficiency

slower than the one that does not use such a range. This by no means prohibits use of ##[1:$], but if you can find a better way to solve the problem, you will get better simulation efficiency. The bottom part of shows the alternate way. It uses the "goto" operator, which models the same behavior, namely, that there will be at least 1 rdy-Ack after a rdy.

17.15 It's a Vacuous World! Huh?

This section could have gone much earlier in the book, but I did not want the reader to get confused from the get-go. Once you go through this example, you will see why the "implication" operator is (almost) a must in an assertion. The example figures below have detailed annotation for ease of understanding.

Here we go.

17.16 Concurrent Assertion—Without—an Implication

Let us examine Fig. 17.23.

There is no implication operator in property pr1. As shown, property "pr1" reads as "@ (posedge clk) req should be true and 2 clocks later gnt should be true." Note

Look! NO IMPLICATION

```
property pr1;
  @(posedge clk) req ##2 gnt;
endproperty

reqGnt: assert property (pr1) $display($stime,,,"\t\t %m PASS"); else
                              $display($stime,,,"\t\t %m FAIL");
```

```
#10  clk=1 req=0 gnt=0
#10 test_basic_property.reqGnt FAIL

#30  clk=1 req=0 gnt=0
#30 test_basic_property.reqGnt FAIL

# 50  clk=1 req=1 gnt=0

#70  clk=1 req=0 gnt=0
#70 test_basic_property.reqGnt FAIL

#90  clk=1 req=0 gnt=1
#90 test_basic_property.reqGnt FAIL
#90 test_basic_property.reqGnt PASS

#110  clk=1 req=0 gnt=0
#110 test_basic_property.reqGnt FAIL
```

Whenever 'req' is Low, the assertion FAILs

That's because, a sequence simply says that 'req' be true at the clock edge and that gnt must be true 2 clocks later.

It does NOT say check the sequence "Only *If* 'req' is true at posedge clk".

But you really don't care for result when 'req' is Low.

That's where an implication operator comes into picture…

Fig. 17.23 Assertion without implication operator

that we have not used implication operator in the property. Hence, read the property carefully. It does –not- say that "*if*" req is true that the property should check for gnt. It simply says that "req" be true at the posedge clk and 2 clks later gnt be true. Hence, every clock that req is *not* true the property FAILs. Is that what we really want? I don't think so. That's where an implication operator comes into picture.

More importantly, do you notice that at time 90, the property PASSes as well as FAILs!! Amazing! The property passes because at time 50, req=1 so the property looks for gnt=1 at 90. It does find gnt=1 at 90, so it PASSes. But since req=0 at 90, it also FAILs. Amazing, again!

My suggestion is to NOT use properties without implication, unless you are absolutely sure of what you are doing. Read On. The story does not quite end with implication either… but there is hope.

17.17 Concurrent Assertion—with—an Implication

Ok, so we decide to add an implication operator as in "@ (posedge clk) req |-> ##2 gnt;" (Fig. 17.24) so that we don't get false failures. But wait! See the simulation log carefully. Now the assertion passes whenever req=0. What's going on?

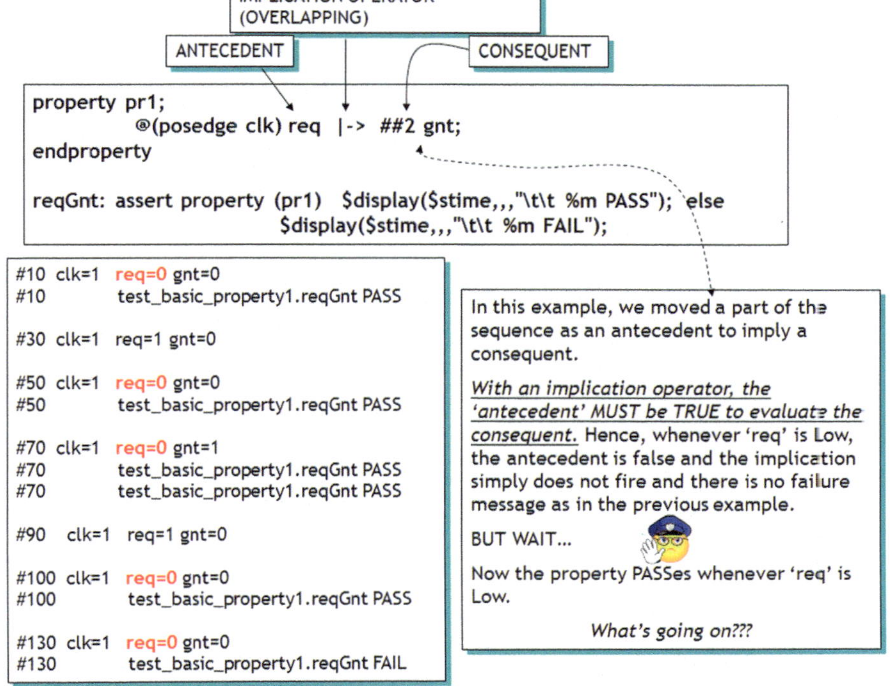

Fig. 17.24 Assertion resulting in vacuous pass

Everything seems Ok. If the antecedent is not true, the consequent won't fire. But when the antecedent is not true, the properties PASS action block triggers and tells us that the property PASSes. Read on.

17.18 Vacuous Pass

The reason we get a PASS on the antecedent not matching is that according to the LRM "If there is no match of the antecedent sequence_expr, then evaluation of the implication succeeds vacuously and returns true." Hence, whenever you see "req" low, you get a "vacuous" pass which triggers the PASS action block and we get the PASS display (Fig. 17.25)

Ok, so what is the solution? Why did we not see this behavior until now with all the examples we have been through? Read on…

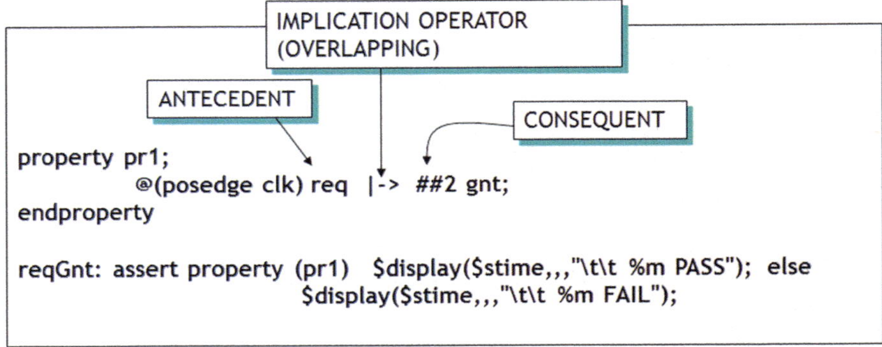

LRM 3.1a (Page 232) ::

"If there is no match of the antecedent sequence_expr, then evaluation of the implication succeeds vacuously and returns true"

A couple of ways to get around this…

One is to simply not use the action_block associated with 'pass' (duh…) of the property, so that you don't get pass indication vacuously

But what if you do want to know when the property passes…

Fig. 17.25 Vacuous Pass

17.19 Concurrent Assertion—With "cover"

There are two ways to overcome the vacuous pass behavior. One is to simply ignore the PASS action block on "assert" of a property, i.e., simply do not have a PASS action block. That way if there is a vacuous pass, our log will not be cluttered with misleading PASS messages. This is the obvious solution.

But what if you do want to know when the property PASSes? That's where "cover" comes into picture.

The solution with "cover" allows us to see if the property is indeed covered (i.e., exercised). "cover" does *not* have the vacuous pass property. It indicates at the *end* of the assertion if it has been covered. When it is covered, it triggers a PASS action block. In this action block you may put a $display statement to indicate that the property has been covered or that it has "passed."

Note that "cover" simply does *not* have a FAIL action block and does not have the vacuous pass property.

This way, with "assert" and "cover," we have a method to code an assertion that gives us the required FAIL and PASS indication without any other message. Please read the simulation log in Fig. 17.26 carefully in to see the behavior of the property.

NEW TO IEEE-1800 2012 LRM. So, for all this "vacuous" pass dilemma, 2012 LRM came up with a clean solution. They have introduced a new System Task called $assertcontrol. I have devoted complete Sect. 20.18 to describe this task. But here's an example of how you can turn OFF the vacuous pass indication.

```
property pr1;
        @(posedge clk) req |-> ##2 gnt;
endproperty

A_reqGnt: assert property (pr1)  else $display($stime,,,"\t\t %m FAIL");

C_reqGnt: cover property (pr1) $display($stime,,,"\t\t %m PASS");
```

No action_block associated with true eval of the property.

You may use a 'cover' statement to cover the same property that is asserted. 'cover' does not report vacuous pass. *Note that 'cover' does not allow an action_block if the property fails.*

```
#10 clk=1  req=0 gnt=0
#30 clk=1  req=1 gnt=0
#50 clk=1  req=0 gnt=0
#70 clk=1  req=0 gnt=1
#70  test_basic_property2.C_reqGnt PASS
#90 clk=1  req=1 gnt=0
#110 clk=1  req=0 gnt=0
#130 clk=1  req=0 gnt=0
#130 test_basic_property2.A_reqGnt FAIL
```

In this example, we removed the action block associated with the true (i.e. pass) evaluation of the property to avoid the vacuous $display.

If you do need an action block for a match (i.e. Pass) of a property, you may use a 'cover' statement to cover the same property that is asserted.

Fig. 17.26 Assertion with "cover" for PASS

```
$assertcontrol (VACUOUSOFF, CONCURRENT | EXPECT);
```

This systasks affect the whole design so no modules are specified. Disable vacuous pass action for all the concurrent asserts and expects in the design.

The 2012 LRM also introduces a specific system task simply to disable vacuous pass indication.

$assertvacuousoff system task turns off the PASS indication based on a vacuous success. An assertion that is already executing is not affected. By default, we get a PASS indication on vacuous pass.

Note also the use of "%m" in the $display task. This good old Verilog feature displays the entire path to the assertion. This is one way to distinguish two properties with same name in two separate scopes.

17.20 Empty Sequence

In Fig. 17.27, we are using the consecutive operator "∗" but with "0" repetition [∗0]. In other words, we are saying "b" should not repeat *ever*. In yet other words, that means "b" simply does not exist (empty) even though it is part of the property. Hence, "b[∗0] ##1 !a" simply means check for empty sequence "b" and 1 clock

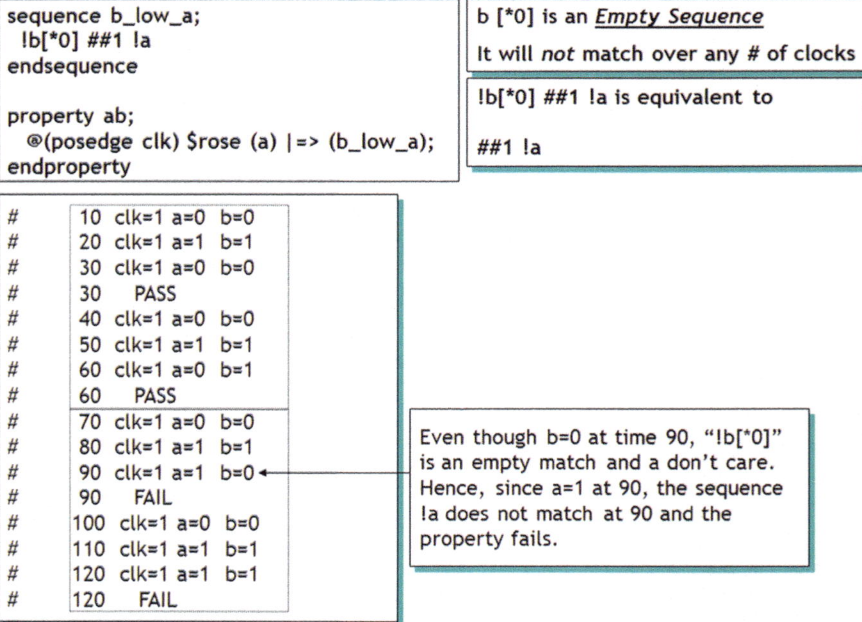

Fig. 17.27 Empty match [∗m] where m=0

later check for "!a." Since, empty sequence does not mean anything, we are basically checking for "!a" 1 clock after $rose(a).

Following examples are given for the sake of completeness. Regard them as Reference Material.

Figure 17.28 is indeed interesting. LRM provides two rules on how to interpret (seq ##n *empty*) and (*empty* ##n seq) both with n>0. The explanation is noted at the top of the figure. To make that clear, let us go through the example.

Property "ab" says that if the antecedent "z" is true that the consequent sequence "sc1" should execute. Sequence "sc1" says that "a" be true when "z" is true; then (according to the LRM rule (seq ##n *empty* == seq##(n-1) `true), the sequence can be read as "a" be true; then "b" may not be true (i.e., empty—does not exist) at all or will continue to repeat forever until c==1.

Let us look at the simulation log to see if the new definition holds.

At 30, z=1 so the property looks for "a"=1 at the same time. "a" = 1 at 30. "b" is also equal to "1"—which does not really matter because "b" can have zero match, as long as there is c==1 at 1 clock after the *last* "b" or "a." In our case which started at 30, we do have c==1 at 50 which is one clock after "a==1" as well as "b==1." Hence the property passes.

At 70, z=1, a=1 but b=0. That's an empty match. Hence, the property looks for c==1 after the last "a." It finds that at 90 and the property passes.

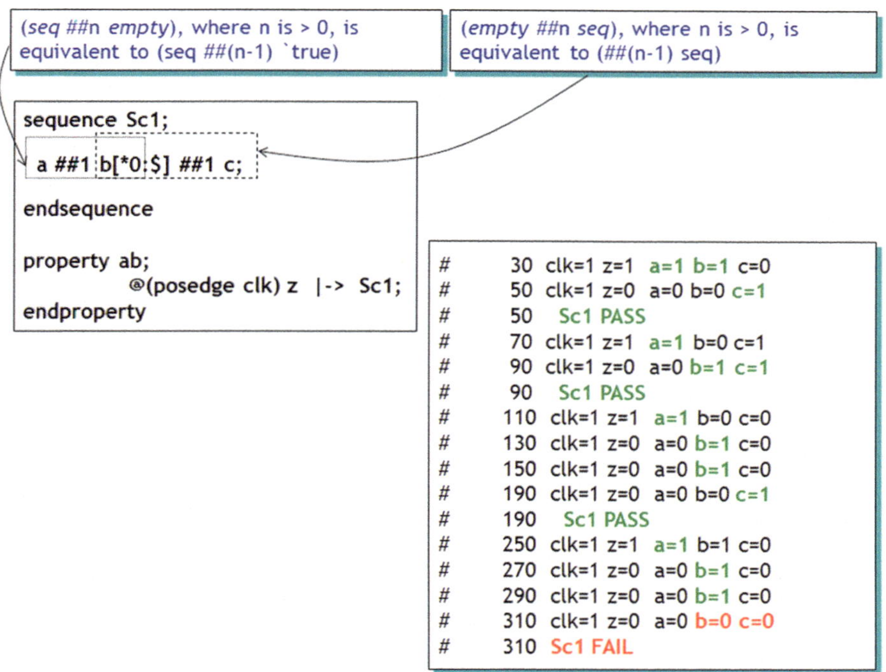

Fig. 17.28 Empty match—example

At 110, z=1, a=1, b=0. Next clock at 130, a=0 and c is also equal to "0"—but "b" = 1. As we saw before, "b" may not match or may continually match forever until "c==1." Since c==0 at time 130, the property continues to look for b==1 until c==1. That happens at time 150 (b==1) and 190 (c==1) and the property passes.

At 250, z=1, a=1 and b=1. The next clock at 270, c is still zero, so the property continues to see that b remains "1." At 290, b==1, so we move on. But at time 310, "b" does not remain asserted and "c" is not equal to "1" either. "c" should have been "1" (to satisfy b[*0:$] ##1 c)—or—"b" should have remain asserted. Neither happens and the property fails.

The example in can be used effectively when you want to check that if a certain sequence takes place that another never takes place. You can do that with non-zero consecutive repetition operator also, but the [*0] or [=0] makes it that much easier.

For example, in the following example (Fig. 17.29),

```
@ (posedge clk) a |=> b[=0];
```

means that on "a" being true, one clock later, "b" should never occur. In other words, "b" should be negated (zero) forever. This is quite straightforward to read/interpret. The behavior is shown in simulation log in.

So, how else would you write this property another way?

```
@ (posedge clk) a |=> !b[*1:$];
```

Same meaning. Once "a" is true that starting next clock "b" should remain "!b" consecutively forever.

Following is purely reference material. Keep it in your back pocket. It will be useful on a rainy day! (Fig. 17.30).

```
property ab;
        @(posedge clk) a |=> b [=0];
endproperty
```

b [=0]

means that the signal 'b' should *never* be true (in other words, it means that the # of non-consecutive occurrences are ZERO).

```
#     5  clk=1 a=1 b=0
#     15 clk=1 a=0 b=0
#     15          property ab PASS

#     35 clk=1 a=1 b=0
#     45 clk=1 a=0 b=0
#     45          property ab PASS
#     55 clk=1 a=0 b=1

#     65 clk=1 a=1 b=0
#     75 clk=1 a=0 b=1
#     75          property ab FAIL
```

Fig. 17.29 Empty match example—II

SystemVerilog 3.1a LRM; Page 210

(*empty* ##0 *seq*) does not result in a match

(*seq* ##0 *empty*) does not result in a match

(*empty* ##n *seq*), where n is > 0, is equivalent to (##(n-1) seq)

(*seq* ##n *empty*), where n is > 0, is equivalent to (seq ##(n-1) `true)

Examples:

b ##1 (a[*0] ##0 c) - will never produce a match

b ##1 a[*0:1] ##2 c is equivalent to
(b ##2 c) or (b ##1 a ##2 c)

Fig. 17.30 Empty sequence. Further rules

Chapter 18
Asynchronous FIFO Assertions

Introduction: An asynchronous FIFO (in contrast to a synchronous FIFO) is a difficult proposition when it comes to writing assertions. The Read and the Write clocks are asynchronous which means the most important property to check for is data transfer from Write to Read clock. Other assertions are to check for fifo_full, fifo_empty, etc. conditions.

A. B. Mehta, *SystemVerilog Assertions and Functional Coverage*,
https://doi.org/10.1007/978-3-030-24737-9_18

First, we present a comprehensive design of the asynchronous fifo. A bit complicated but you don't need to go into its detail. Next, we see a test-bench within which I have designed the assertions. Yes, you can have assertions in the design (RTL) (but not recommended), test-bench (as in this example), in a systemverilog Interface, in a systemverilog Program, and in a file on its own (this is where the "bind" comes into picture (this is highly recommended)).

This FIFO design and test-bench are available on Springer server.

FIFO design uses gray code counters (pointers) for Read and Write pointers for asynchronous transfer of write_data to read_data.

18.1 Asynchronous FIFO Design

```
module asynchronous_fifo (
  // Outputs
  fifo_out, full, empty,
  // Inputs
  wclk, wclk_reset_n, write_en,
  rclk, rclk_reset_n, read_en,
  fifo_in
);

  `define FF_DLY 1'b1
   parameter  D_WIDTH = 20;
   parameter  D_DEPTH = 4;
   parameter  A_WIDTH = 2;

   input                   wclk_reset_n;
   input                   rclk_reset_n;
   input                   wclk;
   input                   rclk;
   input                   write_en;
   input                   read_en;
   input [D_WIDTH-1:0]     fifo_in;

   output [D_WIDTH-1:0]    fifo_out;
   output                  full;
   output                  empty;

   reg [D_WIDTH-1:0]    reg_mem[0:D_DEPTH-1];
   reg [A_WIDTH:0]      wr_ptr;
   reg [A_WIDTH:0]      wr_ptr_gray;
   reg [A_WIDTH:0]      wr_ptr_gray_rclk_q;
   reg [A_WIDTH:0]      wr_ptr_gray_rclk_q2;
   reg [A_WIDTH:0]      rd_ptr;
```

```
reg [A_WIDTH:0]      rd_ptr_gray;
reg [A_WIDTH:0]      rd_ptr_gray_wclk_q;
reg [A_WIDTH:0]      rd_ptr_gray_wclk_q2;

reg                  full;
reg                  empty;

wire [A_WIDTH:0]     nxt_wr_ptr;
wire [A_WIDTH:0]     nxt_rd_ptr;
wire [A_WIDTH:0]     nxt_wr_ptr_gray;
wire [A_WIDTH:0]     nxt_rd_ptr_gray;
wire [A_WIDTH-1:0]   wr_addr;
wire [A_WIDTH-1:0]   rd_addr;
wire                 full_d;
wire                 empty_d;

assign wr_addr = wr_ptr[A_WIDTH-1:0];
assign rd_addr = rd_ptr[A_WIDTH-1:0];

always @ (posedge wclk)
 if (write_en) reg_mem[wr_addr] <= #`FF_DLY fifo_in;

assign fifo_out = reg_mem[rd_addr];


always @ (posedge wclk or negedge wclk_reset_n)
  if (!wclk_reset_n) begin
    wr_ptr <= #`FF_DLY {A_WIDTH+1{1'b0}};
    wr_ptr_gray <= #`FF_DLY {A_WIDTH+1{1'b0}};
  end else begin
    wr_ptr <= #`FF_DLY nxt_wr_ptr;
    wr_ptr_gray <= #`FF_DLY nxt_wr_ptr_gray;
  end

assign nxt_wr_ptr = (write_en) ? wr_ptr+1 : wr_ptr;
assign nxt_wr_ptr_gray = ((nxt_wr_ptr>>1) ^ nxt_wr_ptr ;

always @ (posedge rclk or negedge rclk_reset_n)
  if (!rclk_reset_n) begin
    rd_ptr <= #`FF_DLY {A_WIDTH+1{1'b0}};
    rd_ptr_gray <= #`FF_DLY {A_WIDTH+1{1'b0}};
  end else begin
    rd_ptr <= #`FF_DLY nxt_rd_ptr;
    rd_ptr_gray <= #`FF_DLY nxt_rd_ptr_gray;
  end
```

```
assign nxt_rd_ptr = (read_en) ? rd_ptr+1 : rd_ptr;
assign nxt_rd_ptr_gray = (nxt_rd_ptr>>1) ^ nxt_rd_ptr;

// check full
always @ (posedge wclk or negedge wclk_reset_n)
  if (!wclk_reset_n)
    {rd_ptr_gray_wclk_q2, rd_ptr_gray_wclk_q} <= #`FF_DLY
    {{A_WIDTH+1{1'b0}}, {A_WIDTH+1{1'b0}}};
  else
    {rd_ptr_gray_wclk_q2, rd_ptr_gray_wclk_q} <= #`FF_DLY
    {rd_ptr_gray_wclk_q, rd_ptr_gray};

assign full_d = (nxt_wr_ptr_gray == {~rd_ptr_gray_wclk_q2[A_
WIDTH:A_WIDTH-1], rd_ptr_gray_wclk_q2[A_WIDTH-2:0]});

always @ (posedge wclk or negedge wclk_reset_n)
  if (!wclk_reset_n)
    full <= #`FF_DLY 1'b0;
  else
    full <= #`FF_DLY full_d;

// check empty
always @ (posedge rclk or negedge rclk_reset_n)
  if (!rclk_reset_n)
    {wr_ptr_gray_rclk_q2, wr_ptr_gray_rclk_q} <= #`FF_DLY
    {{A_WIDTH+1{1'b0}}, {A_WIDTH+1{1'b0}}};
  else
    {wr_ptr_gray_rclk_q2, wr_ptr_gray_rclk_q} <= #`FF_DLY
    {wr_ptr_gray_rclk_q, wr_ptr_gray};

  assign empty_d = (nxt_rd_ptr_gray == wr_ptr_gray_rclk_q2);

always @ (posedge rclk or negedge rclk_reset_n)
  if (!rclk_reset_n)
    empty <= #`FF_DLY 1'b1;
  else
    empty <= #`FF_DLY empty_d;

  endmodule
```

18.2 Asynchronous FIFO Test-Bench and Assertions

```
module test_asynchronous_fifo
(
   fifo_out, full, empty,
   wclk, wclk_reset_n, write_en,
   rclk, rclk_reset_n, read_en,
   fifo_in
);

parameter    D_WIDTH = 20;
parameter    D_DEPTH = 4;
parameter    A_WIDTH = 2;

output       wclk_reset_n;
output       rclk_reset_n;
output       wclk;
output       rclk;
output       write_en;
output       read_en;

output [D_WIDTH-1:0] fifo_in;


   logic                 wclk_reset_n;
   logic                 rclk_reset_n;
   logic                 wclk;
   logic                 rclk;
   logic                 write_en;
   logic                 read_en;
   logic [D_WIDTH-1:0]  fifo_in;

   input [D_WIDTH-1:0]      fifo_out;
   input        full;
   input        empty;

asynchronous_fifo aff1
(
       fifo_out, full, empty,
       wclk, wclk_reset_n, write_en,
       rclk, rclk_reset_n, read_en,
       fifo_in
);
/*
```

Following property checks to see if the FIFO is full that the wr_ptr does not change. Note that this assertion needs to be clocked using "wclk" since we are dealing with write operation. This assertion can be written other ways too (for example, try $stable). Please try them out and see if the results match with this assertion.

```
*/
property check_full;
  @ (posedge wclk) disable iff (!wclk_rstn)
      (full) |=> @ (posedge wclk) aff1.wr_ptr == $past (aff1.
      wr_ptr);
endproperty

cfull : assert property (check_full) else $display($stime,,,"%m
Check wr_ptr full FAIL");
cfullc  :  cover property  (check_full)  $display($stime,,,"%m
Check wr_ptr full PASS");

/*
```

Following property checks to see that if the FIFO is empty that the rd_ptr does not change. "empty" means that the rd_ptr remains the same as its value at the last clk—thus guaranteeing that the rd_ptr have not changed. BUT note that we are using !$isunknown and passing it to $past as an expression. Why? If FIFO is empty, the rd_ptr in the past could be "X." So, we make sure that rd_ptr is the same as the past value and that it is not unknown.

This assertion can be written other ways too (using, for example, "full" condition or without the use of $past). Please try them out and see if the results match with this assertion.

```
*/

property check_empty;
  @ (posedge rclk) disable iff (! rclk_rstn)
    (empty) |=> @ (posedge rclk)
      if (!$isunknown( $past (aff1.rd_ptr)))
       (aff1.rd_ptr == $past (aff1.rd_ptr));
endproperty

cempty:assertproperty(check_empty)else$display($stime,,,"%m
Check rd_ptr empty FAIL");
cemptyc : cover property (check_empty) $display($stime,,,"%m
Check rd_ptr empty PASS");

/*-----------------------------------------
ASYNCHRONOUS DATA TRANSFER CHECK
------------------------------------------*/
/*
```

This is a very important assertion for an asynchronous FIFO. Check that the data that is written at a wr_ptr is the same data that is read when rd_ptr reaches that wr_ptr. Simple specification! Let us look at the assertion step by step

*/

/*

In the assertion, data_check property checks to see that FIFO is not full. If so, the assertion saves wr_ptr into the local variable "ptr" and the data from fifo into local variable "data" (and $displays the operation so that we can easily see how the assertion is progressing during simulation).

If the antecedent is true, the consequent says that the first match of rd_ptr being the same as wr_ptr (note wr_ptr was stored in local variable ptr) that the read data is the same as the write data (note write data were stored in local variable data in the antecedent).

Sequence rd_detect(ptr) is used as an expression to first_match. It says that wait from now until forever until you detect a read and its rd_ptr is equal to the wr_ptr (which is stored in the local variable "ptr" in the antecedent).

Note that you need to be careful on the selection of clocks since there are two clocks, "wclk" and "rclk." The "data_check" sequence is based on "wclk" and the rd_detect sequence is based on "rclk." Then we cross from "wclk" to "rclk."

Please note that these are multi-clocked properties since we have a wr_clk and a rd_clk.

So, in this property, we see

- multi-clock property (wclk and rclk)
- use of local variables for storing/comparing
- first_match
- attaching a subroutine (here a $display) to an expression for effective debugging
- effective use of ##0 to create overlapping condition in a sequence.

Try out different ways to write the assertion. Refer to different operators, sampled value functions, etc., and see if you can write an equivalent assertion. There isn't just one way to write an assertion. But there is indeed a right way and a wrong way. You will get that through practice.

*/

```
sequence rd_detect(ptr);
  ##[0:$] (read_en && !empty && (aff1.rd_ptr == ptr));
endsequence

property data_check(wr_ptr);
integer ptr, data;
  @ (posedge wclk) disable iff (!wclk_reset_n || !rclk_reset_n)
       (write_en && !full, ptr=wr_ptr, data=fifo_in,
 $display($stime,"\t Assertion Disp wr_ptr=%h data=%h", wr_
 ptr, fifo_in))

|=>
```

```
  @ (negedge rclk) first_match(rd_detect(ptr),
    $display($stime,,," Assertion Disp FIRST_MATCH ptr=%h
    Compare data=%h fifo_out=%h", ptr, data, fifo_out) )
          ##0 (fifo_out === data);
endproperty

dcheck : assert property (data_check(aff1.wr_ptr)) else
$display($stime,,,"FAIL: DATA CHECK");
dcheckc:coverproperty(data_check(aff1.wr_ptr))$display($stime,,,
"PASS: DATA CHECK");
```

/*---
If FULL -> NOT EMPTY Check
---*/

/*

Following property is quite self-explanatory. But note that this is also a multi-clocked property, and since the write and read clocks are different clocks, we must use the nonoverlapping operator |=> .

This is a mutex property. What are the different ways you can write this?

*/

```
property full_empty;
@ (posedge wclk) disable iff (!wclk_reset_n)
      @ (posedge wclk) (full) |=> @ (posedge rclk) (!empty);
endproperty

few: assert property (full_empty) else $display($stime,," FAIL:
Full and Empty BOTH asserted");
cfew: cover property (full_empty) $display($stime,," PASS: Full
and Empty check ");
```

/*---
If EMPTY -> NOT FULL Check
---*/

/*

Following property is quite self-explanatory. But note that this is also a multi-clocked property, and since the write and read clocks are different clocks, we must use the nonoverlapping operator |=> .

This is a mutex property. What are the different ways you can write this?

*/

```
property empty_full;
@ (posedge wclk) disable iff (!wclk_reset_n)
      @ (posedge rclk) (empty) |=> @ (posedge wclk) (!full);
endproperty
```

```
efw: assert property (empty_full) else $display($stime,,,
"FAIL: Full and Empty BOTH asserted");
```

/*--
rclk_reset_n Check on rclk
--*/
/*
Following property checks to see that empty pointer is high (i.e., empty) when you reset the FIFO
*/

```
property reset_n_rclk;
  @ (posedge rclk) !rclk_reset_n |-> empty;
endproperty

reset_nrclkA:assertproperty(reset_n_rclk)else$display($stime,,,
"FAIL: FIFO not empty during rclk_reset_n");
reset_nrclkC: cover property (reset_n_rclk) $display($stime,,,
"PASS: FIFO empty during rclk_reset_n");
```

/*--
wclk_reset_n Check on wclk
--*/
/*
Following property checks to see that the FIFO is not full when you reset the FIFO. FIFO can only go Empty during reset, not Full.
*/

```
property reset_n_wclk;
@ (posedge wclk)  !wclk_reset_n |-> !full;
endproperty

reset_nwclkA:assertproperty(reset_n_wclk)else$display($stime,,,
"FAIL: FIFO FULL during wclk_reset_n");
reset_nwclkC: cover property (reset_n_wclk) $display($stime,,,
"PASS: FIFO FULL during rclk_wstn");

endmodule
```

Chapter 19
Asynchronous Assertions

Introduction: This chapter is solely devoted to Asynchronous Assertions, meaning the sampling edge of the assertion is not a synchronous clock rather an asynchronous edge. Special focus is on pitfalls of using an asynchronous assertion where an expression is used both as a sampling edge and as part of antecedent.

A. B. Mehta, *SystemVerilog Assertions and Functional Coverage*,
https://doi.org/10.1007/978-3-030-24737-9_19

So far in the book we have always used a synchronous clock edge as the sampling edge for the assertion. That is for good reason. The example presented here uses an asynchronous edge (perfectly legal) as the sampling edge. The problem statement goes something like "whenever (i.e., asynchronously) L2TxData == L2ErrorData that L2Abort is asserted." Now that looks very logical to implement without the need for a clock. So, we write a property as shown in Fig. 19.1. We simply say that @ (L2TxData) (i.e., whenever L2TxData changes) that we compare L2TxData == L2ErrorData and if that matches, we imply that L2Abort ==1.

This sounds very logical. What is wrong with it? Hmm... many things. The annotation in Fig. 19.1 takes you systematically on what is going on. Please study it carefully to see why there is a problem. We will not repeat that explanation of the figure again. But here's the high-level hint on the problem. *The "sampling edge" (namely, @ (L2TxData)) is also used in the comparison expression (L2TxData == L2ErrorData)*. Since the value of variables in an expression are always sampled in

EXAMPLE ON ASYNCHRONOUS ASSERTIONS ::

Whenever (i.e. asynchronously) L2TxData == L2ErrorData that L2Abort is asserted (*i.e. you want to check this condition irrespective of the clock*).

You may be tempted to write the property as follows:

```
property CheckData;
        @(L2TxData) (L2TxData == L2ErrorData) |-> (L2Abort == 1);
endproperty
```

:: Let us analyze how this property works ::

• ***First, recall that the values of expression variables in an assertion are evaluated in the preponed region (i.e., variable value is that which existed a delta before the sampling edge (i.e. clock edge))***

• **Now, let us assume that L2TxData changes and is now equal to L2ErrorData. This is what will happen**

@(L2TxData) is triggered

and (L2TxData == L2ErrorData) expression is evaluated.

BUT

• **The value of L2TxData compared in this expression is the value detected -before- the sampling edge L2TxData, meaning the value sampled is the one that became equal to L2ErrorData. So, the expression won't match and implication won't trigger.**

OK, so let's move on

• **Now when L2TxData changes again (i.e. now it is NOT equal to L2ErrorData, assuming L2ErrorData did not change)**

@(L2TxData) is triggered again

• **and again, the value of L2TxData and L2ErrorData used in the expression are the ones -before- L2TxData changed (when they *did* actually match). So now the expression will match and the implication will trigger checking for L2Abort == 1.**

• **So, what have you really proven? Read on ....**

Fig. 19.1 Asynchronous assertion—problem statement

the preponed region that the value of L2TxData in the expression won't be the same as when @ (L2TxData) changed. In other words, @(L2TxData) uses the "current" value of L2TxData, while the L2TxData in the expression (L2TxData == L2ErrorData), is the "sampled" value. When such is the case, I strongly recommend against using an asynchronous assertion.

We continue the analysis in Fig. 19.2. The annotation explains the reasons.

In order to circumvent the problem that we just described in Fig. 19.2, we can continue with the asynchronous sampling edge, only that we put *all* the comparison expressions/variables as part of asynchronous sampling edge. This is shown in Solution 1 that will take care of the problems we first encountered (Fig. 19.3).

Exercise: Think through and determine why Solution 1 works. Please simulate with a simple Verilog test-bench to see the effect of solution 1.

Why do we have "assign #1" in solution 2? That way when L2TxData or L2ErrorDataW or L2ABortW changes that there is a 1-time unit delay which will allow the new value to settle down and "sample" the settled value—before—you check for (L2TxData == L2ErrorData).

This is a (convoluted) way to get around the check of these variables in the preponed region. If all this looks confusing, do not be daunted. *I strongly advice you against using asynchronous edges as sampling edges when the same edge/expression is also used in the antecedent or the consequent.* Again, if you are comfortable with using them and follow the rule I just specified, please do so, but be careful.

```
property CheckData;
        @(L2TxData) (L2TxData == L2ErrorData) |-> (L2Abort == 1);
endproperty
```

:: Continuing with the story ... ::

- You have basically checked for L2Abort == 1 at the *very last temporal moment* when L2TxData == L2ErrorData !

- What' wrong with that? Here...

- What if L2ErrorData changed in the middle before L2TxData changed again?

- What if L2Abort changed (to 0) while you were waiting for L2TxData to change again?

Fig. 19.2 Asynchronous Assertion—problem statement analysis continued

```
Solution 1 ::
property CheckData;
         @(L2TxData or L2ErrorData or L2Abort) (L2TxData == L2ErrorData)
                                        |-> (L2Abort == 1);
endproperty
```

```
Solution 2 ::
wire L2TxDataW, L2ErrorDataW, L2AbortW;

assign #1 L2TxDataW = L2TxData;
assign #1 L2ErrorDataW = L2ErrorData;
assign #1 L2AbortW = L2Abort;

property CheckData;
         @(L2TxDataW or L2ErrorDataW or L2AbortW) (L2TxData ==
                                        L2ErrorData) |-> (L2Abort == 1);
endproperty
```

```
Solution 3 :: Good old Verilog comes to rescue ...
always @(L2TxData or L2ErrorData or L2Abort)
begin
  if (L2TxData == L2ErrorData)
  begin
    if (L2Abort) $display("L2Abort Check PASS")
    else $display("L2Abort Check FAIL");
  end
end
```

Fig. 19.3 Asynchronous assertion—Solution

Refer to the example above to help you with the behavior of asynchronous sampling edge. Note that I have shown all three solutions with asynchronous assertion (so much for my opposition to it!).

Exercise: How would you model this as a synchronous assertion? Please try and see if you succeed. Assume "posedge clk" as your sampling edge.

Solution 3 uses a procedural block to determine when you do the check. Notice that I have not used assertions in this solution. Point being, sometimes it is better and ok to simply use Verilog which will be more intuitive and give the results you desire.

19.1 Glitch Detection

Ok, so I've made a case that one has to be careful when one uses an asynchronous signal both as a sampling edge and in an antecedent expression at the same time.

There are indeed many uses of asynchronous assertions keeping the above in mind. One question that comes up often is how you detect a glitch on a signal. Asynchronous assertion comes into picture. Here's a complete example. A simple test-bench is presented as well. Do simulate this to see the results.

The explanation of how the code works is embedded directly with the code.

```
module glitch_detect_SVA( );
parameter duration = 3ns;
logic glitch;

//The following property checks to see that signal 'glitch'
stays high for greater than (or equal to) //the 'duration'
(i.e. the glitch width). If so, the property PASSes or else
it FAILs. In other words, if //the width of signal 'glitch'
is less than the duration, it's a glitch.
//You can parametrize the property 'detect_glitch' to make it
generic for different glitch widths.

property detect_glitch;
 time leading; //local variable 'leading' of type 'time'.

@(glitch) //Sampling edge is asynchronous
//At asynchronous edge of signal 'glitch', store the time at
which it changed in the local variable //'leading'.
   (1'b1,leading=$time)
        |=>  //Non-overlapping signifies @ the next asynchronous
        change in signal 'glitch'
        (
         ($time - leading) >= duration
//At the next change in 'glitch', subtract the 'leading' time
from the current time. That gives us the //width of the sig-
nal. Check to see that it's >= the signal 'glitch' duration.
If so, the property passes.
        );
```

```
endproperty: detect_glitch

//The following property is simply to showcase how you would
display the Glitch Width that //results in a glitch. Note the
use of $display used as a subroutine attached to always true
1'b1.
property display_glitch_width;
  time leading;
  @(glitch) (1'b1,leading=$time) |=>
           (1'b1, $display("%0t Glitch Width = %0t",$time,
           ($time-leading)));
endproperty

//Assert both properties
glitchD: assert property (detect_glitch) else $display("At
%0tns ERROR: Glitch Detected\n",$time);
glitchW: assert property (display_glitch_width);

initial
  begin
    glitch = 1'b0;
    #5ns glitch = !glitch;
    #5ns glitch = !glitch;
    #2ns glitch = !glitch; //12ns - Glitch
    #5ns glitch = !glitch;
    #2ns glitch = !glitch; //19ns - Glitch
    #2ns glitch = !glitch; //21ns - Glitch

  #100 $finish(2);
  end
endmodule
```

Chapter 20
IEEE-1800-2009/2012 Features

Introduction: This chapter describes the new features of the 2009/2012 LRM. In that sense, it is a long chapter. It describes features such as "strong" and "weak" properties, abort system tasks, deferred immediate assertions, and past and future global clock-based sampling functions such as $rose_gclk, $fell_gclk, $rising_gclk, and $falling_gclk. It further covers "followed by" property operators and "always," "eventually," "until," "nexttime," "case," as well as $inferred_clock and $inferred_disable. Also covered are "restrict" operator for formal verification, "reject"/"accept" properties, and assertion control tasks.

A. B. Mehta, *SystemVerilog Assertions and Functional Coverage*,
https://doi.org/10.1007/978-3-030-24737-9_20

20.1 Strong and Weak Sequences

IEEE-1880-2009/2012 adds the notion of a strong and weak operator applied to sequence expressions. The idea behind these "strengths" is very simple.

They are declared as:

```
strong (sequence_expression)
OR
weak (sequence_expression)
```

Here's an example

```
sequence a_wait_b;
       @ (posedge clk) A |-> (A ##[1:$] B);
endsequence
awb: assert property (strong(a_wait_b)) else $display($stime,
"a_wait_b FAIL"); //default 'weak'
awbc: cover property (weak(a_wait_b)) $display($stime,"a_wait_b
PASS"); //default 'strong'
```

"strong" sequence means that if you run out of simulation ticks (at the end of simulation, for example), the "strong" sequence will FAIL. In other words, "strong" will evaluate to true only if there is a non-empty match of the sequence expression. And in yet other words, the "strong" operator requires "enough" ticks to witness a success. In our example, if "B" never arrives until the end of simulation, the property will FAIL. By default, an "assert" (or "assume") is "weak" and as we have seen so far, if you run out of simulation ticks the sequence will *not* fail (a simulator may still give an indication/warning of an incomplete sequence or simply given an indication of PASS.

Disclaimer: The simulators that the author tried (at the time of writing) gave different results (incomplete or PASS) for "weak" properties. Even for strong properties, the simulation results were not conclusive.

On the other hand, "cover" is strong by default. Analogous to "assert," if the property does not complete, the evaluation of sequence expression does not succeed, and the "cover" will be considered to FAIL (i.e., not covered). In other words, an incomplete "cover" sequence will not give us a "PASS" or "cover" indication, because there haven't been enough ticks to reach a "success" state. That is exactly what we want because we do not want an incorrect "cover" of a sequence that never completes. On the other hand, if you use "weak" operator with "cover" and the sequence never completes, the "cover" will be considered to have completed or covered (*this is simulator dependent from author's experience, so take the description of cover with a "weak" operator with a grain of salt*).

In short, by default, a property is weak in the context of an "assert" (or an "assume") and is strong in the context of a "cover."

20.2 $changed

SystemVerilog 2009/2012 adds $changed sampled value function in addition to the ones we have already seen such as $past, $rose, $fell, and $stable. Refer to Figs. 20.1 and 20.2.

Here's a simple example of where $changed is helpful.

Specification: Make sure that "toggleSig" toggles every clock. In other words, see that "toggleSig" follows the pattern 101010… or 010101….

Solution: First inclination will be to write the assertion as follows.

```
tP: assert property (@(posedge clk) toggleSig ##1 !toggleSig);
```

But will this work? No. This property simply states that toggleSig be true every clock that it is false the next clock. What that also means is that the next clock, we are checking for toggleSig to be both true and false at the same time! Totally contradictory. The assertion (most likely) will fail at this next clock since at this clock toggleSig could be inverted. That's not what we are checking for.

Here's where $changed comes to rescue. Following property will verify the toggle specification.

```
tP: assert property (@(posedge clk) ##1 $changed(toggleSig );
```

$changed(expression [, clocking event]);	*Returns True if the expression changed from the previous tick of the clocking event. Otherwise it returns False.*

Notes:

- *The [, clocking event] is optional and usually derived from the clocking event of the assertion or from the inferred clock of the procedural block where the function is used*
- *When this function is called at or before the simulation time step in which the first clocking event occurs, the results are computed by comparing the sampled value of the expression with its default sampled value*
- *This function can be used in property/sequence as well as in procedural code as expression*
- *$changed(expr) is true if the sampled value of 'expr' in the pre-poned region of current time stamp changed from the sampled value in the pre-poned region of the previous time stamp.*

Fig. 20.1 $changed

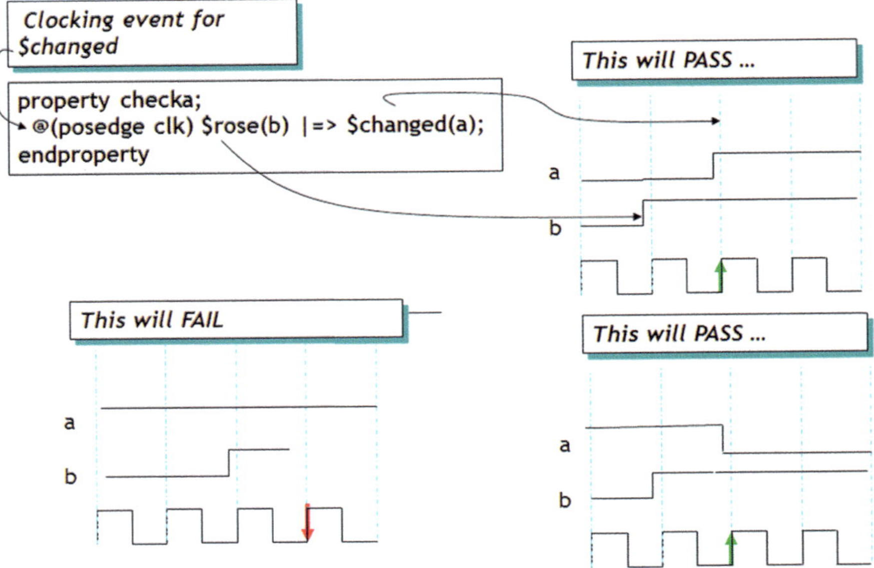

Fig. 20.2 $changed

20.3 $sampled

$sampled simply does—explicitly—what we have seen assertions do. In other words, the expressions in an assertion are always *sampled* in the preponed region of a time stamp. $sampled does exactly the same. It returns the value of an expression sampled in the preponed region of the simulation time stamp in which the function is called.

So, *for concurrent assertions $sampled function is redundant*. Following two are equivalent.

```
z1 : assert property (@ (posedge clk) $sampled(a) == $sampled(b));
z1 : assert property (@ (posedge clk) a == b));
```

The reason they are the same is that in the concurrent assertion (as we have seen throughout the book) the expressions are always sampled in the preponed region of the time stamp in which they are sampled. $sampled function also returns value of expression in the preponed region.

However, there are places where $sampled can be useful for debug purpose. For example, in a simultaneously changing event situation, you can find out the sampled value of an expression (i.e., the value in the preponed region). Let us say, you have an assertion where you want to make sure that "gnt" is asserted on a posedge clk. If "gnt" and posedge clk went high at the same time, the property will fail. That's because the sampled value of “gnt” is 0 in the preponed region. In such a case you can have a $display(“gnt=%b”, $sampled(gnt)). This will tell you right away that the sampled value was “0” which is why the assertion failed.

Important note: The assertion system function $sampled does *not* use a clocking event (note that in the 2005 standard, the explicit clocking event was indeed required but that requirement was removed from the 2012 standard). And as we know, for a sampled value function other than $sampled, the clocking event will be explicitly specified as an argument or inferred from the code where the function is called.

From LRM:

IEEE Std 1800-2005 required that an explicit or inferred clocking event argument be provided for the $sampled assertion system function. In the 2012 version of the standard, the semantics of $sampled have been changed to a form that does not depend on a clocking event. Therefore, the syntax for defining the clocking event argument to $sampled is deprecated and does not appear in this version of the standard.

Also, the use of $sampled in a “disable iff” clause is meaningful since *the disable condition by default is not sampled.* “disable iff” by default is asynchronous. So, what if you want to make it synchronous, i.e., sampled on a clock edge? Here’s how

```
property syncIff;
      (@posedge clk) disable iff ($sampled(rst)) a |=> b;
endproperty
```

Important Note on how variables are treated (displayed or otherwise) in an Action Block:

For debug purposes, it will be *necessary* to $display (or any other system task for that matter) a variable using $sampled function.

For example,

```
z1 : assert property  (@ (posedge clk) a == b)); $display(" a=%b
b=%b", a, b);
```

Let’s say “@posedge clk,” “a” falls (1=> 0) and “b” remains constant “1” from the previous clock. The “sampled” value of “a” will be 1 and the sampled value of “b” will be “1.” The assertion will PASS at that posedge clk. And the pass action block (which is simply a $display) will display the following:

```
a=0 b=1
```

So, your first reaction would be why did the assertion PASS when the values of “a” and “b” are different? There is a discrepancy between the sampled values and the values you see in $display. That’s because in the assert statement the values are *sampled* in the preponed region while since the Action Block is executed in the Reactive Region the values of “a” and “b” would have settled down and you see those values in $display.

Hence, in your Action Block use system function $sampled to present the same value as that sampled in the preponed region. Change your $display as follows,

```
z1 : assert property (@ (posedge clk) a == b));
 $display(" a=%b b=%b", $sampled (a) , $sampled (b));
```

With this $display, you will see the same values that the simulator saw in preponed region which were used for evaluation of the assertion. $display will now display the following and you'll know why the assertion passed.

```
a=1 b=1
```

20.4 Global Clocking past and future Sampled Value Functions

Global clocking sampled value functions may be used only if global clocking has been defined. The global clocking past and future sampled value functions are:

```
$past_gclk (expression)
$rose_gclk (expression)
$fell_gclk (expression)
$stable_gclk (expression)
$changed_gclk (expression)
$future_gclk (expression)
$rising_gclk (expression)
$falling_gclk (expression)
$steady_gclk (expression)
$changing_gclk (expression)
```

As the name suggests, these sampled value functions work off a global clock. Before we investigate these functions, let us see what a global clock is.

```
module top_pd;
  logic clk;
  global clocking sys_clk @ (clk); endclocking
  ....
endmodule
```

"sys_clk" is now considered a global clocking event. It is defined to occur (trigger) if there is a change in "clk." You can access global clock using the system function $global_clock. This system function does not take any arguments but returns the event expression specified in the global clocking declaration. Note that the specification of the name sys_clk in the global clocking declaration is optional since the global clocking event may be referenced by $global_clock. The $global_clock system function will be used to explicitly refer to the event expression in the effective global clocking declaration.

A reference to $global_clock is understood to be a reference to a *clocking_event* defined in a global clocking declaration. A global clock behaves just as any other clocking event. Thus, in the following example:

```
global clocking @clk; endclocking
...
assert property(@$global_clock a);
```

the assertion states that "a" is true at each tick of the global clock. This assertion is logically equivalent to:

```
assert property(@clk a);
```

Global clocking is a SystemVerilog feature and a detailed explanation is beyond the scope of this book. However, here are some high-level points.

- A clocking block may be declared as global clocking for all or part of the design hierarchy. In other words, such a specification may be done for a whole design, or separately for different subsystems in a design.
- Global clocking may be declared in a module, an interface, a checker, or a program. A given module, interface, checker, or program can contain at most one global clocking declaration.
- Although more than one global clocking declaration may appear in different parts of the design hierarchy, at most one global clocking declaration is effective at each point in the elaborated design hierarchy.

Note that *these functions can only be used if there is a global clock defined in your test-bench* (hence the suffix _gclk). They are sampled value functions as we have seen before, and they sample their expression value at the global clock tick that you have defined.

Simulation efficiency tip: Note that the future or past global clocking sampled value functions are *more expensive* than their non-global clocking counter parts. Global clocking functions *take up more simulation time.* This is based on my experience with at least one EDA vendor simulator.

These sampled value functions are divided into two groups. One group looks in the past (these are identical in functionality with the sampled value functions we have seen previously only that the _gclk sampled value functions work off a global clock event). The other group has "future" sampled value functions. They sample the value of the expression in the time step that subsequently (and immediately) follows the time step in which the function is called. Note that the "past" sampled value functions have a non-global clock counterpart as we have seen. *However, for the "future" sampled value functions, there is no non-global clock counterpart.* They work only if you have defined a global clock.

20.5 future Global Clocking Sampled Value Functions

The ***future*** *sampled value functions* are

```
$future_gclk (expression)
$rising_gclk (expression)
$falling_gclk (expression)
$steady_gclk (expression)
$changing_gclk (expression)
```

The future sampled value functions use the subsequent (future) value of the expression. Note that the global clocking *future* sampled value functions may be invoked only in *property_expr* or in *sequence_expr*. This implies that *they cannot be used in assertion action blocks*.

$future_gclk(expression) returns the sampled value of expression at the next global clocking tick.
$rising_gclk(expression) returns a Boolean True if the sampled value of the *least significant bit* of the expression changes to 1 at the next global clocking event. Else it returns false.
$falling_gclk(expression) returns a Boolean True if the sampled value of the *least significant bit* of the expression changes to 0 at the next global clocking event. Else it returns false.
$steady_gclk(expression) returns a Boolean True if the sampled value of the *least significant bit* of expression does not change at the next global clocking event. Else it returns false.
$changing_gclk (expression) returns a Boolean True if the sampled value of the *least significant bit* of expression changes at the next global clocking event. Else it returns false.

Execution of the action block of an assertion containing global clocking future sampled value functions is *delayed* until the global clocking tick follows the last tick of the assertion clock for the attempt.

Some examples.

Example 1:

```
a1: assert property (@$global_clock $changing_gclk(req) |->
$falling_gclk(req))
else $error("req is not stable");
```

In the above example, a future (1 global clock tick away) changing "req" implies that it is a falling "req" at that same global clock tick (since there is an overlapping operator).

Note that the following are illegal assertions using future sampled value functions.

```
always @(posedge_clk) a <= $future_gclk(b) && c;
//ILLEGAL - cannot use in procedural assignment

a1_illegal: assert final (a -> $future_gclk(b));
//ILLEGAL - can't use in immediate assertion

a2_illegal: assert property (@(posedge clk) disable iff (rst
 | $rising_gclk(interrupt)) req |=> gnt);
//ILLEGAL - can't use in disable iff

a3_illegal: assert property (@(posedge clk) req |-> $future_
gclk (ack && $rising_gclk(gnt));
//ILLEGAL - can't have nested sampled value functions
```

20.6 past Global Clocking Sampled Value Functions

The ***past*** sampled value functions are

```
$past_gclk (expression)
$rose_gclk (expression)
$fell_gclk (expression)
$stable_gclk(expression)
$changed_gclk(expression)
```

The globally clocked *past* sampled value functions work the same way as the non-global clocking sampled value function. If you recall, these past sampled value functions take an explicit clocking event. So, $rose_gclk (expr) is equivalent to $rose(expr, @ $global_clock). Please refer to the non-global clocking past sampled value functions to understand how these functions work (Chap. 7)

The global clocking *past* sampled value functions are a special case of the sampled value functions, and therefore the regular restrictions imposed on the sampled value functions and their arguments apply. In particular, the global clocking past sampled value functions are usable in general procedural code and action blocks.

20.7 Illegal Use of Global Clocking Future Sampled Value Functions

Restrictions are imposed on the usage of the global clocking future sampled value functions: they cannot be nested, and they cannot be used in assertions containing sequence match items.

Here are a couple of examples of illegal uses of global clocking functions.

```
Illegal_assert: assert property (@clk $future_gclk(a || $ris-
ing_gclk(b)); //ILLEGAL nesting
```

In this example, $future_glck nests (a || $rising_glck(b) as its arguments. In other words, $future_gclk nests $rising_gclk and that is illegal.

```
a2: assert property (@clk gnt |=> $future_gclk(req)); //ILLEGAL
```

In this example, a global clocking *future* sampled value function is used in an assertion containing sequence match items. In other words, $future_gclk is used with a nonoverlapping operator. So, you are looking for the future value of "req" after 1 clock from when "gnt" is asserted. This is illegal.

Even though global clocking future sampled value functions depend on future values of their arguments, the interval of simulation time steps for an evaluation attempt of an assertion containing global clocking future sampled value functions is defined as though the future sampled values were known in advance.

The end of the evaluation attempt is defined to be the last tick of the assertion clock and is not delayed any additional time steps up to the next global clocking tick.

Execution of the action block of an assertion containing global clocking future sampled value functions will be delayed until the global clocking tick that follows the last tick of the assertion clock for the attempt. If the evaluation attempt fails and $error is called by default, then $error will be called at the global clocking tick that follows the last tick of the assertion clock.

To summarize, following are illegal conditions for the global clocking *future* sampled value functions:

1. *The "future" sampled value functions cannot be used outside of concurrent assertions.*
2. They cannot be nested (for example, $future_gclk ($falling_gclk(gnt_) && req)). But do not confuse this with the following which is legal.

```
F1: assert property (@$ global _clock $rising_gclk(sig1) |->
$falling_gclk(sig2)); //LEGAL
```

3. They *cannot* be used in a "reset" condition (for example, "disable_iff ($falling_gclk (reset_)").
4. The global clocking future sampling functions cannot be used in an assertion *action block (pass or fail).*

20.8 "*followed by*" Properties #-# and #=#

The *followed by* properties has the following form.

```
sequence_expression # - # property_expression
sequence_expression # = # property_expression
```

- # is the overlapped property and # = # is the nonoverlapped, just as in |-> and |=>. But there are differences between the *implication* operators and the *followed by* operators.

For "*followed by*" to succeed *both* the antecedent sequence_expression and the consequent property_expression must be true. *If the antecedent sequence_expression does not have any match, then the property fails.* If the sequence_expression has a match, then the consequent property_expression must match.

This is the fundamental difference between the implication operators (|-> and |=>) and the followed by operators. Recall that with implication operators, if the antecedent does not match, you get a vacuous pass and not a fail.

For overlapped followed-by, there must be a match for the antecedent sequence_expr, where the end point of this match is the start point of the evaluation of the consequent property_expr. For nonoverlapped followed-by, the start point of the evaluation of the consequent property_expr is the clock tick after the end point of the sequence_expr match.

Obviously, # - # being an overlapped operator, it starts the consequent evaluation the same time that the antecedent match ends (and succeeds). Consequently, the # = # nonoverlapped operator will start the consequent evaluation the clock after the antecedent match ends and succeeds.

Here's a simple example

```
property p(a, b, c)
  @ (posedge clk) c |-> a #=# b;
endproperty
assert property (p(req[*5], gnt, c ));
```

Request need to remain asserted (high) for 5 consecutive clocks. One clock later gnt must be asserted (high). If request does *not* remain asserted for 5 consecutive clocks, the assertion will fail. If it does remain asserted for 5 clocks and the next clock gnt is not asserted, the assertion will fail. If both the antecedent and consequent match in the required temporal domain, the property will pass.

Let's look at the simulation log for above property "p."

```
run 200
#        0  clk=0 c=0 req=0 gnt=0
#        5  clk=1 c=0 req=0 gnt=0
#       10  clk=0 c=1 req=1 gnt=0
```

```
#      15  clk=1 c=1 req=1 gnt=0
#      20  clk=0 c=0 req=1 gnt=0
#      25  clk=1 c=0 req=1 gnt=0
#      30  clk=0 c=0 req=1 gnt=0
#      35  clk=1 c=0 req=1 gnt=0
#      40  clk=0 c=0 req=1 gnt=0
#      45  clk=1 c=0 req=1 gnt=0
#      50  clk=0 c=0 req=1 gnt=0
#      55  clk=1 c=0 req=1 gnt=0
#      60  clk=0 c=0 req=0 gnt=1
#      65  clk=1 c=0 req=0 gnt=1
# At 65ns 'followed by' PASS
#      70  clk=0 c=0 req=0 gnt=0
#      75  clk=1 c=0 req=0 gnt=0
#      80  clk=0 c=1 req=1 gnt=0
#      85  clk=1 c=1 req=1 gnt=0
#      90  clk=0 c=0 req=1 gnt=0
#      95  clk=1 c=0 req=1 gnt=0
#      100  clk=0 c=0 req=1 gnt=0
#      105  clk=1 c=0 req=0 gnt=0
#      110  clk=0 c=0 req=0 gnt=0
#      115  clk=1 c=0 req=0 gnt=0
# At 115ns 'followed by' FAIL
#      120  clk=0 c=0 req=0 gnt=0
#      125  clk=1 c=0 req=0 gnt=0
#      130  clk=0 c=0 req=0 gnt=0
#        $finish    : followedby.sv(35)
```

Simulation log reads this way.

At time 10, “c” (the antecedent) is 1 and hence the consequent starts evaluation. After that, “req” is high for 5 clocks consecutively. Then 1 clock later “gnt” is high and the property passes as expected.

At time 80, the property fires again since “c” is 1. But this time “req” remains high only for 2 clocks. So, the property fails. If we had used an implication operator instead of the followed by operator, the property would have passed vacuously. So, this example clearly shows the difference between followed by and implication operator.

Ok, since you asked, one more example !

```
assert property (rst[*1:$] ##1 !rst |=> always !rst);
```

This property will vacuously PASS if “rst” is always High.

```
assert property (rst[*1:$] ##1 !rst #=# always !rst);
```

This property will FAIL if “rst” is always High.

Explanation of "always" is right after this section.

Finally, if you think about it, a # - # b behaves pretty much like a ##0 b. So, what's the difference? a ##0 b requires that "b" is a sequence while a# - #b allows "b" as a property. The same discussion applies to a # = # b, which is equivalent to a ##1 b, except that "b" can be a property.

20.9 "always" and "s_always" Property

"always" property behaves exactly as you would expect. The syntax for "always" (and its variations) is

1. **always property_expression** (weak form)
2. **always [cycle_delay constant range expression] property_expression** (weak form with *unbounded* range)
3. **s_always [constant_range] property_expression** (strong form with *bounded* range)

So, let us see how "always" works. As LRM puts it "A property 'always property_expression' evaluates to true if and only if the property_expression holds at every current and future clock tick." Rather self-explaining. Here's a simple example.

```
property reset_always;
   @ (posedge clk) POR[*5] |=> always !reset;
endproperty
```

The property says that once POR (power on reset) signal has remained high for 5 consecutive clocks that starting next clock, reset would remain de-asserted "always" (forever).

"always" makes it simple to specify the continuous longevity of an assertion.

Next, let us see how always [cycle delay constant range] works.

```
property p1;
  @ (posedge clk) a |-> always [3:$] b;
endproperty
```

property p1 says that if "a" is true that "b" will be true 3 clocks *after* "a" and will remain true "always" (forever) after the 3 clocks. Note that "always[n:m]" allows an unbounded range.

In contrast "s_always" allows only bounded range.

So, let us see what s_always does.

```
property p2;
   @ (posedge clk) a |-> s_always [3:10] b;
endproperty
```

The property says that if "a" is true that "b" remains true from 3rd clock to 10th clock after "a" was detected true. This is a "strong" property. Recall strong property that we discussed earlier. This "s_" property also works the same way. In other words, if you run out of simulation ticks for "s_always," the property will indeed fail (at least with the simulators I have tried—LRM explanation on this is not quite clear).

BUT, why do we need "**always**"? Don't the concurrent assertions always execute at every clock tick? The answer is yes which means we do not always need an "**always**" operator with a concurrent assertion. It is redundant. For example, in the following, "**always**" is redundant.

```
P1: assert property p1p (@ (posedge clk) always bstrap1==0);
```

There is no reason for an "always" in the above concurrent assertion. It is the same as

```
P1: assert property p1p (@ (posedge clk) bstrap1==0);
```

"always" can be useful in "initial" block, however. See the example below.

```
initial
begin
     P1: assert p1p (@ (posedge clk) always bstrap1==0);
end
```

Note that the immediate assertion noted above is slated to execute only once. In our case though, once it is asserted, it will then look for bstrap1==0 at every posedge clock.

Couple more examples,

```
'always (p) and always (q)' is the same as 'always (p and q).
BUT
'always (p) or always (q)' is not the same as 'always (p or q).
```

Consider a case when p holds in all odd clock ticks and q holds in all even ticks. Then always (p or q) is true, whereas (always p) or (always q) is false.

One more:

```
property xx_chk (logic aStrobe, logic data);
@(posedge clk) disable iff(rst)
$rose(aStrobe)[->1]|=>always(!$isunknown(data)&&$stable(data));
endproperty
```

If the aStrobe goes high at least once, the data cannot be unknown and must be stable. So once, aStrobe goes high, the property will "always" check for data integrity at all future posedge clocks.

Exercise:
How would you write this property—without—the "always" operator?
Hint:
Use consecutive repetition operator.

20.10 "eventually,", "s_eventually"

There are two types of this operator, the "weak" kind ("**eventually**") and the "strong" kind ("**s_eventually**"). Here are three forms of these two properties.

s_eventually property_expr (*strong* property without range)

s_eventually [cycle_delay_constant_range] property_expr (*strong* property with range)

- the constant_range can be unbounded

eventually [constant_range] property_expr (*weak* property with range)

- the constant_range must be bounded

Some examples:

```
property p1;
      s_eventually $fell(frame_);
endproperty
```

Eventually PCI cycle will start with assertion of frame_ (frame_ goes low). If frame_ does not assert until the end of simulation time, the property will fail since this is a *strong* property. Note that frame_ can be true in current clock tick or any future clock tick.

```
property p2;
     s_eventually [2:5] $fell(frame_);
endproperty p2;
```

A new PCI cycle must start (frame_ goes low) within the range of 2 clocks from now and eventually by 5th clock (2nd and 5th clock inclusive). Note that as with any *strong* property, s_eventually[n : m] property_expr evaluates to true if, and only if, there exist at least n+1 ticks of the clock of the eventually property, including the current time step, and property_expr evaluates to true beginning in one of the n+1 to m+1 clock ticks, where counting starts at the current time step.

Let's look at a simple test-bench and simulation log to see how "s_eventually" works.

Test-bench:

```
module eventuallyP ( );
bit clk, frame, b, c;

property checkAB;
  @(posedge clk)
  c |-> s_eventually [2:5] $fell(frame_);
endproperty

delayD1: assert property (checkAB) else $display("At %0tns
eventually FAIL \n",$stime);
delayD1c: cover property (checkAB) $display("At %0tns eventually
PASS \n",$stime);

always begin #5ns clk=!clk; end

initial
  begin
    clk=0;
        frame_=1; c=1;
        @(negedge clk) c=0;
        repeat (3) @(posedge clk);
        frame_=0;
        repeat (2) @(posedge clk);

    frame_=1; c=1;
        repeat (1) @(posedge clk);
    frame_=1; c=0;
    repeat (6) @(posedge clk);
    frame_=0;

    repeat (2) @(posedge clk); $finish(2);
  end

initial $monitor($stime,,,"clk=%b frame_=%b c=%b",clk,frame_,c);
endmodule
```

Simulation Log:

```
run 200
#       0  clk=0 frame_=1 c=1
#       5  clk=1 frame_=1 c=1  //antecedent fires
#      10  clk=0 frame_=1 c=0
#      15  clk=1 frame_=1 c=0
```

```
#       20  clk=0 frame_=1 c=0
#       25  clk=1 frame_=1 c=0
#       30  clk=0 frame_=1 c=0
#       35  clk=1 frame_=0 c=0 //frame_ goes low after 3 clocks
#       40  clk=0 frame_=0 c=0
#       45  clk=1 frame_=0 c=0
# At 45ns eventually PASS

#       50  clk=0 frame_=0 c=0
#       55  clk=1 frame_=1 c=1  //antecedent fires again
#       60  clk=0 frame_=1 c=1
#       65  clk=1 frame_=1 c=0
#       70  clk=0 frame_=1 c=0
#       75  clk=1 frame_=1 c=0
#       80  clk=0 frame_=1 c=0
#       85  clk=1 frame_=1 c=0
#       90  clk=0 frame_=1 c=0
#       95  clk=1 frame_=1 c=0
#      100  clk=0 frame_=1 c=0
#      105  clk=1 frame_=1 c=0
#      110  clk=0 frame_=1 c=0
#      115  clk=1 frame_=1 c=0
# At 115ns eventually FAIL

#      120  clk=0 frame_=1 c=0
#      125  clk=1 frame_=0 c=0  //frame_ goes low after 5 clocks
#      130  clk=0 frame_=0 c=0
#      135  clk=1 frame_=0 c=0
#      140  clk=0 frame_=0 c=0
#       $finish  : eventuallyP_book.sv(32)
```

At time, 5, antecedent "c" fires and the property evaluation begins. frame_ goes low after 3 clocks which is within the range of [2:5]. In other words, frame_ goes low eventually within [2:5] clock range and the property passes.

At time 55, antecedent "c" fires again. But this time frame_ does not go low within the [2:5] clock range and the property fails.

Exercise: Are the following properties "p3" and "p2" equivalent? Hint: Simulate from "initial" condition to know the subtle difference.

```
property p2;
    s_eventually [2:5] $fell(frame_);
endproperty p2;
property p3;
     frame_ |-> ##[2:5] $fell(frame_);
endproperty
```

Following is s_eventually with unbounded range.

```
property p4;
    s_eventually [2:$] $fell(frame_);
endproperty
```

A new PCI cycle must start (from the current clock tick) 2 clocks from now or any time after that.

```
property p5;
    eventually [2:$] $fell(frame_); // ILLEGAL. Weak property
    must be bound.
endproperty
property p6;
    s_eventually always a;
endproperty
```

"a" will eventually (starting current clock tick) go high and then remain high at every clock tick after that until the end of simulation.

Simulation Performance efficiency tips:

1. *Checking property "s_eventually always p" in simulation may be costly, especially if it is not in the scope of an initial procedure.*
2. *Consider the following assertions.*

```
a1: assert property ( req |-> ##[1:10000] gnt);   //Inefficient
```

This assertion should be written as

```
a2: assert property (req |-> s_eventually gnt); //efficient
```

Although assertions a1 and a2 are equivalent, simulation performance of a1 may be worse. Avoid using large temporal domain delays.

20.11 "until,", "s_until,", "until_with," and "s_until_with"

There are four forms of "until" property.

1. property_expression1 **until** property_expression2 (weak form—nonoverlapping)
2. property_expression1 **s_until** property_expression2 (strong form—nonoverlapping)
3. property_expression1 **until_with** property_expression2 (weak form—overlapping)
4. property_expression1 **s_until_with** property_expression2 (strong form—overlapping)

Let us start with "until."

```
property p1;
      req until gnt;
endproperty
```

property p1 is true if "req" is true until "gnt" is true. In other words, "req" must remain true as long as "gnt" is false. "req" need *not* be true at the clock tick when "gnt" is found to be true. In other word, **until** is nonoverlapping. An **until** property of the nonoverlapping form evaluates to true if "req" evaluates to true at every clock tick beginning with the starting clock tick of the evaluation attempt and continuing until at least one tick *before* a clock tick where "gnt" evaluates to true. If "gnt" is never true, "req" will remain true at every current and future clock tick. Based on the simulators that I have tried, since **until** is of weak form, if this property never completes (i.e., "gnt" is never true), the property will *not* fail.

Let's look at a simple test-bench and log file to better understand "until" operator.

Test-bench:

```
module untilP ( );
bit clk, req, gnt, antecedent;

property checkAB;
  @(posedge clk)
  antecedent |-> req until gnt;
endproperty

delayD1: assert property (checkAB) else
 $display("At %0tns until FAIL \n", $stime);

delayD1c: cover property (checkAB)
 $display("At %0tns until PASS \n", $stime);

always begin #5ns clk=!clk; end

initial
  begin
      clk=0; c=0;

          @(posedge clk); req=1; gnt=0;
          @(negedge clk); antecedent=1; //fire thread
          @(posedge clk); req=1; gnt=0;
          @(negedge clk); antecedent=0; //turn off thread
          @(posedge clk); req=0; gnt=1;
```

```
            @(posedge clk); req=1; gnt=0;
            @(negedge clk); antecedent=1;
            @(posedge clk); req=1; gnt=0; //fire thread
            @(negedge clk); antecedent=0;
            @(posedge clk); req=0; gnt=0; //turn off thread
            @(posedge clk); req=0; gnt=0;

            @(posedge clk); $finish;
   end

initial $monitor($stime,,,"clk=%b req=%b gnt=%b antecedent=%
b",clk,req,gnt,antecedent);
endmodule
```

Simulation Log:

```
run 200
#     0  clk=0 req=0 gnt=0 antecedent=0
#     5  clk=1 req=1 gnt=0 antecedent=0
#    10  clk=0 req=1 gnt=0 antecedent=1 //fire antecedent
#    15  clk=1 req=1 gnt=0 antecedent=1
#    20  clk=0 req=1 gnt=0 antecedent=0 //turn off antecedent
#    25  clk=1 req=1 gnt=1 antecedent=0
#    30  clk=0 req=0 gnt=1 antecedent=0
#    35  clk=1 req=0 gnt=1 antecedent=0  //'until' PASS
# At 35ns until PASS

#    40  clk=0 req=1 gnt=0 antecedent=1 //fire antecedent
#    45  clk=1 req=1 gnt=0 antecedent=1
#    50  clk=0 req=1 gnt=0 antecedent=0 //turn off antecedent
#    55  clk=1 req=0 gnt=0 antecedent=0
#    60  clk=0 req=0 gnt=0 antecedent=0
#    65  clk=1 req=0 gnt=0 antecedent=0 //'until' FAIL
# At 65ns until FAIL

#    70  clk=0 req=0 gnt=0 antecedent=0
# ** Note: $finish    : untilP_book.sv(34)
```

Here's the explanation of the simulation log.

For the PASS case: At time 10, the "antecedent" is true, so the assertion fires. And "req" is true at that time. It needs to remain true at least 1 clock prior to "gnt" being true. So, when "gnt" goes true at time 25, at the next clock (at time 35), "gnt" is sampled to be true. "req" was true until at least 1 clock before the "gnt" was true and property PASSes.

For the FAIL case: At time 40, "antecedent" fires again and "req" is true at that time. But "req" goes low at time 55 and is sampled low at time 65. BUT gnt is still low at time 65 and since "req" is supposed to remain true until gnt is true the condition is violated, and the property FAILs at time 65.

```
property p1;
   req s_until gnt;
endproperty
```

s_until is identical to **until** except that if "gnt" never arrives and you run out of simulation time, the property will fail (based on the simulators I've tried).

To reiterate the difference between strong and weak properties, an "until" property of one of the strong forms requires that a current or future clock tick *exists* at which "gnt" evaluates to true, while an "until" property of one of the weak forms does not make this requirement. Strong properties require that some terminating condition happen in the future, and this includes the requirement that the property clock ticks enough time to enable the condition to happen. Weak properties do not impose any requirement on the terminating condition, and do not require the clock to tick.

```
property p1;
  req until_with gnt;
endproperty
```

property p1 is true if "req" is true until and *including* a clock tick when "gnt" is true. In other words, "req" must remain true as long as "gnt" is false. "req" must be true at the *same* clock tick when "gnt" is found to be true. If "gnt" is never true, "req" will remain true at every current and future clock tick. In other words, **until_with** is an overlapping property. Since "until_with" is of weak form, if this property never completes (i.e., "gnt" is never true), the property will *not* fail.

Let's look at a simulation log for "until_with."

Simulation Log:

```
run 200
#      0  clk=0 req=0 gnt=0 antecedent=0
#      5  clk=1 req=1 gnt=0 antecedent=0
#     10  clk=0 req=1 gnt=0 antecedent=1 //fire antecedent
#     15  clk=1 req=1 gnt=0 antecedent=1
#     20  clk=0 req=1 gnt=0 antecedent=0 //turn off antecedent
#     25  clk=1 req=1 gnt=0 antecedent=0
#     30  clk=0 req=0 gnt=1 antecedent=0
#     35  clk=1 req=0 gnt=1 antecedent=0  //'until_with' FAIL
# At 35ns until FAIL
```

```
#     40  clk=0 req=1 gnt=0 antecedent=1 //fire antecedent
#     45  clk=1 req=1 gnt=0 antecedent=1
#     50  clk=0 req=0 gnt=0 antecedent=0 //turn off antecedent
#     55  clk=1 req=0 gnt=0 antecedent=0
#     60  clk=0 req=1 gnt=1 antecedent=0
#     65  clk=1 req=1 gnt=1 antecedent=0 //'until_with' PASS
# At 65ns until PASS

#     70  clk=0 req=0 gnt=0 antecedent=0
# ** Note: $finish    : untilP_book.sv(34)
```

Here's the explanation of the simulation log.

At time 35, "gnt" is high but "req" is low. And "until_with" requires that "req" be true until "gnt" is true and including the clock when "gnt" is true. "req" was not low at the clock when "gnt" is true and the property fails.

The counter example is at time 65. Here "req" is true the same clock as "gnt" is true and the property passes.

In short, property "**until_with**" requires "req" and "gnt" to be true at the *same* clock tick when "gnt" is found to be true. "**until**" does not have this requirement.

```
property p1;
  req s_until_with gnt;
endproperty
```

Same as **until_with** but if you run out of simulation tick (end of simulation, for example), and if "gnt" is never found to be true, this property will fail (simulator results on this may vary).

20.12 "nexttime" and "s_nexttime"

"nexttime" property_expression evaluates to true, if property_expression is true at time t+1 clock tick.

There are four forms of "nexttime."

```
nexttime property_expression (weak form)
```

The weak nexttime property **nexttime** *property_expr* evaluates to true if, and only if, either the *property_expr* evaluates to true beginning at the *next clock tick* or there is no further clock tick.

```
s_nexttime property_expression (strong form)
```

The strong nexttime property **s_nexttime** *property_expr* evaluates to true if, and only if, there exists a next clock tick and *property_expr* evaluates to true beginning at that clock tick.

```
nexttime [constant_expression] property_expression (weak form)
```

The indexed weak nexttime property **nexttime** [*constant_expression*] *property_expr* evaluates to true if, and only if, either there are not *constant_expression* clock ticks or *property_expr* evaluates to true beginning at the last of the next *constant_expression* clock ticks.

The "constant_expression" is useful as shown in the following example.

Let's say you want to write a property that looks for "req" to be true after 3 clocks,

```
nexttime nexttime nexttime req;
```

This is cumbersome and difficult to write. You can write the same property as follows:

```
nexttime [3] req;
```

NOTE: **nexttime** with a big "constant_expression" is inefficient both in simulation and in Formal Verification. Try to keep the constant number small, especially in complex assertions. In simulation, it is recommended not to exceed several hundred for the "constant_expression," and in Formal Verification not to exceed a couple of tens. The common rule is the smaller the better.

```
s_nexttime [constant_expression] property_expression (strong form)
```

The indexed strong nexttime property **s_nexttime** [*constant_expression*] *property_expr* evaluates to true if, and only if, there exist *constant_expression* clock ticks and *property_expr* evaluates to true beginning at the last of the next *constant_expression* clock ticks.

Let us examine the following simple example.

```
property p1;
  @ (posedge clk) nexttime req;
endproperty
```

The above property says that the property will pass, if the clock ticks once more and "req" is true at the next clock tick (t+1). In addition, since this is the weak form, if you run out of simulation ticks (i.e., there is no t+1), this property will not fail.

Some examples.

What if you want to check to see that "req" remains asserted for all the clocks following the next clock? Following will do the trick.

```
property p1;
  @ (posedge clk) nexttime always req;
endproperty
```

Or what if you want to see that starting next clock, “req” will eventually become true? Following will do the trick.

```
property p1;
  @ (posedge clk) nexttime eventually req;
endproperty
```

What if you want to see that “req” is true after a certain exact # of clocks? Following will do the trick.

```
property p1;
  @ (posedge clk) nexttime[5] req;
endproperty
```

This property says that “req” shall be true at the fifth future clock tick (provided that there are indeed 5 clock ticks in future, of course).

```
property p1;
  @ (posedge clk) s_nexttime req;
endproperty
```

Same as “nexttime” except that if you run out of simulation ticks after the property is triggered (i.e., there is no (t+1), the property will fail. Other way to look at this is that there exists a next clock and “req” should be true at that next clock, else the property will fail.

Similarly, the following property says that there must be at least 5 clock ticks and that “req” will be true at the fifth future clock tick.

```
property p1;
  @ (posedge clk) s_nexttime [5] req;
endproperty
```

Following example: When “seq1” ends(matches) at “t” that the next time tick (“t+1”) “seq_expr” must be equal to ‘hff.

```
property
  @ (posedge clk) (seq1.matched nexttime seq_expr == 'hff);
endproperty
```

One more real-life issue we face that can be solved with **nexttime**. Initial “x” condition can always give us false failures. This can be avoided with the use of **nexttime**. For example,

Let us say you are using $past to do a simple "xor" of past value and present value (Gray encoding).

```
property
@(posedge clk)
    $onehot (fifocntr ^ $past (fifocntr);
endproperty
```

At time "initial" $past (fifocntr) will return "x" (unknown) and the "xor" would fail right away. This is a false failure and you may spend unnecessary time debugging it. Here's how **nexttime** can solve that problem.

```
property
@(posedge clk)
    nexttime $onehot (fifocntr ^ $past (fifocntr);
endproperty
```

nexttime will avoid the initial $past value of fifocntr and move the comparison to the next clock tick when (hopefully) you have cleared the fifocntr and the comparison will not fail due to the initial "x."

Similarly, if you want to know that a signal stays stable forever (e.g., bootstrap signals). You may write a property as follows

```
property
@(posedge clk)
    $stable (bstrap);
endproperty
```

BUT this will sample the value "x" (e.g., for a "logic" type which has not been explicitly initialized) at time tick 0 and then continue to check to see that it stays at "x." You end up checking for a stable "x." Completely opposite of what you want to accomplish. Again, **nexttime** comes to rescue.

```
property
@(posedge clk)
    nexttime $stable (bstrap);
endproperty
```

This will ensure that you start comparing the previous value of bstrap with the current value, starting *next* clock tick. Obvious, but easy to miss.

We discussed multi-clock properties in Chap. 10. Here's an example of how **nexttime** can be used in a multi-clock property.

N1: **assert property**

```
@(posedge clk1) x |-> nexttime @(posedge clk2) z;
```

Important Note: It is very important to understand how this property works. The "posedge clk1" flows through to "nexttime"—in other words, "nexttime" does not use "@(posedge clk2)" to advance time to next tick. So, when "x" is true at ("posedge clk1"), the "nexttime" causes advance to the next occurrence of "posedge clk1" strictly after when "x" was detected true before looking for a concurrent or subsequent occurrence of "posedge clk2" at which to evaluate "z."

Exercise: How would the following property behave in contrast with the one above?

```
N1: assert property
    @(posedge clk1) x |-> @(posedge clk2) nexttime z;
```

20.13 "*case*" Statement

The *case* statement in assertions is the same as the one we use in systemverilog language. There is no difference, only that in systemverilog assertions, you use the *case* statement in a property. The *case* property statement is a multi-way decision-making mechanism that tests a Boolean expression and sees if it matches one of a number of Boolean expressions. On a match, it will take action specified for that case statement. We are all familiar with this functionality of *case*. The "default" statement is optional.

Here is a simple example.

```
property CycleCase (logic [1:0] CycleType);
      case (CycleType)
        2'b00: $fell(frame_) ##1 (cmd==READ);
        2'b01: $fell(frame_) ##1 (cmd==WRITE);
        2'b10: $fell(frame_) ##1 (cmd==TABORT);
        2'b11: $fell(frame_) ##1 (cmd==MABORT);
       default: $fell(frame_) ##1 (cmd==ILLEGAL); //default is
       optional
endcase
endproperty
```

Note that if the default statement is not given and all of the comparisons fail, then none of the case item property statements are evaluated. In addition, as we know if "assert property ( )" antecedent does not evaluate to true that we get a vacuous pass. The same applies here.

If there is no default and no case branch match, you get a vacuous pass.

20.14 $inferred_clock and $inferred_disable

Many times, while developing assertion logic, we define default blocks for clock and reset. These default blocks-based clock and reset are then available in properties that follow. Please refer to Sect. 6.4 on Default Clocking for further understanding of a default-clocking block.

The inferred clocking event expression is the current resolved event expression that can be used in a clocking event definition. It is obtained by applying clock flow rules to the point where $inferred_clock is called. If there is no current resolved event expression when $inferred_clock is encountered, then an error is issued.

The inferred disable expression is the disable condition from the default disable declaration whose scope includes the call to $inferred_disable. If the call to $inferred_disable is not within the scope of any default disable declaration, then the call to $inferred_disable returns 1'b0 (false).

- $inferred_clock returns the expression of the inferred clocking event.
- $inferred_disable returns the inferred disable expression.

Let us say you have the following default blocks:

```
module (...,clk, rst,...);
default clocking @ (negedge clk); endclocking
default disable iff rst;
```

One of the ways to use this default clocking and reset blocks is as follows.

```
property inferB(a, b, c, clk=$inferred_clock, reset=$inferred_
disable) ;
   @ (clk) disable iff (reset) a |=> b || c;
endproperty
assert property (inferB(x, y, z));
```

The formal parameters of property inferB uses default clocking and reset from their respective default blocks. In other words, "@ (clk)" is now "@ (negedge clk)". Similarly, "disable iff (reset)" is now "disable iff (rst)".

Note that if the property "inferB" is invoked as follows, the $inferred_clock will not take effect—but the actual clocking event "posedge clk" will take effect.

```
assert property inferB (a, b, c, posedge clk, reset);
```

@ (clk) in property inferB will be "@ (posedge clk)" and *not* "@ (negedge clk)" as in the default clocking block. In other words, the actual overwrites the formal argument, as always.

From the above we can see that the inferred clocking event expression is the current resolved event expression that can be used in a clocking event. Of course, if you use $inferred_clock and there is no default clocking block defined, you will get an Error.

Here's simple Verilog code that exemplifies above description.

```
module defaultP ( );
bit clk, x, y, z, rst, reset;

default clocking negclock @ (negedge clk); endclocking
default disable iff rst;

property inferB (a, b, c, clk=$inferred_clock, reset=$inferred_
disable);
 @ (clk) disable iff (reset) a |=> b || c;
endproperty

assert property (inferB (x, y, z )) else $display ($stime,,,
"FAIL");
cover property (inferB (x, y, z )) $display ($stime,,,"PASS");

initial
begin
    clk=0; rst=0; x=1; y=0; z=0;
    #80; rst=1;
    #80; y=1; rst=0;
    #80 $finish;
end

always #10 clk=!clk;

initial $monitor($stime,,,"rst=%b clk=%b x=%b y=%b z=%b",rst,
clk,x,y,z);

endmodule
```

Simulation Log:

```
run 200
#       0  rst=0 clk=0 x=1 y=0 z=0
#      10  rst=0 clk=1 x=1 y=0 z=0
#      20  rst=0 clk=0 x=1 y=0 z=0
#      20  FAIL
#      30  rst=0 clk=1 x=1 y=0 z=0
#      40  rst=0 clk=0 x=1 y=0 z=0
#      40  FAIL //Failure at negedge of clock

#      50  rst=0 clk=1 x=1 y=0 z=0
#      60  rst=0 clk=0 x=1 y=0 z=0
#      60  FAIL //Failure at negedge of clock
```

```
#        70   rst=0 clk=1 x=1 y=0 z=0
#        80   rst=1 clk=0 x=1 y=0 z=0 //rst=1, so disable the
              property
#        90   rst=1 clk=1 x=1 y=0 z=0
#       100   rst=1 clk=0 x=1 y=0 z=0
#       110   rst=1 clk=1 x=1 y=0 z=0
#       120   rst=1 clk=0 x=1 y=0 z=0
#       130   rst=1 clk=1 x=1 y=0 z=0
#       140   rst=1 clk=0 x=1 y=0 z=0
#       150   rst=1 clk=1 x=1 y=0 z=0
#       160   rst=0 clk=0 x=1 y=1 z=0 //rst=0, so enable the
          property
#       170   rst=0 clk=1 x=1 y=1 z=0
#       180   rst=0 clk=0 x=1 y=1 z=0
#       180   PASS //Pass at negedge of clock
```

Here are the nuances:

In the above example, we are not providing clk and reset to the property inferB. This way the property will adopt the default clocking and disable conditions declared in default blocks. Note that the FAIL and PASS occurs at the negedge of clk, since that clk is inferred from the default clocking block and "disable iff (reset)" takes on the definition declared in the default rst block.

A call to an inferred expression function may only be used as the entire default value expression for a formal argument to a property or sequence declaration.

A call to an inferred expression function cannot appear within the body expression of a property or sequence declaration.

If a call to an inferred expression function is used as the entire default value expression for a formal argument to a property or sequence declaration, then it is replaced by the inferred expression as determined at the point where the property or sequence is instantiated. Therefore, if the property or sequence instance is the top-level property expression in an assertion statement, the event expression that is used to replace the default argument $inferred_clock is that as determined at the location of the assertion statement. If the property or sequence instance is not the top-level property expression in the assertion statement, then the event expression determined by clock flow rules up to the instance location in the property expression is used as the default value of the argument.

20.15 "restrict" for Formal Verification

The "restrict" property is strictly for Formal (static functional) verification. *Simulators do not check this property*. Since we are not covering Formal Verification in this book, this property is noted here for the sake of completeness. Note that we have immediate "assert," "cover," and "assume" but *there is no immediate "restrict" assertion statement*. As we saw with the "assume" property, formal verification requires some assumption or restriction in order for it to restrict the logic cones to process and not explode in the state space. "restrict" has the same semantics as "assume," only that "*restrict*" *does not have an action block*. Here is the syntax.

```
restrict property (property_spec);    //Note, no action block.
```

For example, for Formal Verification, if you need to restrict the checking of an assertion which has 5 inputs (a, b, c, d, e) and 2 control bits (x, y). Only if {x, y}=2'b00, that the inputs a, b, and c are of use for the static formal check. So, we restrict the property as follows:

```
restrict property (@ (posedge clk) {x, y} == 2'b00);
```

Note again that the property is ignored by simulation. In other words, {x, y} == 2'b00 is not enforced during simulation.

Please refer to Chap. 15 on "assume" to further understand usage of "restrict."

20.16 Abort Properties: reject_on, accept_on, sync_reject_on, sync_accept_on

Recall "disable_iff" disable condition (6.8), which preempts the entire assertion, if true. "disable_iff" is an asynchronous abort (or reset) condition for the entire assertion. It is also asynchronous in that it's expression is *not* sampled in the preponed region but the expression is evaluated at every time stamp (i.e., in-between clock ticks as well as at the clock ticks) and whenever the "disable_iff" expression turns true that the entire assertion will be abandoned (no pass or fail).

With that background, 1800-2009/2012 adds four more abort conditions. "reject_on" and "accept_on" are asynchronous abort conditions (as in disable_iff) and "sync_reject_on" and "sync_accept_on" are synchronous (i.e., sampled) abort condition. Note that "accept_on" is an abort condition for PASS, even though that may seem a bit counterintuitive at first. In other words, if "accept_on" aborts an evaluation, the result is a PASS. If "reject_on" aborts an evaluation, the result is FAIL.

The syntax for all four is the same.

```
accept_on (abort condition expression) property_expression
sync_accept_on (abort condition expression) property_expression
reject_on (abort condition expression) property_expression
sync_reject_on (abort condition expression) property_expression
```

Before we see examples, here are high-level points to note.

1. One note off the bat to distinguish "disable_iff" from the abort properties is that "disable_iff " works at the "entire concurrent assertion" level while these abort properties work at the "property" level. Only the property_expression associated

with the abort property will get "aborted"—not the entire assertion as with "disable_iff." More on this later.
2. The operators "accept_on" and "reject_on" work at the granularity of simulation time step (i.e., asynchronously).
3. In contrast, the operators "sync_accept_on" and "sync_reject_on" do *not* work at the granularity of simulation time-step. They are sampled at the simulation time step of the clocking event (i.e., the sampling event).
4. You can nest the four abort operators "accept_on," "reject_on," "sync_accept_on," "sync_reject_on." Note that nested operators are in the lexical order "accept_on," "reject_on," "sync_accept_on," and "sync_reject_on" (from left to right). While evaluating the inner abort property, the outer abort property takes precedence over the inner abort condition in case both conditions occur at the same time tick.
5. Abort condition cannot contain any reference to local variables or the sequence methods .triggered and .matched.
6. An abort is a property, so the result of an evaluation is either pass or fail. An aborted evaluation results in pass for the "accept" operators and fail for the "reject" operators.
7. There are no default abort conditions.

Now let us look at some examples to nail down the concepts.

```
property p1;
@ (posedge clk) $fell(bMode) |-> reject_on(bMode)
data_transfer[*4];
endproperty
assert property (p1);
```

The above example specifies that on the falling edge of burst Mode (bMode), data_transfer should remain asserted for 4 consecutive clocks and that the bMode should—not—go high during those 4 data transfers. The way the property reads is; look for the falling edge of bMode and starting that clock *reject* (fail) the property "(data_transfer[*4])" if at any time (i.e., asynchronously—even between clock ticks) it sees bMode going high. As noted before, "reject_on" abort means failure. Hence consequent will FAIL and so will the property p1.

The important thing to note here is that the evaluation of the abort property namely "data_transfer[*4]" and the reject condition reject_on(bMode) start at the *same* time. In other words (as shown below) this is like a "throughout" operator where the LHS is checked to see if it holds for the entire duration of RHS. Similarly, here we check to see that while we are monitoring "data_transfer[*4]" to hold that "bMode" should not go high. If it does go high at any time during "data_transfer[*4]" the property will be rejected, i.e., it will fail.

The same property can be written using sync_reject_on, only that the "bMode" will not be evaluated asynchronously (any time including in-between clock ticks) but will be sampled only at the sampling edge, clock tick.

Note that the above property can be written using "throughout" as well. Please refer to Sect. 8.16 on "throughout" operator to see a similar example.

```
property p1;
  @ (posedge clk) $fell(bMode) |-> !(bMode) throughout data_
  transfer[*4];
endproperty
assert property (p1);
```

Let us look at an example of "accept_on"

```
property reqack;
  @ (posedge clk) accept_on(cycle_end) req |-> ##5 ack;
endproperty
assert property (reqack);
```

This property uses "accept_on(cycle_end)" as the abort condition on the property "req |-> ##5 ack". When "req" is sampled high on a posedge clk, the property "req |-> ##5 ack" starts evaluating waiting for ack to arrive after 5 clocks. At the same time, "cycle_end" is also monitored to see if it goes high. Here are the scenarios that take place.

"cycle_end" arrives within the 5 clocks that the property is waiting for "ack." The accept_on condition will be true in that case and the property will be considered to pass. The next evaluation will again start the next time "req" is sampled high on posedge clk.

"cycle_end" does not arrive within 5 clocks when the property is waiting for "ack." The property will evaluate as with any concurrent assertion and if "ack" does not come in high at 5th clock, the property will fail. If "ack" does come in high at 5th clock, the property will pass.

"cycle_end" arrives exactly the same time as "ack" at the 5 clocks. The abort condition takes precedence. Since in this case, both "ack" arrived and the accept_on were triggered at the same time, the accept_on aborts the evaluation with a pass and so the assertion will pass. What if we used "reject_on" instead of "accept_on" in such a scenario?

In short, the property evaluation aborts on "accept_on" (and passes) or "reject_on" (and fails) OR it will finish on its own (and pass/fail) if the abort condition does not arrive.

Here are some more examples courtesy 1800-2009/2012 LRM.

```
property p; (accept_on(a) p1) and (reject_on(b) p2);
endproperty
```

Note that we are using an "and" operator here between two properties p1 and p2. Note that for an "and" to succeed both the LHS and RHS of "and" must complete and pass. If "a" becomes true, then p1 will abort and pass. But since there is an "and" we will wait for the second property to complete as well. If "b" becomes true during the evaluation of p2, p2 will be aborted and considered to fail and since this is an "and," the "property p" will fail. What if we used an "or" instead?

```
property p;  (accept_on(a) p1) or (reject_on(b) p2);
endproperty
```

Recall that "or" requires either the LHS or the RHS to complete and pass. In the same scenario as above, if "a" becomes true first during the evaluation of p1, p1 is aborted and will be pass (i.e., accepted) and the property "p" will pass. Similarly, if "b" becomes true first, then p2 is aborted and property p will fail (i.e., rejected).

Note that nested operators are in the lexical order "accept_on," "reject_on," "sync_accept_on," and "sync_reject_on" (from left to right). If two nested operator conditions become true in the same time tick during the evaluation of the property, then the outermost operator takes precedence.

```
property p; accept_on(a) reject_on(b) p1; endproperty
```

Note there is no operator between accept_on and reject_on.

If "a" goes high first, the property is aborted on accept_on and will pass. If "b" goes high first, the reject_on succeeds and the property p will fail. If both "a" and "b" go high at the same time during the evaluation of p1, then accept_on takes precedence and "p" will pass.

Another simple example.

```
Frame_accept_reject: assert property (
    @(posedge clk)
        accept_on (Frame_)
        Cycle_start |=> reject_on(Tabort)
    )
    else $display ("Frame_ FAIL);
```

This is another example of nested asynchronous aborts. The outer abort (accept_on (Frame_)) has the scope of the entire property of the concurrent assertion. The inner abort (reject_on(Tabort)) has the scope of the consequent of |=>. The inner abort does not start evaluation until Cycle_start is true. Note that the outer abort ("accept_on") takes precedence over the inner abort ("reject_on").

Exercise: Try the same property with s_accept_on and s_reject_on and note the differences. Try both examples with different triggers of Frame_ and Tabort and see when/how the property passes and fails. I'll leave this for you as an exercise.

Finally, note the following points.

- A disable condition (disable iff) may use the method .triggered (attached to the sequence used in disable condition). But an abort condition (the ones described above) cannot use .triggered method

- Neither the disable nor the abort properties can refer to local variables
- Neither of the reset conditions may use the method .matched attached to the sequence used in reset conditions.

20.17 $assertpassoff, $assertpasson, $assertfailoff, $assertfailon, $assertnonvacuouson, $assertvacuousoff

These system tasks add further control over assertion execution during simulation. We have seen $asserton, $assertoff and $assertkill (refer to Sect. 9.6) before. Here's a brief explanation of what these new system tasks do.

$**assertpassoff**: This system task turns off the action block associated with PASS of an assertion. This includes PASS indication because of both the vacuous and non-vacuous success. To re-enable the PASS action block, use $**assertpasson**. It will turn on the PASS action of both the vacuous and non-vacuous successes. If you want to turn on only the non-vacuous PASS, then use $**assertnonvacuouson** system task. Note that these system tasks do not affect an assertion that is already executing.

$**assertfailoff**: This system task turns off the action block associated with the FAIL of an assertion. In order to turn it on, use $**assertfailon**. Here also, these system tasks do not affect an assertion that is already executing.

$**assertvacuousoff** system task turns off the PASS indication based on a vacuous success (Sect. 17.18). An assertion that is already executing is not affected. By default, we get a PASS indication on vacuous pass.

All the system tasks take arguments, as we have seen before with $asserton, $assertoff, and $assertkill (refer to Sect. 9.6). The first argument indicates how many level of hierarchy below each specified module instance to apply the system tasks. The subsequent arguments specify which scopes of the model to act upon (entire modules or instances).

20.18 $assertcontrol

LRM IEEE-1800 (2012) introduces a new system task $assertcontrol. This system task can be used in lieu of above-mentioned individual system tasks.

The $assertcontrol system task controls the evaluation of assertions. The $assertcontrol system task can also be used to control the execution of assertion action blocks associated with assertions and expect statements.

This system task provides the capability to enable/disable/kill the assertions based on assertion type or directive type. Similarly, this task also provides the capability to enable/disable action block execution of assertions and expect statements based on assertion type or directive type.

The violation reporting for *unique*, *unique0* and *priority if* and *case* constructs can also be controlled using these tasks. I will refrain from explain *unique*, *unique0* and *priority if* and *case* since these are SystemVerilog constructs and not assertions constructs. Please refer to the SystemVerilog LRM for their description.

Here's the syntax:

```
$assertioncontrol ( control_type [ , [ assertion_type ] [ , [
directive_type ] [ , [ levels ]
[ , list_of_scopes_or_assertions ] ] ] ] ) ;
```

where

– **control_type**: This argument controls the effect of the $assertcontrol system task. This argument is an integer expression. The valid values for this argument are

control_type Value	Effect
1	Lock
2	Unlock
3	On
4	Off
5	Kill
6	PassOn
7	PassOff
8	FailOn
9	FailOff
10	NonvacuousOn
11	VacuousOff

Now let us see what each of the "effect" means (LRM):

- Lock: A value of 1 for this argument *prevents status change* of all specified assertions, expect statements, and violation reports until they are unlocked. Once an $assertcontrol with control_type of value 1 (Lock) is applied to an assertion, expect statement, or violation report, it becomes locked and no $assertcontrol will affect it until the locked state is removed by a subsequent $assertcontrol with a control_type value of 2 (Unlock).
- Unlock: A value of 2 for this argument will remove the locked status of all specified assertions, expect statements, and violation reports.
- On: A value of 3 for this argument will re-enable the execution of all specified assertions. A value of 3 for this argument will also re-enable violation reporting from all the specified violation report types. This control_type value does not affect expect statements.
- Off: A value of 4 for this argument will stop the checking of all specified assertions until a subsequent $assertcontrol with a control_type of 3 (On). No new attempts will be started.

- Attempts that are already executing for the assertions, and their pass or fail statements, are *not* affected. In the case of a deferred assertion (Sect. 1.1), currently queued reports are not flushed and may still mature, though further checking is prevented until a subsequent $assertcontrol with a control_type of 3 (On). In the case of a pending procedural assertion instance, currently queued instances are not flushed and may still mature, though no new instances may be queued until a subsequent $assertcontrol with a control_type of 3 (On). A value of 4 for this argument will also disable the violation reporting from all the specified violation report types.
- Currently queued violation reports are not flushed and may still mature, though no new violation reports will be added to the pending violation report queue until a subsequent $assertcontrol with a control_type value of 3 (On). The violation reporting can be re-enabled subsequently by $assertcontrol with a control_type value of 3 (On). This control_type value does not affect expect statements.
- Kill: A value of 5 for this argument will abort execution of any currently executing attempts for the specified assertions and then stop the checking of all specified assertions until a subsequent $assertcontrol with a control_type of 3 (On). This also flushes any queued pending reports of deferred assertions or pending procedural assertion instances that have not yet matured. A value of 5 for this argument will also abort violation reporting from all the specified violation report types. Currently queued violation reports that have not yet matured are also flushed, and no new violation reports shall be added to the pending violation report queue until a subsequent $assertcontrol with a control_type value of 3 (On). This control_type value does not affect expect statements.
- PassOn: A value of 6 for this argument will enable execution of the pass action for vacuous and nonvacuous success of all the specified assertions and expect statements. An assertion that is already executing, including execution of the pass or fail action, is not affected. This control_type value does not affect violation report types.
- PassOff: A value of 7 for this argument will stop execution of the *pass action* for vacuous and nonvacuous success of all the specified assertions and expect statements. Execution of the pass action for both vacuous and nonvacuous successes can be re-enabled subsequently by $assertcontrol with a control_type value of 6 (PassOn), while the execution of the pass action for only nonvacuous successes can be enabled subsequently by $assertcontrol with a control_type value of 10 (NonvacuousOn). An assertion that is already executing, including execution of the pass or fail action, is not affected. This control_type value does not affect violation report types.
- FailOn: A value of 8 for this argument will enable execution of the fail action of all the specified assertions and expect statements. An assertion that is already executing, including execution of the pass or fail action, is not affected. This task also affects the execution of the default fail action block. This control_type value does not affect violation report types.

- FailOff: A value of 9 for this argument will stop execution of the fail action of all the specified assertions and expect statements until a subsequent $assertcontrol with a control_type value of 8 (FailOn). An assertion that is already executing, including execution of the pass or fail action, is not affected. By default, the fail action is executed. This task also affects the execution of default fail action block, i.e., $error, which is called in case no else clause is specified for the assertion. This control_type value does not affect violation report types.
- NonvacuousOn: A value of 10 for this argument will enable execution of the pass action of all the specified assertions and expect statements on nonvacuous success. An assertion that is already executing, including execution of the pass or fail action, is not affected. This control_type value does not affect violation report types.
- VacuousOff: A value of 11 for this argument will stop execution of the pass action of all the specified assertions and expect statements on vacuous success until a subsequent $assertcontrol with a control_type value of 6 (PassOn). An assertion that is already executing, including execution of the pass or fail action, is not affected. By default, the pass action is executed on vacuous success. This control_type value does not affect violation report types.

  The assertion action control tasks or $assertcontrol with control_type values of 6 (PassOn) to 11 (VacuousOff) do not affect statistics counters for the assertions. The details related to the behavior of $assertcontrol for assertions referring to global clocking future sampled value functions are explained in Sect. 20.4.

– **assertion_type**: This argument selects the assertion types and violation report types that are affected by the $assertcontrol system task. This argument shall be an integer expression. The valid values for this argument are

assertion_type value	Types of assertions affected
1	Concurrent
2	Simple Immediate
4	Observed Differed Immediate
8	Final Differed Immediate
16	Expect
32	Unique
64	Unique0
128	Priority

Note that multiple assertion_type values can be specified at a time by OR-ing different values. For example, a task with assertion_type value of 3 (which is the same as Concurrent | SimpleImmediate) will apply to concurrent and simple immediate assertions. Similarly, a task with assertion_type value of 96 (which is same as Unique|Unique0) will apply to unique and unique0 if and case constructs.

If assertion_type is not specified, then it defaults to 255 and the system task applies to all types of assertions, expect statements, and violation reports.

- **directive_type**: This argument selects the directive types that are affected by the $assertcontrol system task. This argument will be an integer expression.
  The valid values for this argument are

directive_type Values	Type of Directives Affected
1	Assert directives
2	Cover directives
3	Assume directives

Multiple directive_type values can be specified at a time by OR-ing different values. For example, a task with directive_type value of 3 (which is same as Assert | Cover) shall apply to assert and cover directives.

If directive_type is not specified, then it defaults to 7 (Assert|Cover|Assume) and the system task applies to all types of directives.

- **levels**: This argument specifies the levels of hierarchy, consistent with the corresponding argument to SystemVerilog $dumpvars system task. If this argument is not specified, it defaults to 0. This argument will be an integer expression.
- **list_of_scopes_or_assertions**: This argument specifies which scopes of the model to control. These arguments can specify any scopes or individual assertions.

Now, let's see how $assertcontrol maps to the individual controls that were described in the previous section. Note that these individual system tasks are still supported in 2012 LRM for backward compatibility.

- **$asserton**[(levels[, list])] is equivalent to $assertcontrol (3, 15, 7, levels [,list])
- **$assertoff**[(levels[, list])] is equivalent to $assertcontrol (4, 15, 7, levels [,list])
- **$assertkill**[(levels[, list])] is equivalent to $assertcontrol (5, 15, 7, levels [,list])

And as I just mentioned, assertion action control tasks $assertpasson, $assertpassoff, $assertfailon, $assertfailoff, $assertvacuousoff, and $assertnonvacuouson are also provided for backward compatibility. Following example from LRM.

These tasks can be defined as follows:

- **$assertpasson**[(levels[, list])] is equivalent to $assertcontrol (6, 31, 7, levels [,list])
- **$assertpassoff**[(levels[, list])] is equivalent to $assertcontrol (7, 31, 7, levels [,list])
- **$assertfailon**[(levels[, list])] is equivalent to $assertcontrol (8, 31, 7, levels [,list])
- **$assertfailoff**[(levels[, list])] is equivalent to $assertcontrol (9, 31, 7, levels [,list])

- **$assertnonvacuouson**[(levels[, list])] is equivalent to $assertcontrol (10, 31, 7, levels [,list])
- **$assertvacuousoff**[(levels[, list])] is equivalent to $assertcontrol (11, 31, 7, levels [,list])

Examples:

```
$assertcontrol (VACUOUSOFF, CONCURRENT | EXPECT);
```

This systasks affects the whole design, so no modules are specified. Disable vacuous pass action for all the concurrent asserts, covers, and assumes in the design. Also disable vacuous pass action for expect statements.

```
$assertcontrol (OFF);
```

Disable concurrent and immediate asserts and covers. This system task does not affect *expect* statements as control type is Off using default values of all the arguments after first argument. This will also disable violation reporting.

```
$assertcontrol (ON, CONCURRENT|S_IMMEDIATE|D_IMMEDIATE, ASSERT|
COVER|ASSUME, 0);
```

This system task enables assertions. This will not enable violation reporting.

```
$assertcontrol (KILL, CONCURRENT, ASSERT, 0);
```

Kill currently executing concurrent assertions but do not kill concurrent covers/assumes and immediate/deferred asserts/covers/assumes using appropriate values of second and third arguments.

```
$assertcontrol (LOCK, ALL_ASSERTS, ALL_DIRECTIVES, 0, a1);
$assertcontrol (ON); // enable all the assertions except a1
$assertcontrol (UNLOCK, ALL_ASSERTS, ALL_DIRECTIVES, 0, a1);
```

Enable all the assertions except a1. To accomplish this, first lock a1 and then enable all the assertions and then unlock a1 as we want future assertion control tasks to affect a1.

Chapter 21
"*let*" Declarations

Introduction: This chapter delves into the detail of "let" declaration. "let" declarations have local scope in contrast with `define which has global scope. A "let" declaration defines a template expression (a let body), customized by its ports (aka parameters).

A. B. Mehta, *SystemVerilog Assertions and Functional Coverage*,
https://doi.org/10.1007/978-3-030-24737-9_21

We have all used the compiler directive `define (as a global text substitution macro). Note the word *global*—it is truly global spanning across *all* scopes of your design modules and files. For example, `define intr 3'b111 will substitute `intr with 3'b111 wherever it sees `intr either within the local scope or global scope. This can be good and bad. Good is that you have to define it only once and it will span across module/file boundaries. Bad is you cannot redefine `intr (well actually you can, but with consequences). For example, if you change the definition of `intr in a package, you will get a warning and also the new definition will overwrite all the previous ones.

That is where "let" comes into picture. "let" allows local scope and allows parameterization (or as LRM puts it, it has "ports") (as in a sequence or a property).

To reiterate, let declarations can be used for customization and can replace the text macros in many cases. The "let" construct is safer because it has a local scope, while the scope of compiler directives is global within the compilation unit. A "let" declaration defines a template expression (a let body), customized by its ports (aka parameters). A "let" construct may be instantiated in other expressions.

The syntax for "let" is

```
let_declaration ::= let let_identifier [ ( [let_port_list] ) =
expression;
```

Let us see each feature of "let" one by one.

21.1 let: Local Scope

First let us see an example using "let."

```
module example;
logic r1,r2, r3,r4,clk,clk1;
let exDefLet = r1 || r2;
always @ (posedge clk) begin: ablock
    let exDefLet = r1 & r2; //exDefLet has a local scope of
    'ablock'
    r3=exDefLet;
end
always @ (posedge clk1) begin: bblock
    r4=exDefLet; // exDefLet will take the definition from the
    scope that is visible to it. Here
                //it is the outer most scope definition of
                (r1 || r2);
end
endmodule
```

We have defined "exDefLet" in two different scopes. One in the always block "ablock" and another at the outermost scope "module example." Note that their definition (expression) is different in each block. Since "let" can have local scope, each of the definition of "let" will be preserved in its local block. The above code will look like the following after "let" substitutions take place.

```
module example;
logic r1,r2, r3,r4,clk,clk1;
always @ (posedge clk) begin :ablock
  r3=r1 & r2;
end
always @ (posedge clk1) begin: bblock
  r4=r1 || r2 ;
end
endmodule
```

If the same design was modeled using `define, here's how the code would look like.

```
module example;
logic r1,r2, r3,r4,clk,clk1;
`define exDefLet r1 || r2;
always @ (posedge clk) begin :ablock
  `define exDefLet r1 & r2;
  r3=`exDefLet;
end
always @ (posedge clk1) begin: bblock
  r4=`exDefLet;
end
endmodule
```

In this example, since there are two `define for the same variable, the compiler will complain right off the bat and use the second definition r1&r2 (it's the latest in lexical order) as the global definition of exDefLet. The above code will look like the following after `define substitutions.

```
module example;
logic r1,r2, r3,r4,clk,clk1;
always @ (posedge clk) begin: ablock
  r3=r1 & r2;
end
always @ (posedge clk1) begin: bblock
  r4=r1&r2;
end
endmodule
```

As you see "let" is very useful from scoping point of view. It follows the normal scoping rules. You can parameterize it (or as LRM puts it, it can have "ports") and reuse the "let" expression repeatedly with different parameters.""

21.2 "let": With Parameters

As mentioned before, "let" can be parameterized . Note that instantiation of "let" is quite different from a "parameterized function." With "let" you replace the instance with *entire* "let" body. With "function" you simply pass the parameters and the function executes using those parameters. Function does not replace the instance of function call.

Ok, let us see a simple example.

```
module abc;
logic clk, x, y, j;
logic [7:0] r1;
let lxor (p, q=1'b0) = p^q;
always @ (posedge clk) begin
  for (i = 0; i <= 256; i++) begin
       r1 = lxor( i ); //After expanding the 'let' instance,
       this will be r1 = i ^ 1'b0;
    end
end
endmodule
```

For each value of "i" r1 will get the "xor" of "i" and "q=1'b0." Note that the formal parameter "q" is assigned a default value of "1'b0." That being the case, when "r1=lxor(i)" is executed, the actual "i" replaces the formal "p" in lxor and "q" takes on its assigned default value of "1'b0." You could have also specified "r1=lxor(i, j)" and the formal "q" will now take the value of "j" (whatever "j" is).

Some rules apply to the formal arguments.

1. Note again that the "let" body gets expanded with the actual arguments (which replace the formal arguments) and the body (RHS of "let") will replace the instance of "let." That being the case, once the body of "let" replaces the instance of "let," all required semantic checks will take place to see that the expanded "let" body with the actual arguments is legal.
2. The formal arguments can have a default value (as we saw in the example above).
3. The formal arguments can be typed or untyped. The typed arguments will force type compatibility between formal and actual (cast compatibility). In other words, the actual argument will be cast to the type of the formal argument before being substituted. Untyped formal in that case is more flexible.

4. If the formal argument is of "event" type, then the actual argument must be an event_expression. Each reference to the formal argument shall be in a place where an event_expression may be written.
5. The self-determined result type of the actual argument must be cast compatible with the type of the formal argument. The actual argument must be cast to the type of the formal argument before being substituted for a reference to the formal argument.
6. If the variables used in "let" are not formal arguments to the "let" declaration, they will be resolved according to the scoping rules of the scope in which "let" is declared.

21.3 "let": In Immediate and Concurrent Assertions

Yes, "let" can be used in an immediate ("assert") as well as concurrent ("assert property") assertions in a procedural block.

Let us start with a very simple example of "let" usage in a sequence. *Note that "let" expression can only be structural or with sampled value function (as in $past).* We will see "let" with sampled value function in the next section.

```
module abc;
logic req, gnt;
let reqack = !req && gnt;
      sequence reqGnt;
        reqack;
      endsequence
endmodule
```

After expanding the 'let' instance

```
module abc;
logic req, gnt;
      sequence reqGnt;
        !req && gnt;
      endsequence
endmodule
```

Here's another example.

```
module abc;
logic clk, r1,r2,req,gnt;
let xxory (x, y) = x ^ y; //bit wise xor
let rorg = req || gnt;
....
```

```
P1: assert property (@ (posedge clk) (rorg));//concurrent
assertion
always_comb begin
     a1: assert (xxory (r1,r2));  //immediate assertion
     a2: assert (rorg);
end
endmodule
```

After expansion,

```
module abc;
logic clk, r1,r2,req,gnt;
P1: assert property (@ (posedge clk) (req || gnt)); //concurrent
assertion
always_comb begin
     a1: assert (r1 ^ r2));  //immediate assertion
     a2: assert (req || gnt);
end
endmodule
```

Now, here's an example that uses the sampled value functions $(rose) and $(fell).

```
module abc;
logic clk,r1,r2,req,gnt,ack,start;
let arose(x) = $rose( x );
let afell(y) = $fell ( y );
always_comb begin
     if (ack) s1: assert(arose(gnt));
     if (start) s2: assert(afell(req));
end
```

Another intended use of let is to provide shortcuts for identifiers or subexpressions. For example, (LRM):

```
task write_value;
input logic [31:0] addr;
input logic [31:0] value;
...
endtask
...
let addr = top.block1.unit1.base + top.block1.unit2.displ;
...
write_value(addr, 0);
```

But note that hierarchical references to “let” expressions are *not* allowed. For example, following is illegal. Assuming “my_let” to be a *let* declaration, following is illegal.

```
assign e = Top.CPU.my_let(a)); // ILLEGAL
```

Also, Recursive “let” instantiations are not permitted.
Here’s an example of how the “let” arguments bind in the declarative context

```
module sys;
logic req = 1'b1;
logic a, b;
let y = req;
...
always_comb begin
        req = 1'b0;
        b = a | y;
end
endmodule: sys
```

The effective code after expanding let expressions:

```
module sys;
logic req = 1'b1;
logic a, b;
...
always_comb begin
        req = 1'b0;
        b = a | (sys.req); //NOTE: y binds to preceding
        definition of 'req' in the declarative context
                       //of 'let'
end
endmodule : top
```

Similarly, here’s another example.

```
module CPU;
logic snoop, cache;
let data = snoop || cache;
sequence s;
     data ##1 cache
endsequence : s
...
endmodule : top
```

The effective code after expanding let expressions:

```
module CPU;
logic snoop, cache;
sequence s;
     (CPU.snoop || CPU.cache) ##1 cache;
endsequence : s
...
endmodule : top
```

Following is an example of "let" with typed formal arguments. The example shows how type conversion works when the type of a formal is different from the type of an actual.

First, module "m" with "let" declarations. Note the typed formals in "let" declarations and the mismatch between "let" instance "actuals" and "formals."

```
module m(input clock);
logic [15:0] a, b;
logic c, d;
typedef bit [15:0] bits;
...
let ones_match(bits x, y) = x == y;
let same(logic x, y) = x === y;

always_comb
a1: assert(ones_match(a, b));
//Note: the actuals 'a' and 'b' are of type 'logic', while the
formals 'x', 'y' are of type 'bits'

property toggles(bit x, y);
     same(x, y) |=> !same(x, y);
     //Note: the actuals 'x', 'y' are of type 'bit', while the
     formals 'x', 'y' are of type 'logic'
endproperty

a2: assert property (@(posedge clock) toggles(c, d));
endmodule : m
```

After expanding the "let" macro, the code looks as follows:

```
module m(input clock);
logic [15:0] a, b;
logic c, d;
typedef bit [15:0] bits;
...
// let ones_match(bits x, y) = x == y;
// let same(logic x, y) = x === y;

always_comb
a1:assert((bits'(a) == bits'(b)));

property toggles(bit x, y);
      (logic'(x) === logic'(y)) |=> ! (logic'(x) === logic'(y));
endproperty

a2: assert property (@(posedge clock) toggles(c, d));
endmodule : m
```

Finally, here's where a "let" can be declared:

- A module
- An interface
- A program
- A checker
- A clocking block
- A package
- A compilation-unit scope
- A generate block
- A sequential or parallel block
- A subroutine

Chapter 22
Checkers

Introduction: In this chapter we explore the construct of a "checker" which allows you to group several assertions in a bigger block with its well-defined functionality and interfaces providing modularity and reusability. The "checker" is a powerful way to design modular and reusable code.

A. B. Mehta, *SystemVerilog Assertions and Functional Coverage*,
https://doi.org/10.1007/978-3-030-24737-9_22

Checkers provide a way to group several assertions together into a bigger block which acts with its well-defined functionality and interfaces providing modularity and reusability. In addition to bundling assertions, you may also put modeling code in these blocks that the assertions or covergroups need. A checker allows you to place all such logic in a well-defined block. One of the intended use of checkers is to serve as verification library units.

But wait. Don't we have "modules" and "interfaces" that do the same thing? Sure, you can have a "module" or "interface" which can keep assertions separate from RTL code and bind them "externally." But there are significant advantages to keeping assertions grouped into a checker.

1. A checker can be instantiated from a procedural block as well as from outside procedural code as with concurrent assertions. On the other hand, and as we are familiar with, a module cannot be instantiated in a procedural block. It can only be instantiated outside of a procedural block.
2. The formal arguments (ports) of a checker can be sequences, properties, or other edge sensitive events. Module I/O ports do not allow this.
3. Synthesis tools normally ignore the entire checker block while in a module you have to use conditional compile if you have synthesizable code mixed with assertions.

Here's the syntax for a checker. A checker is declared using the keyword "*checker*" followed by a name and optional port list and ending with the keyword "*endchecker*."

```
checker checker_identifier [([checker_port_list])];
   {checker_or_generate_item}
endchecker [: checker_identifier]
```

Let us start with a simple example where we show the advantages of grouping assertions in a "checker" vs. a "module."

1. module: First we'll define a "*module*" which holds properties and sequences for a simple bus protocol.
2. module test_bench: Then we'll define a test-bench module that instantiates this "*module*."
3. Checker: Then we'll see how to put all these properties in a "*checker*" (instead of a "*module*").
4. Checker Test-bench: And finally, we'll see how the test-bench "instantiates" this "*checker*" from procedural code.

ONE: Assertions in a "*module*"

```
module checkerModule #(burstSize =4) ( dack_ , oe_, bMode,
bMode_in, clk, rst);
input dack_, oe_, bMode , bMode_in, clk, rst;
```

```
        sequence data_transfer;
            ##2 ((dack_==0) && (oe_==0)) [*burstSize];
        endsequence
        sequence checkbMode;
            (!bMode) throughout  data_transfer;
        endsequence
        property pbrule1;
         @ (posedge clk) disable iff (rst) bMode_in |-> checkbMode;
        endproperty
        checkBurst:assertproperty(pbrule1) else$display($stime,,,
        "Burst Rule Violated");

endmodule
```

module checkerModule is a simple bus protocol checker that is fashioned on the PCI bus. When bMode(burst mode) is asserted (active low) for 2 clocks consecutively that we need to make sure that it remains low *throughout* data_transfer which is 4 clocks long. We have seen a very similar model in Sect. 8.16 while studying throughout. Please refer to the AC specs (timing diagrams) for this *module* in that section. Note that we have parameterized the "burstSize" which is a practical way to model a property that can be reused for different burst lengths.

Now let us see how we instantiate this checkerModule *module* from our test-bench.

TWO: Test-bench for *module* "checkerModule" (Step One above)

```
module test_checkerModule;
logic dataAck_, outputEn_;
logic bMode, bMode_send, rst, clk;
..............
always @ (posedge clk or negedge rst) begin
  if (!rst) begin
    dataAck_=1'b0; outputEn_=0; bMode=0;
end

 /*Following block generates a 'bMode' that is Low for 2 clocks
 consecutively. If so, we send 'bMode_send' to the checker-
 Module module. */
always @ (posedge clk && rst) begin
    if (!bMode) begin
             @ (posedge clk);
            If (!bMode) bMode_send=bMode;
   end
end
```

```
//Now let us instantiate module 'checkerModule'

checkerModule  (#8)
     ck1(.dack_(dataAck_),.oe_(outputEn_),.bMode(bMode),.
     bMode_in(bMode_send), .clk(clk), .rst(rst));
endmodule
```

A few things to note here.

1. We had to explicitly create procedural code using an "always" block (to check that "bMode" is Low for 2 consecutive clocks) in behavioral code since we cannot pass sequences to a module. We could have created a "sequence" to do the same but then you can't pass a sequence to a module port.
2. We must explicitly pass clk and rst to the module checkerModule since a module instance won't infer clk or rst from its context. In other words, clk and rst cannot be inferred from the module test_checkerModule.
3. You cannot pass an edge control to a module. Since an edge cannot be passed to a port, you have to make sure that you send the right polarity on these ports (clk for posedge clk) and (!clk for negedge clk)

Now let us model the same checkerModule "*module*" as a "*checker*"
THREE: Assertions in a *checker*

```
checker checkerM #(burstSize =4) (dack_ , oe_, bMode, bMode_in,
rst, event clk=$inferred_clock);
input dack_, oe_, bMode ,bMode_in, rst;

     sequence data_transfer;
       ##2 ((dack_==0) && (oe_==0)) [*burstSize];
     endsequence

     sequence checkbMode;
       (!bMode)  throughout  data_transfer;
     endsequence

     property pbrule1;
       @ (clk) disable iff (rst) bMode_in |-> checkbMode;
     endproperty

     checkBurst:assertproperty(pbrule1)else$display($stime,,,
     "Burst Rule Violated");
endchecker
```

Note that "clk" is now inferred from the context from which checkerM is instantiated (see the next module test_checkerM). Also, a sequence "bMode_Sequence" will be explicitly assigned to port "bMode_in" from the test_checkerM module. Neither of these two features are possible if we model our assertions in a Verilog *module*.

Here's the test_checkerM module that calls the "checker checkerM" module.

FOUR: Test-bench for *checker* "checkerM" (Step Three above)

```
module test_checkerM;
logic dataAck_, outputEn_, bMode;
logic cycle_start, rst, clk;
..............
/*Following block generates a bMode that is Low for 2 consecu-
tive clocks. */
sequence bMode_Sequence;
     !bMode[*2]
endsequence
.....................
//Now let us call the checker 'checkerM' from a procedural
block
always @ (posedge clk or negedge rst) begin
  if (!rst) begin
    dataAck_=1'b0; outputEn_=0;bMode=0;
end
else
    checkerM (#8) ck1 (.dack_(dataAck_), .oe_(outputEn_),
    .bMode(bMode), .bMode_in(bMode_Sequence) .rst(rst));

endmodule
```

Note

1. We did not explicitly pass "clk" to the *checker* checkerM. The clk was inferred from the context of the procedural block from which it was called (just as in concurrent assertion that is called from a procedural block). We could have done the same for "rst."
2. We passed a sequence bMode_Sequence to checkerM on bMode_in port.

As you noticed, it is much more practical, modular, and easier to code and bundle assertions in a *checker* than in a *module*.

Now let us study further language features and nuances of a "checker."

Once again, the clock and reset (disable iff) contexts are inherited from the scope of the checker instantiation. Here is another simple example.

```
module test;
default clocking @ clk; endclocking
default disable iff reset;

checker test_bMode;
    //directly inherits @ clk and 'reset' from the higher-
    level context of module test
endchecker

checker test_cMode; //Note this is a new checker
//Redefines the default blocks. Point is that you can infer/
inherit or redefine what is
//inherited
     default clocking @ clk1; endclocking //Note that the
     default clocking block is for @ clk1
     default disable iff reset_system; //The default disable
     iff condition is 'reset_system'
endchecker

endmodule
```

The example shows a test-bench called "module test" which defines a clk and a reset at "module test" level. The first checker "test_bMode" inherits the clk and reset from its higher-level scope (which is "module test"). The second checker "test_cMode" defines its own clk1 and reset_system. This enables it to have its own local definition of clk1 and reset_system. It will not inherit the default clk and reset from the top-level module "module test."

22.1 Nested Checkers

As mentioned earlier, a checker can embed another checker, thus making checkers nested. Here is an example that follows the examples above.

```
checker ck1(irdy, trdy, frame_, event clk=$inferred_clock,
event reset = $inferred_disable);
default clocking @ clk; endclocking
default disable iff reset;

property check1;
    irdy |-> ##2 trdy;
endproperty
```

```
property check2;
  $rose(irdy) |=> frame_;
endproperty

     checker ck2; //nested checker
          property check1; //Redefinition of check1 within the
          local scope of checker ck2
              $rose(trdy) |-> irdy;
          endproperty

          property check3;
             $fell(irdy) |-> !frame_;
          endproperty

          checkp1: assert property check1; //local to checker ck2
          checkp3: assert property check3; //local to checker ck2
          checkp2: assert property check2; //declared in checker ck1
     endchecker : ck2

ck2 ck2i; //instantiate ck2

endchecker : ck1
```

Points to note:

1. Checker ck1 properties are visible to checker ck2. Hence checker ck2 is able to "assert" check2 of checker ck1.
2. Checker ck2 redefines property check1 for its local scope use. Since ck2 is instantiated in ck1, property check1 of checker ck2 is not directly visible to checker ck1.
3. The inferred clk and reset of checker ck1 are visible to checker ck2

22.2 Checker: Legal Conditions

A checker body may contain the following elements:

- Declarations of "let" constructs, sequences, properties, and functions (see Sect. 20.15)
- Deferred assertions (see Sect. 1.1)
- Concurrent assertions (see Sect. 5.1)
- Nested Checker declarations
- Other checker instantiations
- Covergroup declarations and instances

- "always" ("always_comb," "always_latch," "always_ff"), "initial" and "final" procedures.
  - An "initial" procedure in a checker body may contain "let" declarations, immediate, deferred immediate, and concurrent assertions. The procedural timing control statement can be using event control only (edge sensitive).
  - The "always" block (and its variations) allows blocking assignments, non-blocking assignments, loop statements, timing even control, subroutine calls, immediate, deferred immediate, concurrent assertions, and "let" declarations.
- "generate" blocks
- "default clocking" and "default disable iff" statements
- A formal argument of a checker may be optionally preceded by a direction qualifier: input or output.
  - If no direction is specified explicitly, then the direction of the previous argument is inferred. If the direction of the first checker argument is omitted, it will default to input. An input checker formal argument cannot be modified by a checker.
  - The type of an output argument cannot be of untyped, sequence, or property.
  - A checker declaration may also specify an initial value for each singular output port using the same syntax as the default value specification for input arguments.
  - A checker declaration may specify a default value for each singular input port.
- Declarations of let constructs, sequences, properties, and functions
- Checker variable declarations and assignments. Checker variables maybe assigned using blocking and nonblocking procedural assignments or non-procedural continuous assignment. But note that only nonblocking assignment is allowed in an "always_ff" procedure.

22.3 Checker: Illegal Conditions

Following is—*not*—allowed in the checker body (this is as far as the author knowledge permits, since the simulators did not fully support checkers as of this writing to validate the following)

A checker body *cannot* contain the following elements (as of writing of this book—IEEE standard is still evolving):

- "if," "case," "for," "continuous assignment," etc., type of procedural conditional and loop statements are *not* allowed.

- "initial" block can only contain concurrent or deferred assertions and a procedural event control statement "@." All other statements are forbidden in the "initial" block.
- modules, interfaces, programs, and packages cannot be declared inside a checker.
- A checker *cannot* be instantiated in a concurrent procedural construct such as fork..join, fork...join_any, or fork...join_none
- Continuous assignment
- Blocking assignment to free checker variables is illegal.
- Declaring *nets* in a checker body is illegal
- Referencing a checker variable using its hierarchical name in assignments is illegal.
- "initial" procedural block may only contain concurrent, deferred, and event control @. Nothing else.

For example, following is illegal

```
checker myCheck;
bit myBit;
      initial begin myBit=1'b1; //'initial' assignment to a
      variable is ILLEGAL
      end
endchecker
```

We assigned a checker variable in the initial block—that is illegal.
Following is legal

```
checker myCheck;
bit myBit=1'b1; //This is Legal
endchecker
```

Following is illegal as well!

```
checker myCheck (a, b, c);
bit myBit;
....
endchecker

module myMod;
...
mycheck mck1(a, b, c);
      $display(mck1.myBit);//Hierarchical reference to checker
      variable is ILLEGAL
endmodule
```

Following is illegal as well!!

```
checker myCheck(a ,b, c);
logic myBus[7:0];

      always @ (posedge clk) begin
      myBus[1:0] = 2'b0;
      myBus[7:1] = 6'b1;
            // Multiple assignments to the same variable are
            ILLEGAL. Bit
            // myBus[1] is common and assigned twice.
      end
endchecker
```

BUT the following is legal.

```
checker myCheck(a, b, c);
logic myBus[7:0];

      always @ (posedge clk) begin
      myBus[1:0] = 2'b0;
      myBus[7:2] = 6'b1; //Multiple assignments to bits of myBus
      is assigned
                              //only once. LEGAL
      end
endchecker
```

Following is illegal as well!

```
checker myCheck(bus, i);
bit [3:0] bus, i;
      initial begin
      @ (posedge clk) i = 0;
      bus[i]=4'b1111;
      end
endchecker

module myMod
logic [3:0] busIndex;
logic [3:0] datafromBus;
..........
      myCheck m1 (datafromBus,busIndex); //busIndex is non-
      constant ILLEGAL
      myCheck m2(datafromBus,4'b0); //busIndex is constant -
      LEGAL
endmodule
```

Rule of thumb is that it is illegal to pass a non-constant index value to a checker variable which is used on the LHS of an assignment.

Also note that if you instantiate a checker from a "loop," the loop index cannot be used as actual for the checker variable.

So, with all these restrictions on checker variable assignments what is one supposed to do? One of the solution is to use functions, as in the example below.

```
checker myCheck(a, b);
bit a, b;

        initial begin
        @ (posedge clk);
        a = returnAvalue;
        end

        function (bit a) returnAvalue;
         return a+1;
        endfunction

endchecker
```

In this example, since we cannot assign a value to "a" directly in the "initial" block, we called the function "returnAvalue" to accomplish the same. Note that "a" is visible in the function "returnAvalue." Since function "returnAvalue" is within the scope of checker mycheck, all the variables available to "mycheck" are also visible to "returnAvalue." As evident, procedural control statements are allowed in a function.

22.4 Checker: Important Points

1. A checker can be declared in a
   (a) module
   (b) interface
   (c) program
   (d) checker (nested checkers)
   (e) package
   (f) generate block
   (g) compilation unit scope
2. How are variables/arguments in a checker evaluated (i.e., sampled or current)? Note that sampled value means evaluated in the preponed region. Now, this gets a bit tricky.
   Except for the variables used in the event control, all other expressions in always_ff procedures are sampled. This means that the expressions in immediate and deferred assertions instantiated in this procedure are also sampled. However,

expressions in always_comb and always_latch procedures are *not* implicitly sampled and the assignments appearing in these procedures use the *current* values of their expressions.
Some examples,

```
module myMod;
      assert #0 (rdy); //This is a deferred assert in a module.
      The current value of 'rdy' is used
endmodule

checker myCheck (rdy);
      assert #0 (rdy)   //In the checker, sampled value of
      'rdy' is used.
endchecker
```

From LRM (2012):

```
checker check(logic a, b, c, clk, rst);
logic x, y, z, v, t;

assign x = a; // current value of a

always_ff @(posedge clk or negedge rst) // current values of clk
and rst
begin
    a1: assert (b); // sampled value of b
    if (rst) // current value of rst
             z <= b; // sampled value of b
    else z <= !c; // sampled value of c
end

always_comb begin
    a2: assert (b); // current value of b
    if (a) // current value of a
              v = b; // current value of b
    else v = !b; // current value of b
end

always_latch begin
    a3: assert (b); // current value of b
    if (clk) // current value of clk
    t <= b; // current value of b
end
// ...
endchecker : check
```

3. "type" and "data" declarations within the checker are local to the checker scope and are static.
4. Clock and "disable iff" contexts are inherited from the scope of the checker declaration.
5. You *can* modify/access DUT variables from a "checker"! But my suggestion is to not overdo it. Checker code will not be portable and may result in spaghetti code. Try to keep a checker modular and reusable.
6. Checker formal arguments cannot be of type "local."
7. Checker formal argument cannot be an "interface."
8. The connectivity between the actual arguments and formal arguments of a checker follow exactly the same rules as those for modules, namely,

   (a) positional association
   (b) explicit named association
   (c) implicit named association
   (d) wildcard name associations.

   Author leaves it to the reader to know of these techniques since they are age old Verilog.
9. A checker body may contain the following elements (LRM 1800-2009/2012)

   (a) Declarations of "let" constructs, sequences, properties, and functions
   (b) Deferred assertions
   (c) Concurrent and Deferred assertions

      • A checker can contain only concurrent and deferred assertions. *Immediate assertions are allowed only in the "action" blocks of assertions and in the final procedural blocks.*

   (d) Nested checkers are allowed
   (e) Covergroup declarations and assignments. One or more "covergroup" declarations are permitted in a checker. These declarations and instances cannot appear in any procedural block in a checker. A "covergroup" may reference any variable visible in its scope, including checker formal arguments and checker variables. But it is indeed an Error if a formal argument referenced by a "covergroup" has a "const" actual argument. Please refer to Chap. 26 on Functional Coverage to see how "covergroups" are defined. The same definition can be directly embedded in a checker.
   (f) Default clocking and disable iff declarations are allowed.
   (g) Checker output arguments must be typed, and their type cannot be sequence or property.
   (h) initial, always, and final procedural blocks are allowed in a "checker" body.

      • An "initial" procedure in a checker body may contain *let* declarations, *immediate*, *deferred*, and *concurrent* assertions, and a procedural timing control statement using an event control only. Similarly, an "always" procedural block also may contain concurrent, deferred assertions, variable assignments, and event control "@." Nothing else.

- The following forms of “always” procedures are allowed in checkers: always_comb, always_latch, and always_ff. Checker “always” procedures may contain the following statements:
  - Blocking assignments
  - Nonblocking assignments
  - Loop statements
  - Timing event control
  - subroutine calls
  - “let” declarations

10. Generate blocks, containing any of the above elements are allowed.
11. Checker variables can be “rand” (free variables).
12. The semantics of “checker” formal arguments is similar to the semantics of “property” arguments. Almost all formal argument types allowed in properties are also allowed in “checkers.” BUT they cannot have “local” qualifier.
13. Just as in properties, you can also use inference system functions such as $inferred_clock and $inferred_disable for checker argument initialization. For example,

```
checker checker_args
   (sequence start,
    property end,
    string message = " ",
    event clk = $inferred_clock,
    rst = $inferred_disable
   );
```

14. You can use “let” declarations in a checker (see Sect. 20.15 for “let”).
15. The RHS of a checker variable assignment may contain the sequence method .triggered

```
checker myCheck(a, b, c);
   ...
      sequence busSeq ; ...; endsequence
      always @ (posedge clk) begin a <= busSeq.triggered; end
endchecker
```

22.5 Checker: Instantiation Rules

We have already covered Nested Checker rules. This section provides further guidelines on checker instantiation rules.

As noted, in the previous section, a checker can be instantiated anywhere a concurrent assertion can be, except that

- A checker cannot be instantiated in a procedural construct such as fork...join, fork...join_any, or fork...join_none.
- checkers cannot contain declarations of modules, interfaces, programs, and packages. *Modules, interfaces, and programs cannot be instantiated inside checkers.*

There is a difference between a checker instantiation inside a procedural block or outside. Let us study this via an example

```
`define MAX_SUM 256
checker c1( logic[7:0] a, b);
logic [7:0] add;
     always @ (posedge clk) begin
         add <= a + 1'b1;
     end

  p1: assert property (@ (posedge) clk add < `MAX_SUM);
  p2: assert property (@ (posedge clk) clk a != b);

endchecker

module m(input logic rst, clk, logic en, logic[7:0] in1, in2,
in_array [20:0]);

c1 check_outside(in1, in2); //Concurrent (static) instantiation
of 'c1' checker

always @ (posedge clk) begin
   automatic logic [7:0] v1=0;
        if (en) begin
              c1 check_inside(in1, v1); //Procedural instantiation
              of 'c1'
        end
        for (int i = 0; i < 4; i++) begin
              v1 = v1+5;
        if (i != 2) begin
              c1 check_loop (in1, in_array [v1]); //Procedural
              (Loop) instantiation of 'c1'
          end
        end
end
endmodule: m
```

Points to note in the above example

1. *check_outside* is a static instantiation, while *check_inside* and *check_loop* are procedural. Total of three instantiations of "c1."
2. Each of the three instantiation of "c1" has its own copy of "add"—which is rather obvious because without it, one instance of "add" would clobber the "add" of another instance. This copy of "add" is updated at every positive clock edge, regardless of whether that instance was visited in procedural code. Even in the case of check_loop, there is only one instance of "add," and it will be updated using the sampled value of "in1."
3. For checker instance "check_outside," "p1" and "p2" are checked at every posedge clock. For checker instance "check_inside," "p1" and "p2" are queued for evaluation anytime "en" is true (on posedge clk).
4. For check_loop, three procedural instances of "p1" and "p2" are queued (for I = 0,1,3) and they will evaluate at every posedge clk. For "p1," all three instances are identical using the sampled value of "add"; but for "p2," the three instances compare the sampled value of "in1" to the sampled value of "in_array" indexed by constant "v1" values of 5,10,20, respectively.
5. Since "c1" (check_outside) instance is static (concurrent), the assertion statements in "checker c1" are continually monitored and begin execution on any time step when their sampling edge (clock event) occur.

22.6 Checker: "Formal" and "Actual" Rules

The mechanism for passing actual arguments to the formal arguments of a checker is the same as that for passing actual arguments to a "property." It is important to note that it's the "sampled" value (i.e., the value in the preponed region) of an actual that is assigned to the formal of the checker (this rule is also the same as that for a property).

Here are the rules for "formal" and "actual" of a checker and checker instantiation. Again, they are similar to those applied to a "property" or a "sequence." But some are repeated here for the sake of completeness.

- A "formal" argument of a checker can be optionally preceded by a direction qualifier: "input" or "output."
- If no direction is specified explicitly, then the direction of the previous argument will be inferred.
- If the direction of the first checker argument is omitted, it will default to "input."
- Obviously, an "input" checker formal argument cannot be modified by a checker.

- The legal data types of a checker formal arguments are the same as those legal for a property.
- The type of an "output" argument cannot be of type "untyped," "sequence" or "property."
- You cannot omit the type of a formal argument, if you have assigned an explicit direction qualifier.
- If you do omit the type of a checker formal argument and if it's the first argument of the checker, then it will be assumed to be "input untyped."
- If you do omit the type of a checker formal argument and it is *not* the first argument, then the type of the "previous" formal argument will be inferred.
- A checker declaration may specify a "default" value for each singular input.
- A checker declaration may also specify an initial value for each of its singular output using the same syntax as the default value specification for input arguments.
- As modules, checkers may access elements from their enclosing scope through their hierarchical names, except the following:
  - Automatic and dynamic variables.
  - Elements of fork … join blocks (including join_any and join_none).

Please note: Checkers for Formal verification are not covered in this book since it is beyond the scope of the book. Specifically, "assume property" is not explored beyond its context in simulation.

22.7 Checker: In a Package

For modularity and reusability, it is good to keep checkers (or properties or sequences for that matter) in SystemVerilog "package." In addition, this will keep the scope of the checker to local scope of the package and will avoid name conflicts if someone named their checker (or module for that matter) the same as your checker.

Let us look at a simple example,

```
package project_library;
      checker irdy_trdy (irdy, trdy, event clk = $inferred_clock);
           a1: assert property (@clk) irdy |=> trdy;
      endchecker : irdy_trdy
endpackage: project_library
```

As in SystemVerilog, if you want to instantiate this checker from a module, you'll have to import this package.

```
module PCI(irdy, trdy, clk);
import project_library::*;
     irdy_trdy (irdy, trdy, posedge clk)
endmodule : PCI
```

Points to note in this example,

- The statement project_library::* makes the entire content of the package visible to module PCI.
- You could have imported only the checker of interest rather than importing the entire package, as in,

```
import project_library::irdy_trdy;
```

Chapter 23
SystemVerilog Assertions LABs

Introduction: This chapter goes through six labs with increasing difficulty to solidify the practical features of properties and sequences. The LABs are as follows:

1. "bind" and implication operators
2. Overlap and nonoverlap operators
3. Synchronous FIFO
4. Counter
5. Data Transfer Protocol
6. PCI Read Protocol

A. B. Mehta, *SystemVerilog Assertions and Functional Coverage*,
https://doi.org/10.1007/978-3-030-24737-9_23

Each of these completely self-contained LABs is included on the Springer server whose information is provided with the book. Each LAB includes the DUT/Test-bench models, LAB Questions, "run" scripts for both Linux and Windows, and, of course, the .solution directory with all required models so that you can simply execute the "run" scripts and understand the results and answer the LAB questions.

23.1 LAB1: Assertions With/Without Implication and "bind"

Please note again that everything noted below (and for all the LABs) is provided on the Springer server. You do not need to rewrite any of the following to run the LABs. The overall view of LAB objectives/questions is given here. Required run scripts/logs/etc. are on the server. The answers for each LAB are included in the .solution directory. The answers are included in the book as well.

23.2 LAB1: "bind" DUT Model and Test-Bench

****
LAB1 :: Objective
****
How to bind a design module to a property module that carries assertions for the design module.

And further confirm your understanding of writing a property with/without implication.

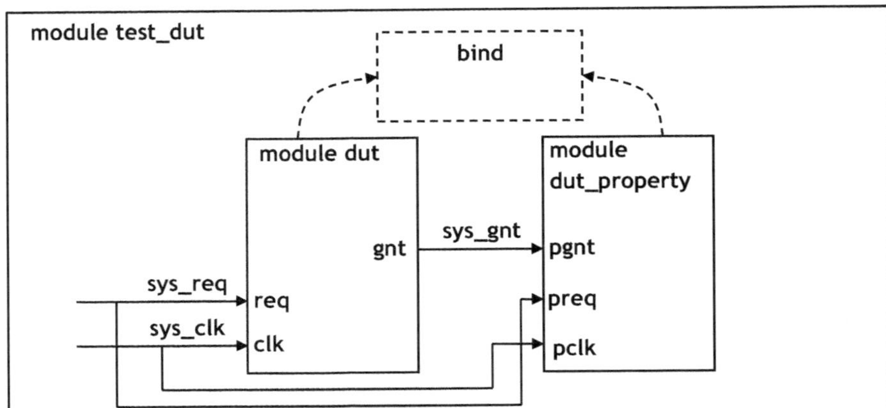

Fig. 23.1 LAB1: "bind" assertions. Problem Definition

****
LAB1 :: What you will do…
****

1. “bind” the design module “dut” with its property module “dut_property.”
2. Compile/simulate according to instructions given in the “run” scripts on the server.
3. Study simulation log.
4. Answer questions embedded in the simulation log file.

****
LAB1 :: Database
****
FILES:

dut.v :: Verilog module that drives a simple req/gnt protocol.
dut_property.sv :: File that contains ‘dut’ properties/assertions.
test_dut.sv :: Test-bench for the ‘dut’.
This is the file in which you’ll “bind” the “dut” with “dut_property.”

```
*********************************************************
dut.v
*********************************************************
/* Behavioral Verilog model that acts as the DUT driving a simple
req/gnt protocol */

     module dut(clk, req, gnt);
     input logic clk, req;
     output logic gnt;

     initial
     begin
       gnt=1'b0;
     end

     initial
     begin
           @ (posedge req);
           @ (negedge clk); gnt=1'b0;
           @ (negedge clk); gnt=1'b1;

           @ (posedge req);
           @ (negedge clk); gnt=1'b0;
           @ (negedge clk); gnt=1'b0;
     end

     endmodule
```

```
**********************************************************
dut_property.sv
**********************************************************
    module dut_property(pclk,preq,pgnt);
    input pclk, preq, pgnt;

    `ifdef no_implication
    property pr1;
      @ (posedge pclk) preq ##2 pgnt;
    endproperty
    preqGnt: assert property (pr1) $display($stime,,,"\t\t %m
    PASS");
        else $display($stime,,,"\t\t %m FAIL");

    `elsif implication
    property pr1;
      @ (posedge pclk) preq |-> ##2 pgnt;
    endproperty

    preqGnt: assert property (pr1) $display($stime,,,"\t\t %m
    PASS");
        else $display($stime,,,"\t\t %m FAIL");

    `elsif implication_novac
    property pr1;
      @ (posedge pclk) preq |-> ##2 pgnt;
    endproperty
    preqGnt: assert property (pr1) else $display($stime,,,"\t\t %m
    FAIL");

    property pr2;
      @ (posedge pclk) preq ##2 pgnt;
    endproperty
    cpreqGnt: cover property (pr2) $display($stime,,,"\t\t %m
    PASS");
    `endif

    endmodule
```

23.3 LAB1: Questions

```
*****************************************************
test_dut.sv
*****************************************************
     module test_dut;
     bit sys_clk, sys_req;
     wire sys_gnt;

     /* Instantiate 'dut' */

     dut dut1 (
             .clk(sys_clk),
             .req(sys_req),
             .gnt(sys_gnt)
             );

     //-----------------------------------
     // LAB EXERCISE - START
     //-----------------------------------
     //
     // Add your code to bind 'dut' with 'dut_property' here.

     // You need to know the names of the ports in the design as
     well as the
     // property module to be able to bind them. So, here they are:

     // ---------------------
     // Design module (dut.v)
     // ---------------------
     // module dut(clk, req, gnt);
     //      input logic clk,req;
     //      output logic gnt;

     // --------------------------------
     // Property module (dut_property.sv)
     // --------------------------------
     //module dut_property(pclk, preq, pgnt);
     //input pclk, preq, pgnt;

     //-----------------------------------
     // LAB EXERCISE - END
     //-----------------------------------
```

```
always @ (posedge sys_clk)
$display($stime,,,"clk=%b req=%b gnt=%b", sys_clk, sys_req,
sys_gnt);

always #10 sys_clk = !sys_clk;

initial
begin
                          sys_req = 1'b0;
  @ (posedge sys_clk) sys_req = 1'b1; //30
  @ (posedge sys_clk) sys_req = 1'b0; //50
  @ (posedge sys_clk) sys_req = 1'b0; //70
  @ (posedge sys_clk) sys_req = 1'b1; //90
  @ (posedge sys_clk) sys_req = 1'b0; //110
  @ (posedge sys_clk) sys_req = 1'b0; //130

@ (posedge sys_clk);
@ (posedge sys_clk); $finish(2);
end

endmodule
```

```
*************************************************************
LAB1-QUESTIONS (based on simulation log)
*************************************************************

/* +define+no_implication

run -all
KERNEL:         10  clk=1 req=0 gnt=0
KERNEL:         10                  test_implication FAIL
KERNEL:         30  clk=1 req=1 gnt=0
KERNEL:         50  clk=1 req=0 gnt=0
KERNEL:         50                  test_implication FAIL
KERNEL:         70  clk=1 req=0 gnt=1
KERNEL:         70                  test_implication FAIL
KERNEL:         70                  test_implication PASS

        Q: WHY DOES THE PROPERTY FAIL -AND- PASS AT TIME (70) ??

KERNEL:         90  clk=1 req=1 gnt=0
KERNEL:        110  clk=1 req=0 gnt=0
KERNEL:        110                  test_implication FAIL
KERNEL:        130  clk=1 req=0 gnt=0
```

```
KERNEL:          130                  test_implication FAIL
KERNEL:          130                  test_implication FAIL

        Q: WHY ARE THERE 2 failures AT TIME (130) ??
*/

/* +define+implication

run -all
KERNEL:          10  clk=1 req=0 gnt=0
KERNEL:          10                   test_implication PASS
KERNEL:          30  clk=1 req=1 gnt=0
KERNEL:          50  clk=1 req=0 gnt=0
KERNEL:          50                   test_implication PASS
KERNEL:          70  clk=1 req=0 gnt=1
KERNEL:          70                   test_implication PASS
KERNEL:          70                   test_implication PASS

        WHY ARE THERE 2 PASSes AT TIME 70 ??


KERNEL:          90  clk=1 req=1 gnt=0
KERNEL:         110  clk=1 req=0 gnt=0
KERNEL:         110                   test_implication PASS
KERNEL:         130  clk=1 req=0 gnt=0
KERNEL:         130                   test_implication FAIL
KERNEL:         130                   test_implication PASS

        WHY IS THERE A PASS -and- a FAIL AT TIME 130 ??


*/

/* +define+implication_novac

run -all
KERNEL:          10  clk=1 req=0 gnt=0
KERNEL:          30  clk=1 req=1 gnt=0
KERNEL:          50  clk=1 req=0 gnt=0
KERNEL:          70  clk=1 req=0 gnt=1
KERNEL:          70                   test_implication PASS
KERNEL:          90  clk=1 req=1 gnt=0
KERNEL:         110  clk=1 req=0 gnt=0
KERNEL:         130  clk=1 req=0 gnt=0
KERNEL:         130                   test_implication FAIL

*/
```

23.4 LAB2: Overlap and Nonoverlap Operators

We will learn how overlap and non-overlap operators work. Pay attention to finer nuances of these two operators.

23.5 LAB2 DUT Model and Test-Bench

****

LAB: Objective

****

1. Learn how overlapping implication operator works.
2. Learn how nonoverlapping implication operator works.
3. Learn how pipelined threads work in SVA.

****

LAB: Database

****

test_overlap_nonoverlap.sv :: This file contains the properties, sequences, and the test-bench required to simulate the DUT.

```
************************************************************
test_overalp_nonoverlap.sv
************************************************************
module test_overlap_nonoverlap;
bit clk, cstart, req, gnt;

always @ (posedge clk)
  $display($stime,,, "clk=%b cstart=%b req=%b gnt=%b", clk,
  cstart, req, gnt);

always #10 clk = !clk;

sequence sr1;
  req ##2 gnt;
endsequence

`ifdef overlap
property pr1;
  @ (posedge clk) cstart |-> sr1;
endproperty

//Note that if a simulator supports filter on vacuous pass for
a 'cover'
```

```
//the following property is not needed. You can simply use
"property pr1"
//for 'cover' as well.

property pr1_for_cover;
  @ (posedge clk) cstart ##0 sr1;
endproperty

`elsif nonoverlap
property pr1;
  @ (posedge clk) cstart |=> sr1;
endproperty

//Note that if a simulator supports filter on vacuous pass for
a 'cover'
//the property pr1_for_cover is not needed. You can simply use
"property pr1"
//for 'cover' as well.

property pr1_for_cover;
  @ (posedge clk) cstart ##1 sr1;
endproperty
`endif

reqGnt: assert property (pr1) else $display($stime,,,"\t\t %m
FAIL");
creqGnt: cover property (pr1_for_cover) $display($stime,,,"\t\t
%m PASS");

initial
begin
  {cstart, req, gnt}=3'b000;
end

initial
begin
  @ (negedge clk); {cstart, req, gnt}=3'b100;
  @ (negedge clk); {cstart, req, gnt}=3'b110;
  @ (negedge clk); {cstart,req,gnt}=3'b000;
  @ (negedge clk); {cstart,req,gnt}=3'b001;

  @ (negedge clk); {cstart,req,gnt}=3'b110;
  @ (negedge clk); {cstart,req,gnt}=3'b110;
  @ (negedge clk); {cstart,req,gnt}=3'b111;
  @ (negedge clk); {cstart,req,gnt}=3'b010;
```

```
    @ (negedge clk); {cstart,req,gnt}=3'b000;
    @ (negedge clk); {cstart,req,gnt}=3'b001;

    @ (negedge clk); $finish(2);
  end

  endmodule
```

23.6 LAB2: Questions

```
*******************************************************
LAB Questions based on simulation log
*******************************************************
*******************************************************
Simulation Log—QUESTIONS
*******************************************************

/* +define+overlap
run -all
KERNEL:         10  clk=1 cstart=0 req=0 gnt=0
KERNEL:         30  clk=1 cstart=1 req=0 gnt=0
KERNEL:         30          test_overlap_nonoverlap FAIL

        Q: WHY DOES THE PROPERTY FAIL at 30?

KERNEL:         50  clk=1 cstart=1 req=1 gnt=0
KERNEL:         70  clk=1 cstart=0 req=0 gnt=0
KERNEL:         90  clk=1 cstart=0 req=0 gnt=1
KERNEL:         90          test_overlap_nonoverlap PASS

        Q: WHY DOES THE PROPERTY PASS at 90?

KERNEL:        110  clk=1 cstart=1 req=1 gnt=0
KERNEL:        130  clk=1 cstart=1 req=1 gnt=0
KERNEL:        150  clk=1 cstart=1 req=1 gnt=1
KERNEL:        150          test_overlap_nonoverlap PASS

        Q: WHY DOES THE PROPERTY PASS at 150?

KERNEL:        170  clk=1 cstart=0 req=1 gnt=0
KERNEL:        170          test_overlap_nonoverlap FAIL

        Q: WHY DOES THE PROPERTY FAIL at 170?
```

```
KERNEL:          190  clk=1 cstart=0 req=0 gnt=0
KERNEL:          190              test_overlap_nonoverlap FAIL

        Q: WHY DOES THE PROPERTY FAIL at 190?

KERNEL:          210  clk=1 cstart=0 req=0 gnt=1
*/

/* +define+nonoverlap

run -all

KERNEL:           10  clk=1 cstart=0 req=0 gnt=0
KERNEL:           30  clk=1 cstart=1 req=0 gnt=0
KERNEL:           50  clk=1 cstart=1 req=1 gnt=0
KERNEL:           70  clk=1 cstart=0 req=0 gnt=0
KERNEL:           70              test_overlap_nonoverlap FAIL

        Q: WHY DOES THE PROPERTY FAIL at 70?

KERNEL:           90  clk=1 cstart=0 req=0 gnt=1
KERNEL:           90              test_overlap_nonoverlap PASS

        Q: WHY DOES THE PROPERTY PASS at 90?

KERNEL:          110  clk=1 cstart=1 req=1 gnt=0
KERNEL:          130  clk=1 cstart=1 req=1 gnt=0
KERNEL:          150  clk=1 cstart=1 req=1 gnt=1
KERNEL:          170  clk=1 cstart=0 req=1 gnt=0
KERNEL:          170              test_overlap_nonoverlap FAIL

        Q: WHY DOES THE PROPERTY FAIL at 170?

KERNEL:          190  clk=1 cstart=0 req=0 gnt=0
KERNEL:          190              test_overlap_nonoverlap FAIL

        Q: WHY DOES THE PROPERTY FAIL at 190?

KERNEL:          210  clk=1 cstart=0 req=0 gnt=1
KERNEL:          210              test_overlap_nonoverlap PASS

        Q: WHY DOES THE PROPERTY PASS at 210?

*/
```

23.7 LAB3: Synchronous FIFO Assertions

We will learn how assertions are applied to test a Synchronous FIFO.

23.8 LAB3: DUT Model and Test-Bench

****

LAB3

****

We saw an example of an asynchronous FIFO in Chap. 18 and assertions thereof. For this LAB, I have chosen a simpler Synchronous FIFO for which you will exercise writing assertions. This way you will be familiar with writing assertions for both styles of

LAB Overview

A simple synchronous FIFO design is presented. FIFOs are some of the most commonly used design elements which require close scrutiny. FIFO assertions deployed directly at the source of a FIFO can greatly reduce the time to debug since these assertions point to the exact instance of fifo where an assertion fires.

LAB Objectives

1. *You will learn how to model various FIFO assertions that will be applicable to most any FIFO.*
2. *You will learn use of boolean expressions and sampled value functions as part of this exercise.*

LAB Design Under Test (DUT)

A simple synchronous FIFO design is presented as the DUT.

- *FIFO is 8 bit wide and 8 deep.*
- *FIFO INPUTS*
  fifo_write, fifo_read, clk, rst_ and fifo_data_in[7:0]
- *FIFO OUTPUTS*
  fifo_full, fifo_empty, fifo_data_out[7:0]

Fig. 23.2 LAB3: Synchronous FIFO: Problem Definition

FIFO. Note that one of the most important set of assertions that you may write for your project are the FIFO assertions. Like it or not, FIFOs always give trouble!.

FIFO Specs

- *fifo maintains a wr_ptr and a rd_ptr*
  - *wr_ptr increments by 1 everytime a write is posted to the fifo on a fifo_write request*
  - *rd_ptr increments by 1 everytime a read is posted to the fifo on a fifo_read request*
- *fifo maintains a 'cnt' that increments on a write and decrements on a read. It is used to signal fifo_full and fifo_empty conditions as follows*
  - *When fifo 'cnt' is >= 7, fifo_full is asserted*
  - *When fifo 'cnt' is 0, fifo_empty is asserted.*

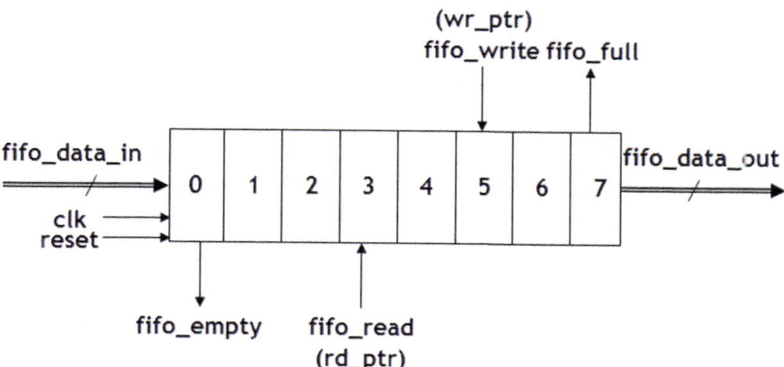

****

LAB3 :: Database

****

LAB: Database

FILES:

1. *fifo.v :: Verilog RTL for 'fifo'*
2. *fifo_property.sv :: SVA file for fifo assertion.s*
   This is the file in which you will add your assertions.
3. *test_fifo.sv :: Testbench for the fifo.*
   Note the use of 'bind' in this testbench.

****

fifo.v

****

```
module  fifo  (fifo_data_out,fifo_full,fifo_empty,fifo_write,
fifo_read,clk,rst_,fifo_data_in);

parameter fifo_depth=8;
parameter fifo_width=8;

output logic [(fifo_width - 1):0] fifo_data_out;
output logic fifo_full, fifo_empty;
input  logic fifo_write, fifo_read, clk, rst_;
input  logic [(fifo_width - 1):0] fifo_data_in;

logic [(fifo_width-1):0] fifomem [0:(fifo_depth-1)];

logic [3:0] wr_ptr, rd_ptr;
logic [3:0] cnt;

always @ (posedge clk or negedge rst_)
  if (!rst_) begin
    rd_ptr  <= 0;
    wr_ptr  <= 0;
    cnt   <= 0;
```

```
`ifndef check1
    fifo_empty <= 1;
`endif
    fifo_full  <= 0;
  end
  else begin
    case ({fifo_write, fifo_read})
      2'b00: ;      // everyone's sleeping!
      2'b01: begin // read
               if (cnt>0) begin
                 rd_ptr <= rd_ptr + 1;
                 cnt   <= cnt - 1;
               end
`ifdef check2
               if (cnt==0) fifo_empty <= 1;
`else
`ifdef check5
               if (cnt==1) fifo_empty <= 1;
           rd_ptr <= rd_ptr+1;
`else
               if (cnt==1) fifo_empty <= 1;
`endif
`endif
               fifo_full <= 0;
             end
      2'b10: begin // write
               if (cnt< fifo_depth) begin
                 fifomem[wr_ptr] <= fifo_data_in;
                 wr_ptr <= wr_ptr + 1;
                 cnt   <= cnt  + 1;
               end
`ifdef check3
               if (cnt>(fifo_depth - 1)) fifo_full <= 1;
`else
`ifdef check4
               if (cnt>=(fifo_depth - 1)) fifo_full <= 1;
           wr_ptr <= wr_ptr+1;
`else
               if (cnt>=(fifo_depth - 1)) fifo_full <= 1;
`endif
`endif
               fifo_empty <= 0;
             end
      2'b11: // write && read
           //You cannot write if cnt is full; so, read only
               if (cnt>(fifo_depth - 1)) begin
                 rd_ptr  <= rd_ptr + 1;
```

```
                  cnt    <= cnt - 1;
                end
              //You cannot read if cnt is empty; so, write only
                else if (cnt<1) begin
                  fifomem[wr_ptr] <= fifo_data_in;
                  wr_ptr  <= wr_ptr + 1;
                  cnt    <= cnt  + 1;
                end
              //else write and read both
                else begin
                  fifomem[wr_ptr] <= fifo_data_in;
                  wr_ptr  <= wr_ptr + 1;
                  rd_ptr  <= rd_ptr + 1;
                end
      endcase
    end

  assign fifo_data_out = fifomem[rd_ptr];

  endmodule
```

23.9 LAB3: Questions

LAB: Assertions to Code

Code assertions to check for the following conditions in the 'fifo' design.

CHECK # 1. Check that on reset
rd_ptr=0; wr_ptr=0; cnt=0; fifo_empty=1 and fifo_full=0

CHECK # 2. Check that fifo_empty is asserted when fifo 'cnt' is 0.
Disable this property 'iff (!rst)'

CHECK # 3. Check that fifo_full is asserted any time fifo 'cnt' is greater than 7.
Disable this property 'iff (!rst)'

CHECK # 4. Check that if fifo is full and you attempt to write (but not read) that the wr_ptr does not change.

CHECK # 5. Check that if fifo is empty and you attempt to read (but not write) that the rd_ptr does not change.

CHECK # 6. Write a property to Warn on write to a full fifo

CHECK # 7. Write a property to Warn on read from an empty fifo

****

LAB3: Questions—Assertion Questions embedded in the fifo_property.sv

****

I have provided the fifo_property.sv file. All you have to do is write your assertions in this file and simulate. I have coded ***"dummy"*** properties so that you can compile the code. The .solution directory contains correct assertions and simulation log against which you can compare your results. Note that there is not just but one way to write an assertion and your assertion could look different from the one you see in the .solution directory. But the results must match with the simulation log in the .solution directory

```
`define rd_ptr test_fifo.fi1.rd_ptr
`define wr_ptr test_fifo.fi1.wr_ptr
`define cnt test_fifo.fi1.cnt

module fifo_property (
    input logic [7:0] fifo_data_out,
    input logic       fifo_full, fifo_empty,
    input logic       fifo_write, fifo_read, clk, rst_,
    input logic [7:0] fifo_data_in
      );

parameter fifo_depth=8;
parameter fifo_width=8;

//   -----------------------------------------
//   1. Check that on reset,
//      rd_ptr=0; wr_ptr=0; cnt=0; fifo_empty=1 and fifo_full=0
//   -----------------------------------------

`ifdef check1
property check_reset;
@ (posedge clk) !rst_ |-> `rd_ptr==0; //DUMMY... remove this
line and
                    //replace it with correct check
endproperty
check_resetP: assert property (check_reset) else $display
($stime,"\t\t FAIL::check_reset\n");
`endif
```

```
//    ------------------------------------------
//    2. Check that fifo_empty is asserted the same clock
         that fifo 'cnt'
//       is 0. Disable this property 'iff (!rst)'
//    ------------------------------------------

`ifdef check2
property fifoempty;
@ (posedge clk) !rst_ |-> `rd_ptr==0; //DUMMY... remove this
line and
                   //replace it with correct check
endproperty
fifoemptyP: assert property (fifoempty) else $display($stime,
"\t\t FAIL::fifo_empty condition\n");
`endif

//    ------------------------------------------
//    3. Check that fifo_full is asserted any time fifo 'cnt'
//       is greater than 7. Disable this property 'iff (!rst)'
//    ------------------------------------------

`ifdef check3
property fifofull;
@ (posedge clk) !rst_ |-> `rd_ptr==0; //DUMMY... remove this
line and
                   //replace it with correct check
endproperty
fifofullP: assert property (fifofull) else $display($stime,"\
t\t FAIL::fifo_full condition\n");
`endif

//    ------------------------------------------
//    4. Check that if fifo is full and you attempt to write
         (but not read)
//       that the wr_ptr does not change.
//    ------------------------------------------

`ifdef check4
property fifo_full_write_stable_wrptr;
@ (posedge clk) !rst_ |-> `rd_ptr==0; //DUMMY... remove this
line and
                   //replace it with correct check
endproperty
```

```
fifo_full_write_stable_wrptrP: assert property (fifo_full_
write_stable_wrptr)
        else $display($stime,"\t\t FAIL::fifo_full_write_stable_
        wrptr condition\n");
`endif

`ifdef check5

//    ------------------------------------------
//    5. Check that if fifo is empty and you attempt to read
         (but not write)
//       that the rd_ptr does not change.
//    ------------------------------------------

property fifo_empty_read_stable_rdptr;
@ (posedge clk) !rst_ |-> `rd_ptr==0; //DUMMY... remove this
line and
                    //replace it with correct check
endproperty
fifo_empty_read_stable_rdptrP: assert property (fifo_empty_read_
stable_rdptr)
    else $display($stime,"\t\t FAIL::fifo_empty_read_stable_
    rdptr condition\n");
`endif

//    ------------------------------------------
//    6. Write a property to Warn on write to a full fifo
//       This property will give Warning with all simulations
//    ------------------------------------------

`ifdef check6
property write_on_full_fifo;
@ (posedge clk) !rst_ |-> `rd_ptr==0; //DUMMY... remove this
line and
                    //replace it with correct check
endproperty
write_on_full_fifoP: assert property (write_on_full_fifo)
    else $display($stime,"\t\t WARNING::write_on_full_fifo\n");
`endif
```

```
//     ------------------------------------------
//     7. Write a property to Warn on read from an empty fifo
//        This property will give Warning with all simulations
//     ------------------------------------------

`ifdef check7
property read_on_empty_fifo;
@ (posedge clk) !rst_ |-> `rd_ptr==0; //DUMMY... remove this
line and
                    //replace it with correct check
endproperty
read_on_empty_fifoP: assert property (read_on_empty_fifo)
       else $display($stime,"\t\t WARNING::read_on_empty_fifo
       condition\n");
`endif

endmodule


*****************************************************
test_fifo.sv
*****************************************************
module test_fifo;

wire  [7:0] fifo_data_out;
wire        fifo_full, fifo_empty;
logic       fifo_write, fifo_read, clk, rst_;
logic [7:0] fifo_data_in;

parameter fifo_depth = 8, fifo_width = 8;

fifo  #(fifo_depth,  fifo_width)  fi1  (fifo_data_out,fifo_full,
fifo_empty,fifo_write,fifo_read,clk,rst_,
       fifo_data_in);

bind fifo fifo_property #(fifo_depth, fifo_width) fi1bind
      (fifo_data_out,fifo_full,fifo_empty,fifo_write,fifo_read,clk,
      rst_, fifo_data_in);

initial
begin
  clk=0;
  fiforeset;
```

```
  fifowrite(10);
  fiforead(9);
  @ (posedge clk);
  @ (posedge clk);
  @ (posedge clk); $finish(2);
end

always #5 clk=!clk;

task fiforeset;
  fifo_write=0; fifo_read=0; rst_=1;
  @ (negedge clk); rst_=0;
  @ (negedge clk);
  @ (negedge clk); rst_=1;
endtask

task fifowrite;
input int nwrite;
  fifo_read=0;
  for (int i=0; i<=nwrite-1; i++)
  begin
    @ (negedge clk); fifo_write=1; fifo_data_in=i;
    //$display($stime,,,"fifo Write Data = %0d",fifo_data_in);
  end
endtask

task fiforead;
input int nread;
  fifo_write=0;
  repeat(nread)
  begin
    @ (negedge clk); fifo_read=1;
    //$display($stime,,,"fifo Read Data = %0d",fifo_data_out);
  end
endtask

always @ (posedge clk)
  $display($stime,,,"rst_=%b clk=%b fifo_write=%b fifo_read=%b
  fifo_full=%b fifo_empty=%b wr_ptr=%0d rd_ptr=%0d cnt=%0d",
    rst_,clk,fifo_write,fifo_read,fifo_full,fifo_empty.fi1.wr_
    ptr,fi1.rd_ptr,fi1.cnt);

endmodule
```

23.10 LAB4: Counter

LAB: Database

FILES:

1. *counter.v :: Verilog RTL for a simple counter.*
2. *counter_property.sv :: SVA file for counter properties*
   This is the file in which you will add your assertions.
3. *test_counter.sv :: Testbench for the counter.*
   Note the use of 'bind' in this testbench.

```
*****************************************************************
counter.v
*****************************************************************
     module counter (
       input clk, rst_, ld_cnt_, updn_cnt, count_enb,
       input [7:0] data_in,
       output logic [7:0] data_out
       );

     always @ (posedge clk or negedge rst_)
     begin

       if (!rst_)
         begin
         `ifndef check1
            data_out <= 0;
         `endif
          end
       else
       begin

         //LOAD DATA
         if (!ld_cnt_)
           data_out <= data_in;

         //HOLD DATA
         `ifndef check2
          else if (!count_enb)
             data_out <= data_out;
         `endif
```

```
        //COUNT DATA
        `ifdef check3
        else
          case (updn_cnt)
            1'b1: data_out <= data_out - 1;
            1'b0: data_out <= data_out + 1;
          endcase
        `else
        else
          case (updn_cnt)
            1'b1: data_out <= data_out + 1;
            1'b0: data_out <= data_out - 1;
          endcase
        `endif

      end
    end
    endmodule
```

LAB: Assertions to Code

Code assertions to check for the following conditions in the 'counter' design.

CHECK # 1. Check that when 'rst_' is asserted (==0) that data_out == 8'b0

CHECK # 2. Check that if ld_cnt_ is deasserted (==1) and count_enb is not enabled (==0) that data_out HOLDS it's previous value.

Disable this property if rst is low.

CHECK # 3. Check that if ld_cnt_ is deasserted (==1) and count_enb is enabled (==1) that if updn_cnt==1 the count goes UP and if updn_cnt==0 the count goes DOWN.

Disable this property if rst is low.

LAB Overview

A simple UP/DOWN COUNTER design is presented. Counter assertions deployed directly at the source can greatly reduce the time to debug since these assertions will point to the exact cause of a Counter error without the need for extensive back-tracing debug when design fails.

LAB Objectives

1. *You will learn use of sampled value functions.*
2. *Alternate ways of modeling an assertion.*

LAB Design Under Test (DUT)

A simple UP/DOWN COUNTER design is presented as the DUT.

**) The counter has 8 bit data input and 8 bit data output*
**) When ld_cnt_ is asserted (active Low), data_in is loaded and output to data_out*

**) When count_enb (active High) is enabled (high) and*
**) updn_cnt is high, data_out = data_out+1;*
**) updn_cnt is low, data_out = data_out-1;*

**) When count_enb is LOW, data_out = data_out;*

Fig. 23.3 LAB4: Counter: Problem Definition

23.11 LAB4: Questions

```
****************************************************
LAB4: Questions embedded in counter_property.sv
****************************************************
    module counter_property (
      input clk, rst_, ld_cnt_, updn_cnt, count_enb,
      input [7:0] data_in,
      input logic [7:0] data_out
      );

    //-----------------------------------
    //   CHECK # 1. Check that when 'rst_' is asserted (==0) that
    //   data_out == 8'b0
    //-----------------------------------
```

```
`ifdef check1
property counter_reset;
@ (posedge clk) data_in |=> data_out; // DUMMY - REMOVE this
line and code
                   // correct assertion
endproperty

counter_reset_check: assert property(counter_reset)
  else $display($stime,,,
                         "\t\tCOUNTER RESET CHECK FAIL:: rst_=
                         %b data_out=%0d \n",
                         rst_,data_out);
`endif

//------------------------------------
// CHECK # 2.Check that if ld_cnt_ is de-asserted (==1) and
count_enb is not
// enabled (==0) that data_out HOLDS its previous value.
// Disable this property 'iff (!rst)'
//------------------------------------

`ifdef check2
property counter_hold;
@ (posedge clk) data_in |=> data_out; // DUMMY - REMOVE this
line and code
                   //correct assertion
endproperty

counter_hold_check: assert property(counter_hold)
else $display($stime,,,"\t\tCOUNTER HOLD CHECK FAIL:: counter
HOLD \n");
`endif

//------------------------------------
// CHECK # 3. Check that if ld_cnt_ is de-asserted (==1) and
count_enb is
// enabled(==1) that if updn_cnt==1 the count goes UP and if
// updn_cnt==0 the count goes DOWN.
// Disable this property 'iff (!rst)'
//------------------------------------

`ifdef check3
property counter_count;
@ (posedge clk) data_in |=> data_out; // DUMMY - REMOVE this
line and code
```

```
                    //correct assertion
endproperty

counter_count_check: assert property(counter_count)
 else $display($stime,,,
              "\t\tCOUNTER COUNT CHECK FAIL:: UPDOWN COUNT
              using $past \n");
`endif

endmodule


***************************************************
test_counter.sv
***************************************************
module test_counter;

logic clk, rst_, ld_cnt_, updn_cnt, count_enb;
logic [7:0] data_in;
wire [7:0] data_out;

int seed1;
counter  upc(
  clk, rst_, ld_cnt_, updn_cnt, count_enb,
  data_in,
  data_out
  );

bind counter counter_property bind_inst (
  clk, rst_, ld_cnt_, updn_cnt, count_enb,
  data_in,
  data_out
  );

initial
begin
  clk=1'b0;
  counter_init;
      count_up(100,10);
  repeat (2) @ (posedge clk);
      count_down(100,10);
  repeat (2) @ (posedge clk);
  @ (posedge clk); $finish(2);
end
```

```
always @ (posedge clk)
   $display($stime,,,"rst_=%b clk=%b count_enb=%b ld_cnt_=%b
   updn_cnt=%b DIN=%0d DOUT=%0d",
   rst_, clk, count_enb, ld_cnt_, updn_cnt, data_in, data_
   out);

always #5 clk=!clk;
task counter_init;
  rst_=1'b1; ld_cnt_=1'b1; count_enb=1'b0; updn_cnt=1'b1;

  @ (negedge clk); rst_=1'b0;
  @ (negedge clk);
  @ (negedge clk); rst_=1'b1;
  @ (negedge clk); data_in=8'b0; ld_cnt_=1'b0;
  @ (negedge clk);
endtask

task count_up;
input logic [7:0] din;
input int count;
  @ (negedge clk); data_in=din; ld_cnt_=1'b0;
  @ (negedge clk); ld_cnt_=1'b1; count_enb=1'b1; updn_cnt=1'b1;
  repeat (count-1) @ (negedge clk);
  @ (negedge clk); count_enb=1'b0;
endtask

task count_down;
input logic [7:0] din;
input int count;
  @ (negedge clk); data_in=din; ld_cnt_=1'b0;
  @ (negedge clk); ld_cnt_=1'b1; count_enb=1'b1; updn_cnt=1'b0;
  repeat (count-1) @ (negedge clk);
  @ (negedge clk); count_enb=1'b0;
endtask

endmodule
```

23.12 LAB5: Data Transfer Protocol

LAB Overview

Specification for a simple data transfer protocol.

- dValid must remain asserted for minimum of 2 clocks but no more than 4 clocks.

- 'data' must be known when 'dValid' is High.

- 'dack' going high signifies that target have accepted data and that master must de-assert 'dValid' the clock after 'dack' goes high.

  - *Note that since data must be valid for minimum 2 cycles, that 'dack' cannot go High for at least 1 clock after the transfer starts (i.e. after the rising edge of 'dValid') and that it must not remain low for more than 3 clocks (because data must trasnfer in max 4 clocks).*

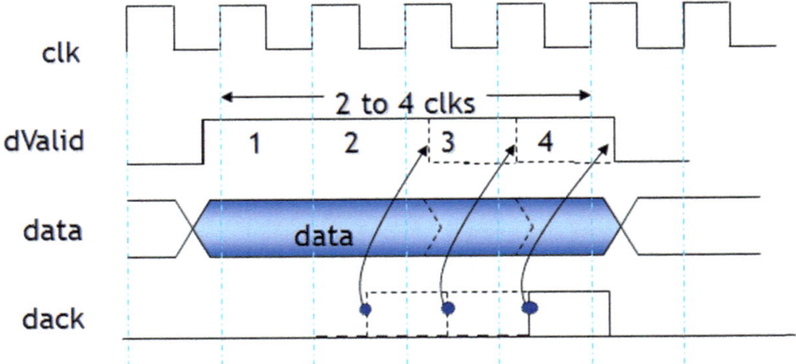

Fig. 23.4 LAB5: Data Transfer Protocol: Problem Definition

LAB: Database

FILES:

1. bus_protocol.v :: bus_protocol module that drive a simple bus protocol
2. bus_protocol_property.sv :: SVA file for bus_protocol assertions.

   Note that this file is only an empty module shell.

   You will add properties that meet the specification described ablve.

3. test_bus_protocol.sv :: Testbench for the bus_protocol module.

   Note the use of 'bind' in this testbench.

LAB Objectives

Bus interfaces are common to any design and this lab will show you how to model assertions for common bus protocol specification.

You will learn

1. Modeling temporal domain assertions for bus interface type logic.
2. Reinforce understanding of Edge sensitive and sampled value functions, consecutive repetition, boolean expressions, etc.

LAB: Assertions to Code

Code assertions to check for the following conditions in the 'bus protocol' design

CHECK # 1. Check that once dValid goes high that it is consecutively asserted (high) for minimum 2 and maximum 4 clocks

CHECK # 2. Check that data is not unknown and remains stable after dValid goes high and until dAck goes high.

CHECK # 3. Check that 'dAck' and 'dValid' relationship is maintained to complete the data transfer.

In other words,

'dack' going high signifies that target have accepted data and that master must de-assert 'dValid' the clock after 'dack' goes high.

Note that since data must be valid for minimum 2 cycles, that 'dack' cannot go High for at least 1 clock after the transfer starts (i.e. after the rising edge of 'dValid') and that it must not remain low for more than 3 clocks (because data must transfer in max 4 clocks).

```
*****************************************************
bus_protocol.sv
*****************************************************
     /* bus_protocol.v module

        This module drives the bus protocol
        timing diagram.

        This module acts as the bus interface unit of
        your design whose protocol you are trying to verify.

     */
     module bus_protocol (input bit clk, reset,
                          output bit dValid, dAck,
                          output logic [7:0] data
     );

     initial
     begin
       $display("SCENARIO 1");
       @(negedge clk); dValid=1'b0; data=8'h0; dAck=1'b0;
       @(negedge clk); dValid=1'b1; data=8'h0; dAck=1'b0;
       @(negedge clk); dValid=1'b1; data=8'h0; dAck=1'b0;
       @(negedge clk); dValid=1'b1; data=8'h0; dAck=1'b1;
       @(negedge clk); dValid=1'b0; data=8'h0; dAck=1'b0;
       @(negedge clk); dValid=1'b0; data=8'h0; dAck=1'b0;
       $display("\n");

       $display("SCENARIO 2");
       @(negedge clk); dValid=1'b1; data=8'h0; dAck=1'b0;
       @(negedge clk); dValid=1'b1; data=8'h0; dAck=1'b0;
       @(negedge clk); dValid=1'b1; data=8'h0; dAck=1'b0;
       @(negedge clk); dValid=1'b1; data=8'h0; dAck=1'b1;
       @(negedge clk); dValid=1'b0; data=8'h0; dAck=1'b0;
       @(negedge clk); dValid=1'b0; data=8'h0; dAck=1'b0;
       $display("\n");

       $display("SCENARIO 3");
       @(negedge clk); dValid=1'b1; data=8'h0; dAck=1'b0;
       @(negedge clk); dValid=1'b1; data=8'h0; dAck=1'b0;
       @(negedge clk); dValid=1'b1; data=8'h0; dAck=1'b0;
       @(negedge clk); dValid=1'b1; data=8'h0; dAck=1'b0;
       @(negedge clk); dValid=1'b1; data=8'h0; dAck=1'b1;
       @(negedge clk); dValid=1'b0; data=8'h0; dAck=1'b0;
       @(negedge clk); dValid=1'b0; data=8'h0; dAck=1'b0;
       $display("\n");
```

```
`ifdef nobugs
  $display("SCENARIO 4");
  @(negedge clk); dValid=1'b1; data=8'h0; dAck=1'b0;
  @(negedge clk); dValid=1'b1; data=8'h0; dAck=1'b0;
  @(negedge clk); dValid=1'b1; data=8'h0; dAck=1'b0;
  @(negedge clk); dValid=1'b1; data=8'h0; dAck=1'b0;
  @(negedge clk); dValid=1'b1; data=8'h0; dAck=1'b1;
  @(negedge clk); dValid=1'b0; data=8'h0; dAck=1'b0;
  @(negedge clk); dValid=1'b0; data=8'h0; dAck=1'b0;
  $display("\n");
`else
  $display("SCENARIO 4");
  @(negedge clk); dValid=1'b1; data=8'h0; dAck=1'b0;
  @(negedge clk); dValid=1'b1; data=8'h0; dAck=1'b0;
  @(negedge clk); dValid=1'b1; data=8'h0; dAck=1'b0;
  @(negedge clk); dValid=1'b1; data=8'h0; dAck=1'b0;
  @(negedge clk); dValid=1'b1; data=8'h0; dAck=1'b0;
  @(negedge clk); dValid=1'b1; data=8'h0; dAck=1'b1;
  @(negedge clk); dValid=1'b0; data=8'h0; dAck=1'b0;
  $display("\n");
`endif

`ifdef nobugs
  $display("SCENARIO 5");
  @(negedge clk); dValid=1'b1; data=8'h0; dAck=1'b0;
  @(negedge clk); dValid=1'b1; data=8'h0; dAck=1'b0;
  @(negedge clk); dValid=1'b1; data=8'h0; dAck=1'b0;
  @(negedge clk); dValid=1'b1; data=8'h0; dAck=1'b0;
  @(negedge clk); dValid=1'b1; data=8'h0; dAck=1'b1;
  @(negedge clk); dValid=1'b0; data=8'h0; dAck=1'b0;
  @(negedge clk); dValid=1'b0; data=8'h0; dAck=1'b0;
  $display("\n");
`else
  $display("SCENARIO 5");
  @(negedge clk); dValid=1'b1; data=8'h0; dAck=1'b0;
  @(negedge clk); dValid=1'b1; data=8'h0; dAck=1'b0;
  @(negedge clk); dValid=1'b1; data=8'h0; dAck=1'b0;
  @(negedge clk); dValid=1'b1; data=8'h0; dAck=1'b0;
  @(negedge clk); dValid=1'b1; data=8'h0; dAck=1'b1;
  @(negedge clk); dValid=1'b1; data=8'h0; dAck=1'b0;
  @(negedge clk); dValid=1'b0; data=8'h0; dAck=1'b0;
  $display("\n");
`endif
```

```
`ifdef nobugs
  $display("SCENARIO 6");
  @(negedge clk); dValid=1'b1; data=8'h0; dAck=1'b0;
  @(negedge clk); dValid=1'b1; data=8'h0; dAck=1'b0;
  @(negedge clk); dValid=1'b1; data=8'h0; dAck=1'b1;
  @(negedge clk); dValid=1'b0; data=8'h0; dAck=1'b0;
  @(negedge clk); dValid=1'b0; data=8'h0; dAck=1'b0;
  $display("\n");
`else
  $display("SCENARIO 6");
  @(negedge clk); dValid=1'b1; data=8'h0; dAck=1'b0;
  @(negedge clk); dValid=1'b0; data=8'h0; dAck=1'b0;
  @(negedge clk); dValid=1'b0; data=8'h0; dAck=1'b0;
  @(negedge clk); dValid=1'b0; data=8'h0; dAck=1'b0;
  @(negedge clk); dValid=1'b0; data=8'h0; dAck=1'b0;
  $display("\n");
`endif

`ifdef nobugs
  $display("SCENARIO 7");
  @(negedge clk); dValid=1'b1; data=8'h0; dAck=1'b0;
  @(negedge clk); dValid=1'b1; data=8'h0; dAck=1'b0;
  @(negedge clk); dValid=1'b1; data=8'h0; dAck=1'b1;
  @(negedge clk); dValid=1'b0; data=8'h0; dAck=1'b0;
  $display("\n");
`else
  $display("SCENARIO 7");
  @(negedge clk); dValid=1'b1; data=8'h0; dAck=1'b0;
  @(negedge clk); dValid=1'b1; data=8'h0; dAck=1'b0;
  @(negedge clk); dValid=1'b1; data=8'hx; dAck=1'b1;
  @(negedge clk); dValid=1'b0; data=8'h0; dAck=1'b0;

  @(negedge clk); dValid=1'b1; data=8'h0; dAck=1'b0;
  @(negedge clk); dValid=1'b1; data=8'h1; dAck=1'b0;
  @(negedge clk); dValid=1'b1; data=8'h0; dAck=1'b1;
  @(negedge clk); dValid=1'b0; data=8'h0; dAck=1'b0;

  @(negedge clk);
  @(negedge clk);
  $display("\n");
`endif

  @(negedge clk);

  $finish(2);
end
endmodule
```

23.13 LAB5: Questions

```
*************************************************************
bus_protocol_property.sv : Questions embedded in the file
*************************************************************
     /* Properties (assertions) for bus_protocol.v
     */
     module bus_protocol_property (input bit clk, dValid, dAck,
     reset,
                  input logic [7:0] data
     );

     /*----------------------------------------------------
     CHECK # 1. Check that once dValid goes high that it is
     consecutively
     asserted (high) for minimum 2 and maximum 4 clocks.
     Check also that once dValid is asserted (high) for 2 to 4
     clocks that
     it does de-assert (low) the very next clock.
     ----------------------------------------------------*/

     `ifdef check1
         property checkValid;
             @ (posedge clk) dValid |-> dValid; //DUMMY - REMOVE
             this line and code
                       //correct assertion
             endproperty
             assert property (checkValid) else
                           $display($stime,,,"checkValid FAIL");
     `endif

        /*----------------------------------------------------
        CHECK # 2. Check that data is not unknown and remains stable
        after
        dValid goes high and until dAck goes high.
        ----------------------------------------------------*/

     `ifdef check2
         property checkdataValid;
             @ (posedge clk) disable iff (reset)
             @ (posedge clk) dValid |-> dValid; //DUMMY - REMOVE
             this line and
                       //code correct assertion
             endproperty
```

```
        assert property (checkdataValid) else
                     $display($stime,,,"checkdataValidFAIL");
`endif

    /*--------------------------------------------------
    CHECK # 3.
    'dack' going high signifies that target have accepted data
    and that master
    must de-assert 'dValid' the clock after 'dack' goes high.

    Note that since data must be valid for minimum 2 cycles,
    that 'dack' cannot
    go High for at least 1 clock after the transfer starts (i.e.
    after the
    rising edge of 'dValid') and that it must not remain low
    for more than 3
    clocks (because data must transfer in max 4 clocks).
    --------------------------------------------------*/

`ifdef check3
        property checkdAck;
        @ (posedge clk) dValid |-> dValid; //DUMMY - REMOVE
        this line and code
                  //correct assertion
        endproperty
        assert property (checkdAck) else $display($stime,,,
        "checkdAck FAIL");
`endif

endmodule


*****************************************************
test_bus_protocol.v
*****************************************************
    module test_bus_protocol (output bit clk, reset,
                              input logic dValid, dAck,
                              input logic [7:0] data);

    bus_protocol bp1(.*);
    bind bus_protocol bus_protocol_property bpb1 (.*);

    initial begin clk=1; reset=1; end
    always #5 clk=!clk;
```

```
initial
begin
  @ (negedge clk); reset=1;
  @ (negedge clk); reset=0;
end

always @ (posedge clk)
  $display($stime,,,"clk=%b dValid=%b data=%h dAck=%b",
                clk, dValid, data, dAck);

endmodule
```

23.14 LAB6: PCI Read Protocol

LAB Overview

A simple system with a PCI Master and PCI Target modules designed to do a simple basic PCI Read operation.

The LAB shows how to derive and write simple but effective assertions for a PCI type bus.

LAB: Database

FILES:

pci_master.v :: A (very) simple PCI Master module driving only a simple Read cycle.
pci_target.v :: A (very) simple PCI Target module responding to a simple Read Cycle.
pci_protocol_property.v :: SVA file for PCI Read cycle assertions.

Note that this file is only an empty module shell. You will add properties that meet the specification described below.

test_pci_protocol.sv :: Testbench for the pci_protocol module.

Fig. 23.5 LAB6: PCI Protocol: Problem Definition

LAB Objectives

1) Learn how to model temporal domain assertions for bus interface type logic.

2) Reinforce understanding of Edge sensitive sampled value functions, consecutive repetition, boolean expressions, etc.

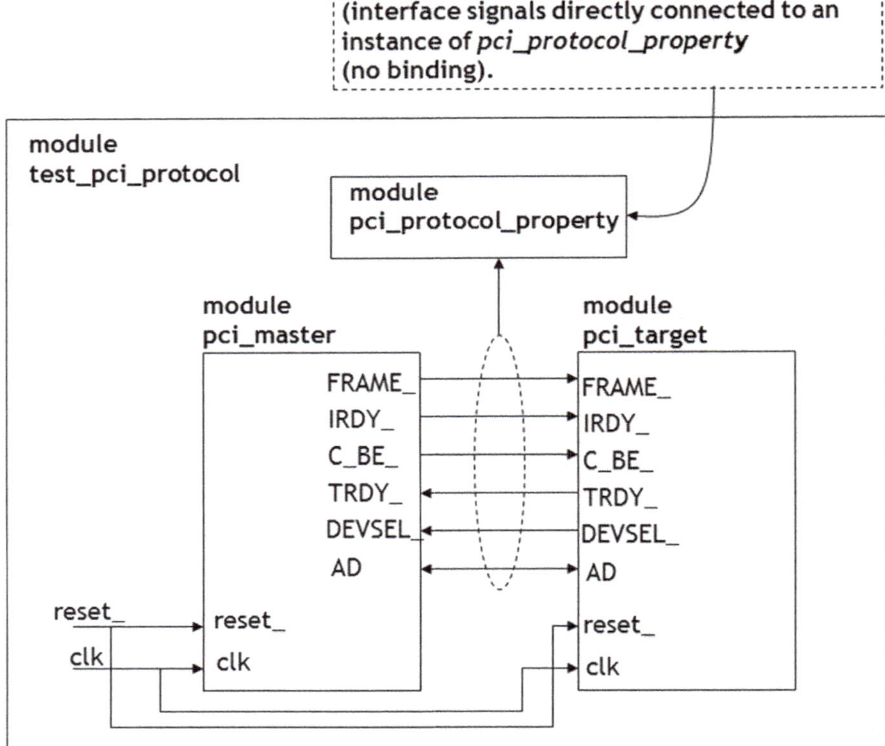

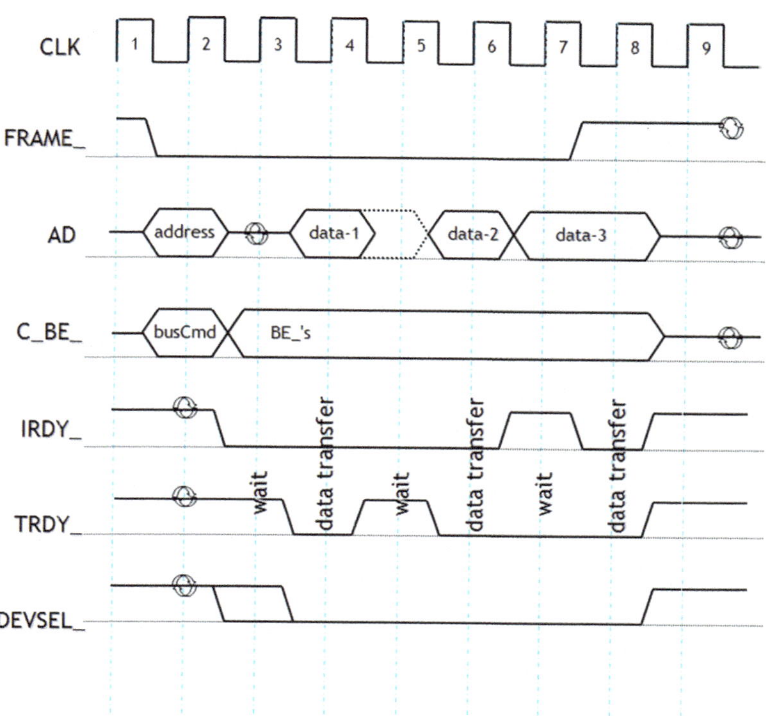

23.15 LAB6: Questions

LAB: Assertions to Code

Property Name	Description
checkPCI_AD_CBE (check1)	On falling edge of FRAME_, AD or C_BE_ bus cannot be unknown
checkPCI_DataPhase (check2)	When both IRDY_ and TRDY_ are asserted, AD or C_BE_ bus cannot be unknown
checkPCI_Frame_Irdy (check3)	FRAME can be de-asserted only if IRDY_ is asserted
checkPCI_trdyDevsel (check4)	TRDY_ can be asserted only if DEVSEL_ is asserted
checkPCI_CBE_during_trx (check5)	Once the cycle starts (i.e. at FRAME_ assertion) C_BE_ cannot float until FRAME_ is de-asserted.

****

pci_master.v (PCI Master DUT)

****

```
/* pci_master.v module
   This is a simple behavioral model that drives only a
   simple (canned!) PCI Read Transaction from the master.
   This is -not- a complete PCI master model.
*/

module pci_master (input bit clk, reset_,
                   input logic TRDY_, DEVSEL_,
             output logic FRAME_, IRDY_,
                   output logic [3:0] C_BE_,
             inout wire [31:0] AD
);

reg [31:0] data [0:7];

bit AD_enb;
reg [31:0] AD_reg;
assign AD = AD_enb ? AD_reg:32'hZ;

initial
begin

    //de-assert control signals
    FRAME_=1'b1; IRDY_=1'b1;

    //Don't drive AD - yet.
    AD_enb=1'b0;

    //On de-assertion of reset_
    @(posedge reset_);

    //Drive FRAME_, AD and C_BE_ (for a memory read)
    @(negedge clk);
          FRAME_ = 1'b0;
    //check1
    `ifdef check1
          AD_reg = 32'h 0000_1234;
    `else
          AD_reg = 32'h 0000_1234; AD_enb=1'b1;
    `endif
          C_BE_ = 4'b 0110;
          $display("\n","\tDrive FRAME_, AD and Read Command");
```

```
//Start the cycle (and drive Byte Enables)
@(negedge clk);
      IRDY_ = 1'b0;
      AD_enb=1'b0;
      C_BE_ = 4'b 1111;
      $display("\n","\tDrive IRDY_ and Byte Enables");

//Wait for TRDY_ to assert
@(negedge TRDY_);

//Read data received
if (! DEVSEL_) data[0] = AD;
        $display("\n","\tData Transfer Phase");

//Wait for the next TRDY_ to assert
@(negedge TRDY_);

//Read data received
if (! DEVSEL_) data[1] = AD;
        $display("\n","\tData Transfer Phase");

//Insert a wait state from the master
@(negedge clk);
      IRDY_ = 1'b1;
      $display("\n","\tMaster Wait Mode");

//Remove wait state
@(negedge clk);
if (! DEVSEL_ && !TRDY_) data[1] = AD;
//check3
`ifdef check3
      IRDY_ = 1'b1;
`else
      IRDY_ = 1'b0;
`endif
      $display("\n","\tData Transfer Phase");

//De-assert FRAME_
@(negedge clk);
      FRAME_ = 1'b1;
      $display("\n","\tFRAME_ De-asserted");

//De-assert C_BE_ just to introduce bug for check#5
`ifdef check5
      C_BE_ = 4'b zzzz;
`endif
```

```
    //De-assert IRDY_
    @(negedge clk);
          IRDY_ = 1'b1;
          $display("\n","\tCYCLE COMPLETE");

end
endmodule
```

```
*****************************************************
```

pci_target.v (PCI Target DUT)

```
*****************************************************
/* pci_target.v module
   This is a simple behavioral model that drives only a simple
   (canned!) PCI Read Transaction from the target. This is
   -not- a complete PCI target model.
*/

module pci_target (input bit clk, reset_,
    output logic TRDY_, DEVSEL_,   input logic FRAME_, IRDY_,
    input logic [3:0] C_BE_,       inout wire [31:0] AD
);

bit AD_enb;
reg [31:0] AD_reg;
assign AD = AD_enb ? AD_reg:32'hZ;

initial
begin

    //Keep AD float...until you want to drive data on it
    //You are not yet selected, so keep DEVSEL_ de-asserted
    AD_enb = 1'b0;
    DEVSEL_ = 1'b1;

    //On assertion of IRDY_
    @(negedge IRDY_);

    //Drive DEVSEL_
    //Check 4
`ifdef check4
          DEVSEL_=1'b1;
    `else
          DEVSEL_=1'b0;
    `endif
    //$display("\tTARGET selected");
```

```
    //Drive TRDY_  and data
    @(negedge clk);
           TRDY_ = 1'b0;

    //For Check2
`ifdef check2
           AD_reg = 32'h CAFE_CAFE;
    `else
           AD_reg = 32'h CAFE_CAFE; AD_enb = 1'b1;
    `endif

    //Insert a WAIT state
    @(negedge clk);
           TRDY_ = 1'b1;
           AD_enb = 1'b0;
    $display("\n","\tTARGET Wait Mode");

    //Drive TRDY_  and second data
    @(negedge clk);
           TRDY_ = 1'b0;
    //For Check2
`ifdef check2
           AD_enb = 1'b0;
    `else
           AD_reg = 32'h FACE_FACE; AD_enb = 1'b1;
    `endif

    //Drive TRDY_  and third data
    @(negedge clk);
           TRDY_ = 1'b0;
    //For Check2
`ifdef check2
           AD_enb = 1'b0;
    `else
           AD_reg = 32'h CAFE_FACE; AD_enb = 1'b1;
    `endif

    @(negedge clk);
    //De-assert TRDY_ and DEVSEL_
    @(negedge clk);
        TRDY_ = 1'b1;
        DEVSEL_ = 1'b1;
        AD_enb = 1'b0;

end
endmodule
```

```
*******************************************************
pci_protocol_property.sv-LAB6 Questions embedded in code
*******************************************************
module pci_protocol_property (input logic clk, reset_, TRDY_,
DEVSEL_, FRAME_, IRDY_,
             input logic [3:0] C_BE_,
             input logic [31:0] AD
);

/*---------------------------------------------------
   CHECK # 1. On falling edge of FRAME_, AD or C_BE_
      cannot be unknown.
---------------------------------------------------*/

`ifdef check1
   property checkPCI_AD_CBE;
   @ (posedge clk) disable iff (!reset_)
            FRAME_ |-> 1'b1; //DUMMY -REMOVE this line and
            code correct
                           //assertion
   endproperty
   assert property (checkPCI_AD_CBE) else $display($stime,,,
   "CHECK1:checkPCI_AD_CBE FAIL\n");
`endif

/*---------------------------------------------------
   CHECK # 2. When IRDY_ and TRDY_ are asserted (low) AD
      or C_BE_ cannot be unknown.
---------------------------------------------------*/

`ifdef check2
   property checkPCI_DataPhase;
   @ (posedge clk) disable iff (!reset_)
            FRAME_ |-> 1'b1; // DUMMY - REMOVE this line and
            code correct
            //assertion
   endproperty
   assert property (checkPCI_DataPhase) else $display($stime
   ,,,"CHECK2:checkPCI_DataPhase FAIL\n");
`endif

/*---------------------------------------------------
CHECK # 3. FRAME_ can go High only if IRDY_ is asserted.
   In other words, master can signify end of cycle
   only if IRDY_ is asserted.
---------------------------------------------------*/
```

```
`ifdef check3
   property checkPCI_Frame_Irdy;
   @ (posedge clk) disable iff (!reset_)
            FRAME_ |-> 1'b1; // DUMMY - REMOVE this line and
            code correct
            //assertion
   endproperty
   assert property (checkPCI_Frame_Irdy) else $display($stim
   e,,,"CHECK3:checkPCI_frmIrdy FAIL\n");
`endif

/*-----------------------------------------------
 CHECK # 4. TRDY_ can be asserted (low) only if DEVSEL_
   is asserted (low)
 ------------------------------------------------*/

`ifdef check4
   property checkPCI_trdyDevsel;
   @ (posedge clk) disable iff (!reset_)
            FRAME_ |-> 1'b1; // DUMMY - REMOVE this line and
            code correct
            //assertion
   endproperty
   assert property (checkPCI_trdyDevsel) else $display($stim
   e,,,"CHECK4:checkPCI_trdyDevsel FAIL\n");
`endif

/*-----------------------------------------------
 CHECK # 5. Once the cycle starts (i.e. at FRAME_ assertion)
   C_BE_ should not float until FRAME_ is de-asserted
 ------------------------------------------------*/

`ifdef check5
   property checkPCI_CBE_during_trx;
   @ (posedge clk) disable iff (!reset_)
            FRAME_ |-> 1'b1; // DUMMY - REMOVE this line and
            code correct
            //assertion
   endproperty
   assert property (checkPCI_CBE_during_trx) else $display($
   stime,,,"CHECK5:checkPCI_CBE_during_trx FAIL\n");
`endif
endmodule
```

Chapter 24
System Verilog Assertions: LAB Answers

Introduction: This chapter provides answers to all the LAB questions posed in previous chapter, namely, answers for the following LABs are presented.

1. "bind" and implication operators
2. Overlap and nonoverlap operators
3. Synchronous FIFO
4. Counter
5. Data Transfer Protocol
6. PCI Read Protocol

A. B. Mehta, *SystemVerilog Assertions and Functional Coverage*,
https://doi.org/10.1007/978-3-030-24737-9_24

24.1 LAB1: Answers: "bind" and Implication Operators

LAB 1 : Code snippet from test_dut.sv showing 'bind' between 'dut' and 'dut_property'

```
bind dut dut_property dut_bind_inst (
            .pclk(clk),
            .preq(req),
            .pgnt(gnt)
      );
```

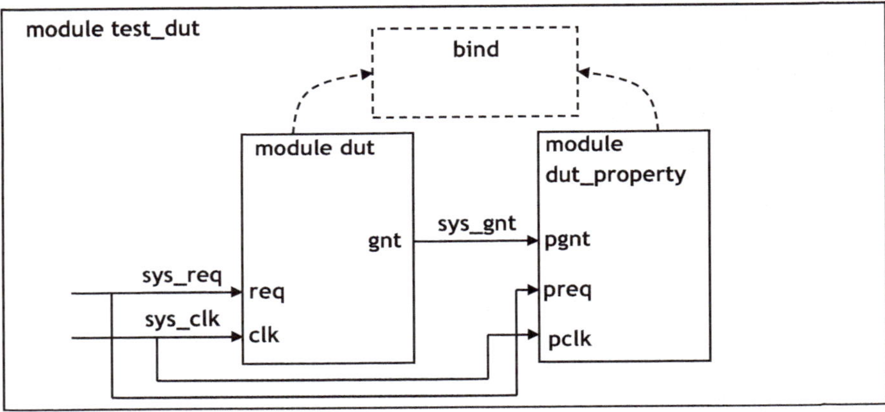

Fig. 24.1 LAB1: "bind" assertions (answers)

LAB 1 : Code snippet for "no_implication"

```
property pr1;
  @(posedge clk) req ##2 gnt;
endproperty
reqGnt: assert property (pr1) $display($stime,,,"\t\t %m PASS");
            else $display($stime,,,"\t\t %m FAIL");
```

LAB 1 : Q&A on "no_implication"

```
/* +define+no_implication

run -all
KERNEL:        10 clk=1 req=0 gnt=0
KERNEL:        10           test_implication FAIL
KERNEL:        30 clk=1 req=1 gnt=0
KERNEL:        50 clk=1 req=0 gnt=0
KERNEL:        50           test_implication FAIL
KERNEL:        70 clk=1 req=0 gnt=1
KERNEL:        70           test_implication FAIL
KERNEL:        70           test_implication PASS
```

Q: WHY IS THERE A FAIL -AND- A PASS AT TIME (70) ??

A: The FAIL at 70 is for the thread starting at time 70.
At 70, req==0 and since there is no implication, the property fails because without an implication there is no antecedent to match before the check begins. Whenever at posedge clk, 'req' is detected low, the property will fail.

The PASS at 70 is for the thread that starts at 30.
At 30, req==1, so property eval proceeds.
At 70 (i.e. 2 clocks later) gnt==1 as required by the property and the property PASSes.

Fig. 24.2 LAB1: Q&A on "no_implication" operator (answers)

LAB 1 : Code snippet for "no_implication"

```
property pr1;
  @(posedge clk) req ##2 gnt;
endproperty
reqGnt: assert property (pr1) $display($stime,,,"\t\t  %m PASS");
                 else $display($stime,,,"\t\t  %m FAIL");
```

LAB 1 : Q&A on "no_implication"

```
KERNEL:     90  clk=1 req=1 gnt=0
KERNEL:    110  clk=1 req=0 gnt=0
KERNEL:    110            test_implication FAIL
KERNEL:    130  clk=1 req=0 gnt=0
KERNEL:    130            test_implication FAIL
KERNEL:    130            test_implication FAIL
```

Q: WHY ARE THERE 2 FAILs AT TIME (130) ??

A: The first failures is for the thread starting at time 90
At 90, req==1, so property eval proceeds.
At 130 (i.e. 2 clocks later) gnt==0 which violates the property and the property FAILs.

The second failure is for the thread starting at time 130.
At 130, req==0 and since there is no implication, the property fails because without an implication there is no antecedent to match before the check begins. Whenever at posedge clk, 'req' detected low, the property will fail.

Fig. 24.2 (continued)

LAB 1 : Code snippet for "implication"

```
property pr1;
  @(posedge clk) req |-> ##2 gnt;
endproperty

reqGnt: assert property (pr1) $display($stime,,,"\t\t %m PASS");
          else $display($stime,,,"\t\t %m FAIL");
```

LAB 1 : Q&A on "implication"

```
run -all
KERNEL:      10 clk=1 req=0 gnt=0
KERNEL:      10          test_implication PASS
KERNEL:      30 clk=1 req=1 gnt=0
KERNEL:      50 clk=1 req=0 gnt=0
KERNEL:      50          test_implication PASS
KERNEL:      70 clk=1 req=0 gnt=1
KERNEL:      70          test_implication PASS
KERNEL:      70          test_implication PASS
```

Q: WHY ARE THERE 2 PASSes AT TIME 70 ??

A: The first pass is for the thread starting at time 30.
At 30, req==1, so property eval proceeds.
At 70 (i.e. 2 clocks later) gnt==1 as required by the property and the property PASSes.

The second pass is for the thread starting at time 70.
At 70, req==0 and since there is implication, the consequent eval won't start. However, there is a PASS action_block associated with the property which triggers because of the vacuous pass phenomenon. In other wods, whenever 'req' is low, the antecedent won't match and the property will pass vacuosly.

Fig. 24.3 LAB1: Q&A on "implication" operator (answers)

LAB 1 : Code snippet for "implication"

```
property pr1;
  @(posedge clk) req |-> ##2 gnt;
endproperty

reqGnt: assert property (pr1) $display($stime,,,"\t\t %m PASS");
              else $display($stime,,,"\t\t %m FAIL");
```

LAB 1 : Q&A on "implication"

```
KERNEL:        90  clk=1 req=1 gnt=0
KERNEL:       110  clk=1 req=0 gnt=0
KERNEL:       110            test_implication PASS
KERNEL:       130  clk=1 req=0 gnt=0
KERNEL:       130            test_implication FAIL
KERNEL:       130            test_implication PASS
```

Q: WHY IS THERE A PASS -and- a FAIL AT TIME 130 ??

A: The failure is for the thread starting at time 90.
At 90, req==1, so property eval proceeds.
At 130 (i.e. 2 clocks later) gnt==0 which violates the property and the property FAILs.

The pass is for the property stating at 130.
At 130, req==0 and since there is implication, the consequent eval won't start. However, there is a PASS action_block associated with the property which triggers because of the vacuous pass phenomenon. In other words, whenever 'req' is low, the antecedent won't match and the property will pass vacuosly.

Fig. 24.3 (continued)

24.2 LAB2: Answers: Overlap and Nonoverlap Operators

LAB 2 : Code snippet with "overlap" operator

```
sequence sr1;
  req ##2 gnt;
endsequence

property pr1;
  @(posedge clk) cstart |-> sr1;
endproperty

property pr1_for_cover;
  @(posedge clk) cstart ##0 sr1;
endproperty
```

LAB 2 : Q&A on "overlap" operator

```
run -all
KERNEL:     10 clk=1 cstart=0 req=0 gnt=0
KERNEL:     30 clk=1 cstart=1 req=0 gnt=0
KERNEL:     30          test_overlap_nonoverlap FAIL

          Q: WHY DOES THE PROPERTY FAIL at 30?

           A: At time 30, cstart=1; so antecedent matches and consequent eval starts
             At time 30, req is NOT equal to 1 as required by overlapping implication
             and the consequent fails right away and the property FAILs.

KERNEL:     50 clk=1 cstart=1 req=1 gnt=0
KERNEL:     70 clk=1 cstart=0 req=0 gnt=0
KERNEL:     90 clk=1 cstart=0 req=0 gnt=1
KERNEL:     90          test_overlap_nonoverlap PASS

          Q: WHY DOES THE PROPERTY PASS at 90?

           A: At time 50, cstart=1; so antecedent matches and consequent eval starts
             At time 50 (i.e, the same clock as required by overlapping implication).
             req=1; so consequent eval continues
             At time 70, gnt=1 as required by the property and the consequent
             matches and the property PASSes.
```

Fig. 24.4 LAB1: Q&A on “overlap” operator (answers)

LAB 2 : Q&A on "overlap" operator

```
KERNEL:     110  clk=1 cstart=1 req=1 gnt=0
KERNEL:     130  clk=1 cstart=1 req=1 gnt=0
KERNEL:     150  clk=1 cstart=1 req=1 gnt=1
KERNEL:     150            test_overlap_nonoverlap PASS
```

Q: WHY DOES THE PROPERTY PASS at 150?

A: At time 110, cstart=1; antecedent matches and consequent eval starts.
At time 110 (i.e, the same clock as required by overlapping implication), req=1; so consequent eval continues
At time 150 (i.e 2 clocks after 110), gnt=1 as required by the property so the consequent matches and the property PASSes

```
KERNEL:     170  clk=1 cstart=0 req=1 gnt=0
KERNEL:     170            test_overlap_nonoverlap FAIL
```

Q: WHY DOES THE PROPERTY FAIL at 170?

A: At time 130, cstart=1;antecedent matches and consequent eval starts
At time 130 (i.e, the same clock as required by overlapping implication), req=1; so consequent eval continues
At time 170 (i.e 2 clocks after 130), gnt is NOT equal to 0 as required by the property so the consequent does not match and the property FAILs

```
KERNEL:     190  clk=1 cstart=0 req=0 gnt=0
KERNEL:     190            test_overlap_nonoverlap FAIL
```

Q: WHY DOES THE PROPERTY FAIL at 190?

A: At time 150, cstart=1; antecedent matches and consequent eval starts
At time 150 (i.e, the same clock as required by overlapping implication), req=1; so consequent eval continues
At time 190 (i.e 2 clocks after 150), gnt is NOT equal 0 as required by the property so the consequent does not match and the property FAILs

Fig. 24.4 (continued)

LAB 2 : Code snippet with "non-overlap" operator

```
sequence sr1;
  req ##2 gnt;
endsequence

property pr1;
  @(posedge clk) cstart |=> sr1;
endproperty

property pr1_for_cover;
  @(posedge clk) cstart ##1 sr1;
endproperty
```

LAB 2 : Q&A on "non-overlap" operator

```
KERNEL:      10  clk=1 cstart=0 req=0 gnt=0
KERNEL:      30  clk=1 cstart=1 req=0 gnt=0
KERNEL:      50  clk=1 cstart=1 req=1 gnt=0
KERNEL:      70  clk=1 cstart=0 req=0 gnt=0
KERNEL:      70            test_overlap_nonoverlap FAIL

         Q: WHY DOES THE PROPERTY FAIL at 70?

          A: This failure is for the thread that started at time 50 (and not 30).
             At time 50, cstart=1; so antecedent matches and consequent eval starts
             At time 70 (i.e, one clock later as required by nonoverlapping
             implication), req is NOT EQUAL to 1; so consequent does not match and
             the property FAILs

KERNEL:      90  clk=1 cstart=0 req=0 gnt=1
KERNEL:      90            test_overlap_nonoverlap PASS

         Q: WHY DOES THE PROPERTY PASS at 90?

          A: This pass is for the thread that started at time 30 (and not 50).
             At time 30, cstart=1; so antecedent matches and consequent eval starts
             At time 50 (i.e, one clock later as required by nonoverlapping
             implication), req == 1; so consequent eval continues
             At time 90 (i.e, two clocks later as required by the property),
             gnt == 1; so consequent matches and the property PASSes.
```

Fig. 24.5 LAB1: Q&A on “nonoverlap” operator (answers)

LAB 2 : Q&A on "non-overlap" operator

```
KERNEL:     110  clk=1 cstart=1 req=1 gnt=0
KERNEL:     130  clk=1 cstart=1 req=1 gnt=0
KERNEL:     150  clk=1 cstart=1 req=1 gnt=1
KERNEL:     170  clk=1 cstart=0 req=1 gnt=0
KERNEL:     170          test_overlap_nonoverlap FAIL

          Q: WHY DOES THE PROPERTY FAIL at 170?

            A: This failure is for the thread that started at time 110
              At time 110, cstart=1; antecedent matches and consequent eval starts
              At time 130 (i.e, one clock later as required by nonoverlapping
              implication), req is EQUAL to 1; so consequent eval continues.
              At time 170 (i.e, two clocks later as required by the property),
              gnt is NOT EQUAL to 1; so consequent does not match and
              the property FAILs.

KERNEL:     190  clk=1 cstart=0 req=0 gnt=0
KERNEL:     190          test_overlap_nonoverlap FAIL

          Q: WHY DOES THE PROPERTY FAIL at 190?

            A: This failure is for the thread that started at time 130
              At time 110, cstart=1; so antecedent matches and consequent eval starts
              At time 150 (i.e, one clock later as required by nonoverlapping
              implication), req is EQUAL to 1; so consequent eval continues.
              At time 190 (i.e, two clocks later as required by the property),
              gnt is NOT EQUAL to 1; so consequent does NOT match and
              the property FAILs.

KERNEL:     210  clk=1 cstart=0 req=0 gnt=1
KERNEL:     210          test_overlap_nonoverlap PASS

          Q: WHY DOES THE PROPERTY PASS at 210?

            A: This pass is for the thread that started at time 150
              At time 150, cstart=1; antecedent matches and consequent eval starts
              At time 170 (i.e, one clock later as required by nonoverlapping
              implication), req is EQUAL to 1; so consequent eval continues.
              At time 210 (i.e, two clocks later as required by the property),
              gnt is EQUAL to 1; so consequent matches and the property PASSes.
```

Fig. 24.5 (continued)

24.3 LAB3: Answers: Synchronous FIFO

LAB 3 : fifo_property.sv

```
//      -----------------------------------------
//      1. Check that on reset,
//                 rd_ptr=0; wr_ptr=0; cnt=0; fifo_empty=1 and fifo_full=0
//      -----------------------------------------
`ifdef check1
property check_reset;
 @(posedge clk)
  (!rst_ |-> (`rd_ptr==0 && `wr_ptr==0 && fifo_empty==1 && fifo_full==0));
endproperty
check_resetP: assert property (check_reset) else $display($stime,"\t\t
FAIL::check_reset\n");
`endif

//      -----------------------------------------
//      2. Check that fifo_empty is asserted the same clock that fifo 'cnt' is 0.
//        Disable this property 'iff (!rst)'
//      -----------------------------------------
`ifdef check2
property fifoempty;
 @(posedge clk) disable iff (!rst_)
  (`cnt==0 |-> fifo_empty);
endproperty
fifoemptyP: assert property (fifoempty) else $display($stime,"\t\t
FAIL::fifo_empty condition\n");
`endif

//      -----------------------------------------
//      3. Check that fifo_full is asserted any time fifo 'cnt' is greater than 7.
//        Disable this property 'iff (!rst)'
//      -----------------------------------------
`ifdef check3
property fifofull;
 @(posedge clk) disable iff (!rst_)
  (`cnt>(fifo_depth-1) |-> fifo_full);
endproperty
fifofullP: assert property (fifofull) else $display($stime,"\t\t FAIL::fifo_full
condition\n");
`endif
```

Fig. 24.6 LAB3: FIFO: answers

LAB 3 : fifo_property.sv

```
//    -----------------------------------------
//     4. Check that if fifo is full and you attempt to write (but not read) that
//       the wr_ptr does not change.
//    -----------------------------------------
`ifdef check4
property fifo_full_write_stable_wrptr;
  @(posedge clk) disable iff (!rst_)
   (fifo_full && fifo_write && !fifo_read |=> $stable(`wr_ptr));
endproperty
fifo_full_write_stable_wrptrP: assert property (fifo_full_write_stable_wrptr)
    else $display($stime,"\t\t FAIL::fifo_full_write_stable_wrptr condition\n");
`endif

`ifdef check5
//    -----------------------------------------
//     5. Check that if fifo is empty and you attempt to read (but not write) that
//       the rd_ptr does not change.
//    -----------------------------------------
property fifo_empty_read_stable_rdptr;
  @(posedge clk) disable iff (!rst_)
   (fifo_empty && fifo_read && !fifo_write |=> $stable(`rd_ptr));
endproperty
fifo_empty_read_stable_rdptrP: assert property (fifo_empty_read_stable_rdptr)
            else $display($stime,"\t\t FAIL::fifo_empty_read_stable_rdptr
condition\n");
`endif

//    -----------------------------------------
//     6. Write a property to Warn on write to a full fifo
//           This property will give Warning with all simulations
//    -----------------------------------------
`ifdef check6
property write_on_full_fifo;
  @(posedge clk) disable iff (!rst_)
   fifo_full |-> !fifo_write;
endproperty
write_on_full_fifoP: assert property (write_on_full_fifo)
            else $display($stime,"\t\t WARNING::write_on_full_fifo\n");
`endif
```

Fig. 24.6 (continued)

LAB 3 : fifo_property.sv

```
//     ----------------------------------------
//      7. Write a property to Warn on read from an empty fifo
//           This property will give Warning with all simulations
//     ----------------------------------------
`ifdef check7
property read_on_empty_fifo;
  @(posedge clk) disable iff (!rst_)
    fifo_empty |-> !fifo_read;
endproperty
read_on_empty_fifoP: assert property (read_on_empty_fifo)
             else $display($stime,"\t\t  WARNING::read_on_empty_fifo  condition\n");
`endif
```

Fig. 24.6 (continued)

24.4 LAB4: Answers: Counter

LAB 4 : counter_property.sv

```
//-----------------------------------
//      CHECK # 1. Check that when 'rst_' is asserted (==0) that data_out == 8'b0
//-----------------------------------
`ifdef check1
property counter_reset;
  @(clk) disable iff (rst_) !rst_ |=> (data_out == 8'b0);
endproperty

counter_reset_check: assert property(counter_reset)
  else $display($stime,,,"\t\tCOUNTER  RESET CHECK FAIL:: rst_=%b data_out=%0d \n",
                                                        rst_,data_out);
`endif

//-----------------------------------
//      CHECK # 2. Check that if ld_cnt_ is deasserted (==1) and count_enb is not enabled
//                 (==0) that data_out HOLDS it's previous value.
//                 Disable this property 'iff (!rst)'
//-----------------------------------
`ifdef check2
property counter_hold;
  @(posedge clk) disable iff (!rst_) (ld_cnt_ & !count_enb) |=> data_out ===
$past(data_out);
endproperty

counter_hold_check: assert property(counter_hold)
  else $display($stime,,,"\t\tCOUNTER  HOLD CHECK FAIL:: counter HOLD \n");
`endif
```

Fig. 24.7 LAB4: Counter: answers

LAB 4 : counter_property.sv

```
//-----------------------------------
// CHECK # 3. Check that if ld_cnt_ is deasserted (==1) and count_enb is
// enabled (==1) that if updn_cnt==1 the count goes UP and if updn_cnt==0 the
// count goes DOWN.
//-----------------------------------

`ifdef check3
property counter_count;
  @(posedge clk) disable iff (!rst_) (ld_cnt_ & count_enb) |->
    if (updn_cnt) ##1 (data_out-8'h01) == $past(data_out)
    else          ##1 (data_out+8'h01) == $past(data_out);
endproperty

counter_count_check: assert property(counter_count)
  else $display($stime,,,"\t\tCOUNTER COUNT CHECK FAIL:: UPDOWN COUNT using
$past \n");
 `endif

//-----------------------------------
// Alternate way of writing assertion for CHECK # 3
// Check for count using local variable
//-----------------------------------
/*
`ifdef check3
property counter_count_local;
logic[7:0] local_data;
  @(posedge clk) disable iff (!rst_) (ld_cnt_ & count_enb, local_data = data_out) |->
    if (updn_cnt) ##1 (data_out == (local_data+8'h01))
    else          ##1 (data_out == (local_data-8'h01));
endproperty

counter_count_check: assert property(counter_count)
  else $display($stime,,,"\t\tCOUNTER COUNT CHECK FAIL:: UPDOWN COUNT using
$past \n");

 `endif
*/
```

Fig. 24.7 (continued)

24.5 LAB5: Answers: Data Transfer Protocol

LAB Overview

Specification for a simple data transfer protocol.

- dValid must remain asserted for minimum of 2 clocks but no more than 4 clocks.
- 'data' must be known when 'dValid' is High.
- 'dack' going high signifies that target have accepted data and that master must de-assert 'dValid' the clock after 'dack' goes high.
    - *Note that since data must be valid for minimum 2 cycles, that 'dack' cannot go High for at least 1 clock after the transfer starts (i.e. after the rising edge of 'dValid') and that it must not remain low for more than 3 clocks (because data must trasnfer in max 4 clocks).*

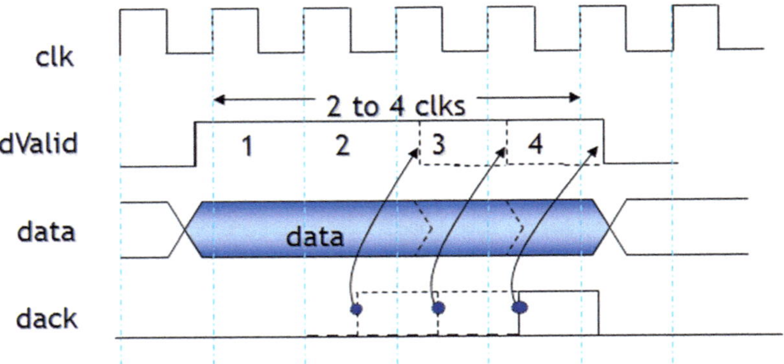

Fig. 24.8 LAB5: Data Transfer Bus Protocol: answers

LAB 5 : bus_protocol_property.sv

```
/*-------------------------------------------------
    CHECK # 1. Check that once dValid goes high that it is consecutively
     asserted (high) for minimum 2 and maximum 4 clocks.
            Check also that once dValid is asserted (high) for 2 to 4 clocks that
            it does de-assert (low) the very next clock.
     -------------------------------------------------*/
`ifdef check1
    property checkValid;
     @(posedge clk) disable iff (reset) $rose(dValid) |=> (dValid) [*2:4] ##1 $fell(dValid);
    endproperty
    assert property (checkValid) else $display($stime,,,"checkValid FAIL");
`endif

    /*-------------------------------------------------
      CHECK # 2. Check that data is not unknow and remains stable after dValid goes
                 high and until dAck goes high.
     -------------------------------------------------*/
`ifdef check2
    property checkdataValid;
     @(posedge clk) disable iff (reset)
       $rose(dValid) |=> (!$isunknown(data) && $stable(data)) [*1:$] ##0 $rose(dAck);
endproperty
    assert property (checkdataValid) else $display($stime,,,"checkdataValid FAIL");
`endif
```

LAB 5 : bus_protocol_property.sv

```
/*-------------------------------------------------
    CHECK # 3. Check that 'dAck' and 'dValid' relationship is maintained to complete the
data transfer. In other words,

    'dack' going high signifies that target have accepted data and that master must de-
assert 'dValid' the clock after 'dack' goes high.

Note that since data must be valid for minimum 2 cycles, that 'dack' cannot go High
for at least 1 clock after the transfer starts (i.e. after the rising edge of 'dValid') and
that it must not remain low for more than 3 clocks (because data must trasnfer in max 4
clocks).
     -------------------------------------------------*/
`ifdef check3
            property checkdAck;
             @(posedge clk) disable iff (reset)
               $rose(dValid) |=> (dValid && !dAck) [*1:3] ##1 $rose (dAck) ##1 $fell (dValid);
            endproperty
            assert property (checkdAck) else $display($stime,,,"checkdAck FAIL");
`endif
```

Fig. 24.8 (continued)

24.6 LAB6: Answers: PCI Read Protocol

LAB 6 : pci_protocol_property.sv

```
/*-------------------------------------------------
      CHECK # 1. On falling edge of FRAME_, AD or C_BE_ cannot be unknown.
        -------------------------------------------------*/
`ifdef check1
      property checkPCI_AD_CBE;
        @(posedge clk) disable iff (!reset_) $fell(FRAME_) |->
                                 !($isunknown(AD) || $isunknown(C_BE_)) ;
      endproperty
      assert property (checkPCI_AD_CBE) else
$display($stime,,,"CHECK1:checkPCI_AD_CBE  FAIL\n");
`endif

      /*-------------------------------------------------
      CHECK # 2. When IRDY_ and TRDY_ are asserted (low) AD or C_BE_ cannot be
unknown.
        -------------------------------------------------*/
`ifdef check2
      property checkPCI_DataPhase;
        @(posedge clk) disable iff (!reset_) (!IRDY_ && !TRDY_) |->
                                 !($isunknown(AD) || $isunknown(C_BE_)) ;
      endproperty
      assert property (checkPCI_DataPhase) else
$display($stime,,,"CHECK2:checkPCI_DataPhase  FAIL\n");
`endif

      /*-------------------------------------------------
      CHECK # 3. FRAME_ can go High only if IRDY_ is asserted.
                 In other words, master can signify end of cycle only if IRDY_ is
asserted.
        -------------------------------------------------*/
`ifdef check3
      property checkPCI_Frame_Irdy;
        @(posedge clk) disable iff (!reset_) $rose(FRAME_) |-> !IRDY_;
      endproperty
      assert property (checkPCI_Frame_Irdy) else
$display($stime,,,"CHECK3:checkPCI_frmIrdy  FAIL\n");
`endif
```

Fig. 24.9 LAB6: PCI Protocol: answers

LAB 6 : pci_protocol_property.sv

```
  /*-----------------------------------------------
      CHECK # 4. TRDY_ can be asserted (low) only if DEVSEL_ is asserted (low)
        -----------------------------------------------*/
`ifdef check4
      property checkPCI_trdyDevsel;
        @(posedge clk) disable iff (!reset_) !TRDY_ |-> !DEVSEL_;
      endproperty
      assert property (checkPCI_trdyDevsel) else
$display($stime,,,"CHECK4:checkPCI_trdyDevsel FAIL\n");
`endif

      /*-----------------------------------------------
      CHECK # 5. Once the cycle starts (i.e. at FRAME_ assertion)
                    C_BE_ should not float until FRAME_ is de-asserted
        -----------------------------------------------*/
`ifdef check5
      property checkPCI_CBE_during_trx;
        @(posedge clk) disable iff (!reset_)
              $fell(FRAME_) |-> !($isunknown(C_BE_)) [*0:$] ##0 $rose(FRAME_);
      endproperty
      assert property (checkPCI_CBE_during_trx) else
$display($stime,,,"CHECK5:checkPCI_CBE_during_trx FAIL\n");
`endif
```

Fig. 24.9 (continued)

24.7 Further PCI Protocol Assertion Examples

Specification:
Once frame_ is de-asserted (high), that the last data phase is completed within 16 cycles. The last data phase is characterized when irdy_, trdy_, and devsel_ are de-asserted (high).

Solution:

```
property check_dataphase;
    @(posedge clk) $rose(frame_) |-> ##[1:16] ($rose (irdy_
    && trdy_ && devsel_));
endproperty
```

Specification:
The PCI Master is required to assert irdy_ (low) within 16 cycles after frame_ is asserted (low).

Solution:

```
`define true 1'b1;
property check_irdy;
    $fell (frame_) |-> `true[*1:16] intersect ($fell (frame_)
    ##[1:$] $fell(irdy_));
endproperty
```

In this property, we check for the sequence $fell(frame_) ##[1:$] $fell(irdy_) to occur (intersect) with `true[*1:16], meaning if the sequence does not occur within 16 clocks, the property will fail. Both the `true[*1:16] and ($fell (frame_) ##[1:$] $fell(irdy_)) start executing at the same time and end within 16 clocks. If irdy_ is asserted within 16 clocks, then both sequences end at the same time, else the property fails.

Specification:
Once PCI master indicates the start of a cycle, it must start a data transfer phase within 16 cycles.

Solution:

```
sequence start_PCI_Cycle;
    $fell (irdy_) and (!devsel_) and (!trdy_abort)
endsequence
sequence dataphase_begin;
    (!irdy_) [*0:16] ##0 $fell(trdy_);
endsequence
property check_dataphase;
    start_PCI_Cycle |-> dataphase_begin;
endproperty
assert property (check_dataphase);
```

As I've mentioned before, always dissect the specification to create distinct smaller sequences and then "connect" them in a property.

In this example, first we characterize the start of a PCI master cycle. This happens when irdy_ falls and at that time the target device is indeed selected (devsel_ low) and also that Target have not indicated an abort. This is modeled in the sequence "start_PCI_Cycle." Another sequence "dataphase_begin" checks to see that once the irdy_ is asserted that it remains asserted for 16 clocks until Target indicates the start of a data phase (trdy_ going low).

Then in the property "check_dataphase" we simply use the sequence "start_PCI_Cycle" as the antecedent to imply the consequent sequence "dataphase_begin." If the consequent does not hold, the property fails.

Specification:

Once PCI starts a Special Cycle (command/bye enable == 4'b 0001) that the parity error cannot be asserted (i.e., going low) until the Special cycle ends, no matter what's on the data bus.

Solution:

```
sequence PCI_Special_Cycle;
     $fell (frame_) && (cbe == 4'b001);
endsequence
sequence parity_error_check;
     perr[*1:$] ##0 ($rose (irdy && trdy));
endsequence
property
     PCI_Special_Cycle |-> parity_error_check;
endproperty
```

Here also, we divide the specification into two sequences. Sequence "PCI_Special_Cycle" detects the start of a special cycle and sequence "parity_error_check" checks to see that "perr" remains asserted until PCI cycle ends. This way the parity on the data bus is ignored.

Part II
System Verilog Functional Coverage (FC)

Chapter 25
Functional Coverage

Introduction: This is the introductory chapter on Functional Coverage which is a distinct language under the SystemVerilog umbrella. We will see different components of functional coverage and methodology.

A. B. Mehta, *SystemVerilog Assertions and Functional Coverage*,
https://doi.org/10.1007/978-3-030-24737-9_25

Ah, so you have done everything to check the design. But what have you done to check your test-bench? How do you know that your test-bench has indeed covered everything that needs to be covered? That's where Functional Coverage comes into picture. But first let us make sure we understand difference between the good old Code Coverage and the new Functional Coverage methodology.

Note that Functional Coverage and Constrained-random verification go hand in hand. You need to know the functional coverage gaps, and based on those gaps, constrain the input stimuli only to target those gaps.

(Mentor) The origin of functional coverage can be traced back to the 1990s with the emergence of constrained-random simulation. Obviously, one of the value propositions of constrained-random stimulus generation is that the simulation environment can automatically generate thousands of tests that would have normally required a significant amount of manual effort to create as directed tests. However, one of the problems with constrained-random stimulus generation is that you never know exactly what functionality has been tested without the tedious effort of examining waveforms after a simulation run. Hence, functional coverage was invented as a measurement to help determine exactly what functionality a simulation regression tested without the need for visual inspection of waveforms.

Today, the adoption of functional coverage is not limited to constrained-random simulation environments. In fact, functional coverage provides an automatic means for performing requirements tracing during simulation, which is often a critical step required for DO-254 compliance checking. For example, functional coverage can be implemented with a mechanism that links to specific requirements defined in a specification. Then, after a simulation run, it is possible to automatically measure which requirements were checked by a specific directed or constrained-random test—as well as automatically determine which requirements were never tested.

25.1 Difference Between Code Coverage and Functional Coverage

25.1.1 Code Coverage

Derived directly from the design code; not user specified. One of the advantages of code coverage is that it automatically describes the degree to which the source code of a program has been activated during testing, thus identifying structures in the source code that have not been activated during testing. One of the key benefits of code coverage, unlike functional coverage, is that creating the structural coverage model is an automatic process. Hence, integrating code coverage into your existing simulation flow is easy and does not require a change to either your current design or verification approach.

- Evaluates to see if design structure has been covered (i.e., line, toggle, assign, branch, expression, states and state transition)

- But does not evaluate the *intent* of the design
  - if the user specified busGnt = busReq && (idle || !(reset));
  - instead of the real *intent* busGnt = busReq && (idle **&&** !(reset));

  Code coverage won't catch it. For intent, you need both a robust test-bench to weed out functional bugs and a way to objectively predict how robust the test-bench is.

Here's a brief snapshot of what code coverage targets (structurally). You may refer to code coverage manuals from EDA vendors for further analysis. Here's a brief description from (Mentor) Mentor's Verification Academy Coverage Cookbook.

Line Coverage

Line coverage is a code coverage metric we use to identify which lines of our source code have been executed during simulation. A line coverage metric report will have a count associated with each line of source code indicating the total number of times the line has executed. The line execution count value is not only useful for identifying lines of source code that have never executed, but also useful when the engineer feels that a minimum line execution threshold is required to achieve sufficient testing.

Line coverage analysis will often reveal that a rare condition required to activate a line of code has not occurred due to missing input stimulus. Alternatively, line coverage analysis might reveal that the data and control flow of the source code prevented it either due to a bug in the code, or dead code that is not currently needed under certain IP configurations. For unused or dead code, you might choose to exclude or filter this code during the coverage recording and reporting steps, which allows you to focus only on the relevant code.

Statement Coverage

Statement coverage is a code coverage metric we use to identify which statements within our source code have been executed during simulation. In general, most engineers find that statement coverage analysis is more useful than line coverage since a statement often spans multiple lines of source code-or multiple statements can occur on a single line of source code.

A metrics report used for statement coverage analysis will have a count associated with each line of source code indicating the total number of times the statement has executed. This statement execution count value is not only useful for identifying lines of source code that have never executed, but also useful when the engineer feels that a minimum statement execution threshold is required to achieve sufficient testing.

Block Coverage

Block coverage is a variant on the statement coverage metric which identifies whether a block of code has been executed or not. A block is defined as a set of statements between conditional statements or within a procedural definition, the key point being that if the block is reached, all the lines within the block will be executed. This metric is used to avoid unscrupulous engineers from achieving a higher statement coverage by simply adding more statements to their code.

Branch Coverage

Branch coverage (also referred to as decision coverage) is a code coverage metric that reports whether Boolean expressions tested in control structures (such as the *if*, *case*, *while*, *repeat*, *forever*, *for* and *loop* statements) evaluated to both true and false. The entire Boolean expression is considered one true-or-false predicate regardless of whether it contains logical-and or logical-or operators.

Expression Coverage

Expression coverage (sometimes referred to as condition coverage) is a code coverage metric used to determine if each condition evaluated both to true and false. A condition is a Boolean operand that does not contain logical operators. Hence, expression coverage measures the Boolean conditions independently of each other.

Finite-State Machine Coverage

Today's code coverage tools can identify finite state machines within the RTL source code. Hence, this makes it possible to automatically extract FSM code coverage metrics to measure conditions. For example, the number of times each state of the state machine was entered, the number of times the FSM transitioned from one state to each of its neighboring states, and even sequential arc coverage to identify state visitation transitions.

25.1.2 Functional Coverage

The idea behind functional coverage is to see that we have covered the intent of the design. Have we covered everything that the design specification requires? For example, you may have 100% code coverage, but your functional coverage could only be 50%. Covering just the structure of the RTL code (code coverage) does not guarantee what that we have functionally covered what RTL actually intended to design. That being the case, it is obvious that the functional coverage matrices cannot be automatically created. One has to meticulously study the design specs and manually create functional coverage matrices using "covergroup," "coverpoint," "bins," etc.

So, functional coverage is,

- User specified; identify features of design specs that require functional coverage. A manual process.
- Based on design specification (as we have already seen with "cover" of an assertion).
- Measures coverage of design *intent.*
- Control-oriented coverage
  - Have I exercised all possible protocols that read cycle supports (burst, non-burst, etc.)?
  - *Transition* coverage

    Did we issue transactions that access Byte followed by Qword followed by multiple Qwords? (use SystemVerilog *transition* coverage).

    A Write to L2 is followed by a Read from the same address (and vice versa). Again, the *transition* coverage will help you determine if you have exercised this condition.
  - *Cross* coverage

    Tag and Data Errors must be injected at the same time (use SystemVerilog *cross* coverage).
- Data-oriented coverage
  - Have we accessed cache lines at all granularity levels (odd bytes, even bytes, word, quad-word, full cache line, etc.)?

25.2 Assertion Based Verification (ABV) and Functional Coverage (FC) Based Methodology

First let us examine the components of SystemVerilog language that contribute to Functional Coverage.

First component is the "cover" statement associated with an assertion This "cover" statement allows us to measure temporal domain functional coverage. Recall that "assert" checks for failures in your design and "cover" sees if the property did get exercised (i.e., got covered). Pure combinatorial coverage is not sufficient. What I call "low level" temporal domain conditions (aka structural Sequential

Conditions) such as every req should be followed by a gnt. If this assertion does not fail, it could be because the logic is correct or *because you never really asserted "req" to start with.* "cover" completes this story. We "cover" exactly the same property that we "assert." In the req/gnt example, if "cover" passes we know that the property did get exercised (i.e., it got covered) and that it did not fail.

Second component is the Functional Coverage language which is the gist of this entire section. Functional coverage allows you to specify the "function" you want to cover via the so-called *coverpoints* and *covergroups*. More importantly, it also allows you to measure *transition* as well as *cross* coverage to see that we have indeed covered finer details of our design. This section will clarify all this.

Figure 25.1 shows the different components of SystemVerilog as well as code coverage that all ties together to determine if a design have indeed been completely verified.

To reiterate, SystemVerilog provides two types of Coverage (*Code coverage is independent of SystemVerilog*). One is the "cover" property feature and the other is the Functional Coverage language features (covergroup, coverpoint, bins, etc.)

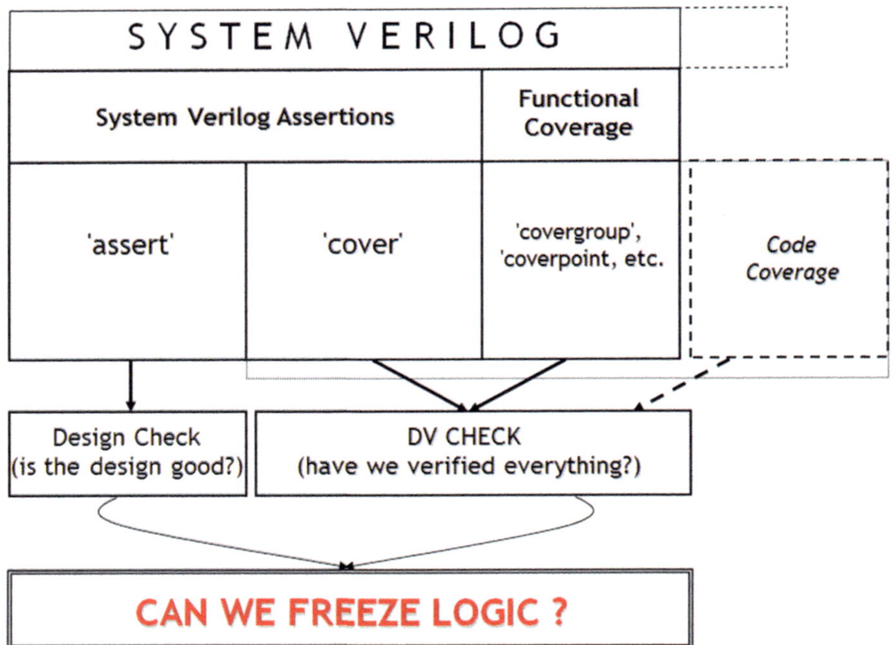

Fig. 25.1 Assertion based verification (ABV) and Functional Coverage (FC) based methodology

25.3 "cover" Property, "cover" Sequence

The first component of functional coverage methodology is "cover" (part of SystemVerilog Assertions language). As we saw in Chap. 3, "cover" uses SVA temporal syntax. Please refer to that section for detailed overview. Following is just a recap.

(a) "cover" is basically a shadow of "assert." In other words, you get double mileage, in that the same property can be used for both assertions and collecting functional coverage at structural level.
(b) "cover" is not accessible from SystemVerilog code.
(c) "cover" works only on structural design.
(d) "cover" provides structural level temporal domain sequential coverage. It can only be placed in modules, programs, and interfaces. Cannot be placed in a "class."
(e) There are two types of "cover." As described in Sects. 3.1 and 3.2, one is **cover property** and the other is **cover sequence.** cover property's body may contain an arbitrary property, as concurrent assertions and assumptions. On the other hand, cover sequence's body is limited to a sequence. See Sect. 6.16 for the difference between a sequence and a property.

25.4 "covergroup" (With Its "coverpoints," "bins," etc.)

The second component of the Functional Coverage methodology is based on covergroup, coverpoint, bins, etc. of the functional coverage language.

(a) Covergroups provides coverage of design variables. They record the number of occurrences of various values specified as coverpoints (of design variables).
(b) These coverpoints can be hierarchically referenced by your testcase or testbench so that you can query whether certain values or scenarios have occurred.
(c) Provides "cross" of coverpoints.
(d) Accessible by SystemVerilog Code and Testcases.
(e) Placeable in a "class-based objects" or structural code.

25.5 Functional Coverage Methodology

Here are some more points from project methodology point of view.

- Your test plan is (obviously) based on what functions you want to test (i.e., cover).
- So, create a Functional Cover Matrix based on your test plan that includes each of the functions (control and data) that you want to test.
  - Identify in this matrix all your functional covergroups/coverpoints (more on that coming soon).

  - Measure their coverage during verification/simulation process.
  - You may even automate updating the matrix directly from the coverage reports. That methodology is depicted in Fig. 25.2.

- Measure effectiveness of your tests from the coverage reports. To reiterate what we just discussed above since the following points are indeed the gist of what functional coverage allows you to accomplish.

  - For example, if your tests are accessing mostly 32-byte granules in your cache line, you will see byte, word, quadword coverage low or not covered. Change or add new tests to hit bytes/words, etc. Use constrained random methodology to narrow down the target of your tests. Constrained random is a very powerful methodology and goes hand in hand with Functional Coverage. Constrained random is beyond the scope of this book.
  - Or that the tests do not fire transactions that access Byte followed by Qword followed by multiple Qwords. Check this with *transition* coverage.
  - Or that Tag and Data Errors must be injected at the same time (*cross* coverage between Tag and Data Errors).

- "cover" temporal domain assertions.
- And add more *coverpoints* for critical functional paths through design.

  - For example, a Write to L2 is followed by a Read from the same address and that this happens from both processors in all possible write/read combinations.

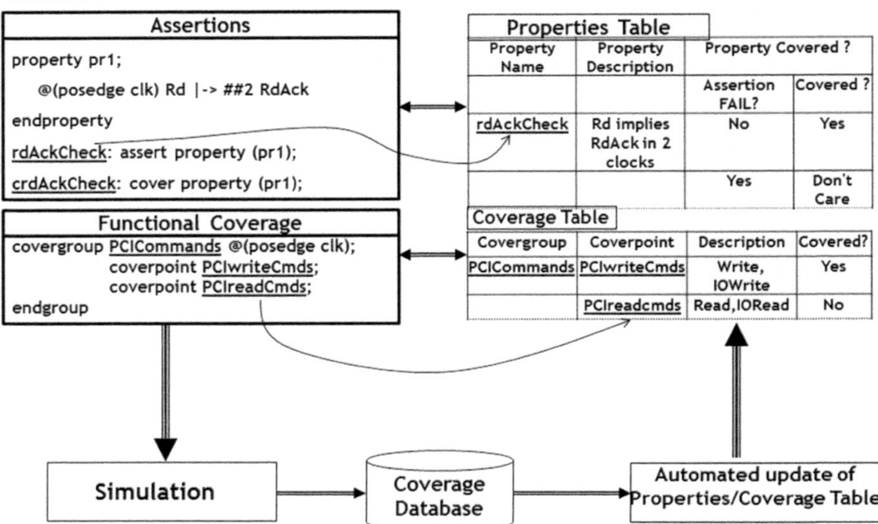

Fig. 25.2 Assertions and Coverage based verification methodology—I

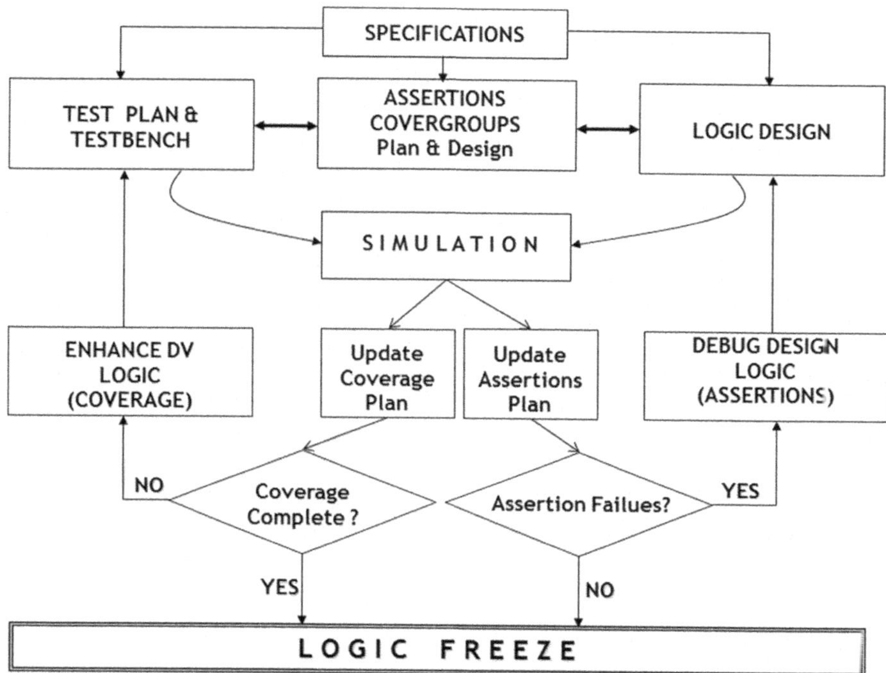

Fig. 25.3 Assertion and Functional Coverage based Verification methodology—II

- Remember to update your Functional Cover plan as verification progresses.
  - Just because you created a plan in the beginning of the project does not mean it's an end in itself.
  - As your knowledge of the design and its corner cases increase, so should the exhaustiveness of your test plan and the functional cover plan.
  - Continue to add *coverpoints* for any function that you didn't think of at the onset.

Figures 25.2 and 25.3 show an assertion and coverage driven methodology.

1. For every "assert" in a property, have an associated "cover." Give meaningful names to the property and assert Labels.
2. Create a Properties Table which automatically reads in your assertions and creates a FAIL/Covered matrix. If the assertion FAILs, well, fill in the FAIL column. If not and if gets covered, fill in the Covered column. How do we fill in this matrix? Read on...
3. Create a Functional Coverage plan with *covergroup* and *coverpoint*. Again, give meaningful names to covergroup and coverpoint(s).

4. Create a Coverage Table that automatically derives the covergroup/coverpoint names from step 3 and creates a matrix for "Covered" results. This matrix is for those functions that are not covered by assertion "cover" nor are they covered by code coverage. So, you need to carefully design your covergroups and coverpoints.
5. Simulate your design with assertions and functional cover groups.
6. Simulation will create a "coverage database." This database has all the information about failed assertions and "cover"ed properties and covered covergroups and coverpoints.
7. Using EDA vendor provided API, shift through this database and update the Properties Table and Coverage Table.
8. Loop back.

Advantage of such methodology is that you continually know if you are spinning the wheel without increasing coverage. Without such continual measure you may keep simulating; bugs don't get reported; you start feeling comfortable only to realize later that the functional coverage was really inadequate. You were basically running the tests that target the same logic over and over again. If you have a methodology as described above, you will have a correct notion of what functional logic to target to increase bug rate.

25.6 Follow the Bugs!!

- So, when do you *start* collecting coverage?
  - Code and Functional Coverage add to simulation overhead.
  - So, don't turn on code/functional coverage at the very "beginning" of the project.
  - But what does "beginning" of the project mean? When does the "beginning" end?
- That's where the bugs come into picture!
  - Create Bug Report charts
  - During the "beginning" time, bug rate will (should) be high. All low hanging fruits are being picked ☺
  - When the Bug Rate starts to drop; the "beginning" has come to an "end"
  - That's when your existing test strategy is running out of steam ☹
  - That's when you start code and functional coverage to determine

    If new tests are simply exercising the same logic repeatedly
    And which part of logic is not yet covered
  - Develop tests for the uncovered functionality. Use constrained random methodology.
  - Your Bug Rate will again go up (guaranteed! ☺).

Chapter 26
Functional Coverage: Language Features

Introduction: This chapter covers the entire "Functional Coverage" language. We will cover the following features in the upcoming sections.

1. covergroups and coverpoints for variables and expressions
2. automatic as well as user-defined coverage bins
3. "bins" for transition coverage
4. "wildcard bins," "illegal_bins," "ignore_bins"
5. Cross Coverage
6. Coverage Options
7. Flexible coverage sample—events, sequences, procedural
8. Directives to control and query coverage
9. SystemVerilog "class" based functional coverage
10. Application: Coverage Methods and procedural activation of coverage methods

A. B. Mehta, *SystemVerilog Assertions and Functional Coverage*,
https://doi.org/10.1007/978-3-030-24737-9_26

26.1 Covergroup/Coverpoint

26.1.1 What Is a Covergroup?

"covergroup" is a user-defined type that allows you to collectively sample all those variables/transitions/cross that are sampled at the same clock (sampling) edge.

- The "covergroup" construct encapsulates the specification of a coverage model.
- A "covergroup" can be defined in a "package," "module," a "program," an "interface," or a "class."
- Accessible by SystemVerilog Code and Testcases.

26.1.2 What Is a Coverpoint?

- A coverpoint is a variable or an expression that functionally covers design parameters (reg, logic, enum, etc.)
- Each coverpoint includes a set of bins associated with its sampled value or its value transition.
- The so-called "bins" can be defined by the user or created automatically by an EDA tool. A bin tells you the actual coverage measure.
- These coverpoints can be hierarchically referenced by your testcase or test-bench so that you can query whether certain values or scenarios have occurred.

Figure 26.1 gives a good basic overview of covergroup and coverpoint.

26.2 System Verilog "covergroup": Basics…

Figure 26.1 is self-explanatory with its annotations. Key syntax of the covergroup and coverpoint is pointed out. A few points to reiterate are as follows.

1. *covergroup* without a coverpoint is useless *and* the compiler won't give a Warning (at least the simulators that the author has tried).
2. *covergroup*, as the name suggests, is a group of coverpoints, meaning you can have multiple coverpoints in a covergroup.
3. You have to instantiate the covergroup with "new" just as you would for a class.
4. You may provide (not mandatory) a sampling edge (e.g., a clock edge) to determine when the *coverpoints* in a *covergroup* get sampled. If the clocking event is omitted, you must procedurally trigger the coverage sample window using a built-in method called sample ( ). We will discuss sample ( ) later in the chapter.
5. The sampling of covergroup takes place at the *instant* the sampling edge occurs. It does *not* take place in the preponed region as with "assert," "cover," "assume," "restrict" semantic of SystemVerilog Assertions.
6. A "covergroup" can specify an optional list of formal arguments (discussed later in the chapter). The actual values are provided when you instantiate the cover-

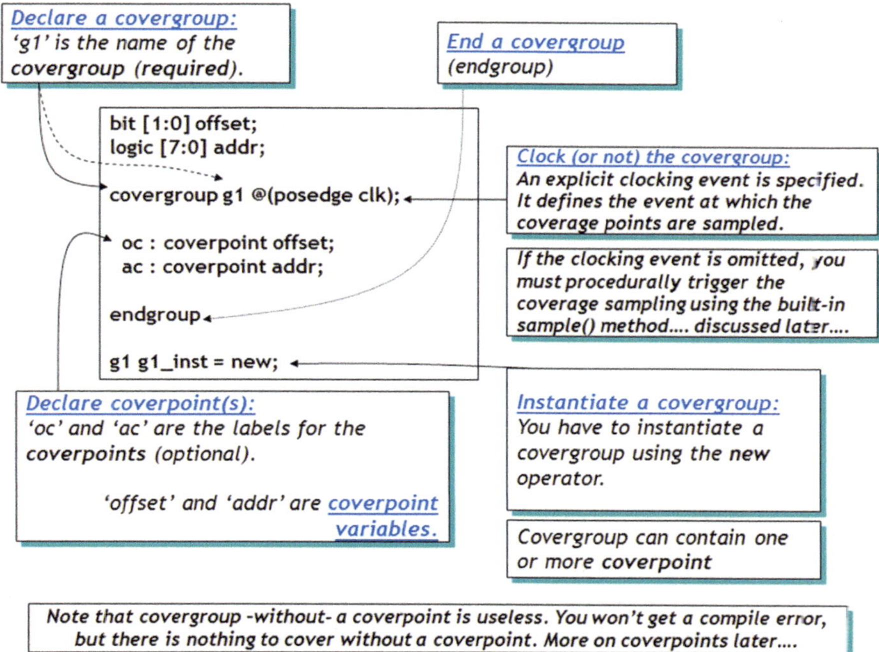

Fig. 26.1 "covergroup" and "coverpoint"—Basics

group (i.e., with the "new" operator). Actual arguments are evaluated when the "new" operator is executed. Note that the formal arguments are only "input" to the covergroup. An output or inout will result in an Error.

7. A "*covergroup*" can be declared in

   (a) package
   (b) interface
   (c) module
   (d) program
   (e) class

Other points are annotated in Fig. 26.1. Carefully study them so that the rest of the chapter is easier to digest.

26.3 SystemVerilog Coverpoint Basics…

Syntax:

Here's the formal syntax from the SystemVerilog LRM.

```
cover_point ::=
[ [ data_type_or_implicit ] cover_point_identifier : ] cover-
point expression [ iff ( expression ) ] bins_or_empty
```

Coverpoint and bins associated with the coverpoint do all the work. The syntax for coverpoint is as shown in Fig. 26.2. "covergroup g1" is sampled at (posedge clk). "oc" is the coverpoint name (or label). This is the name by which simulation log refers to this coverpoint. "oc" covers the 2-bit variable "offset."

We haven't yet covered "bins," so please hang on with the following description for a while. We will cover plenty of "bins" in the upcoming sections. So, in this example, you do not see any "bins" associated with the coverpoint "oc" for variable "offset." Since there are no bins to hold coverage results, simulator will create those for you. In this example, the simulator will create 4 bins because "offset" is a 2-bit variable. If "offset" were a 3-bit vector, there would be 8 bins and so on. We will discuss a lot more on "bins" in upcoming sections and hence I am showing only the so-called auto bins created by the simulator.

I haven't shown the entire test-bench but the simulation log (at the bottom of Fig. 26.2) shows that there are 4 auto-generated bins called "bin auto[0]" … "bin auto[3]." Each of these bins covers 1 value of "offset." For example, auto[0] bin covers "offset == 0." In other words, if "offset==0" has been simulated, then auto[0] will be considered covered. Again, this will become clearer when we go through basics of "bins." Since all 4 bins of coverpoint "oc" have been covered, the coverpoint "oc" is considered 100% covered, as shown in the simulation log. Now let us look at a real-life example of covergroup/coverpoint.

A data type for the coverpoint may also be specified explicitly or implicitly in *data_type_or_implicit*. In either case, it is understood that a data type is specified for the coverpoint. The data type must be an integral type. If a data type is specified, then a *cover_point_identifier* must also be specified.

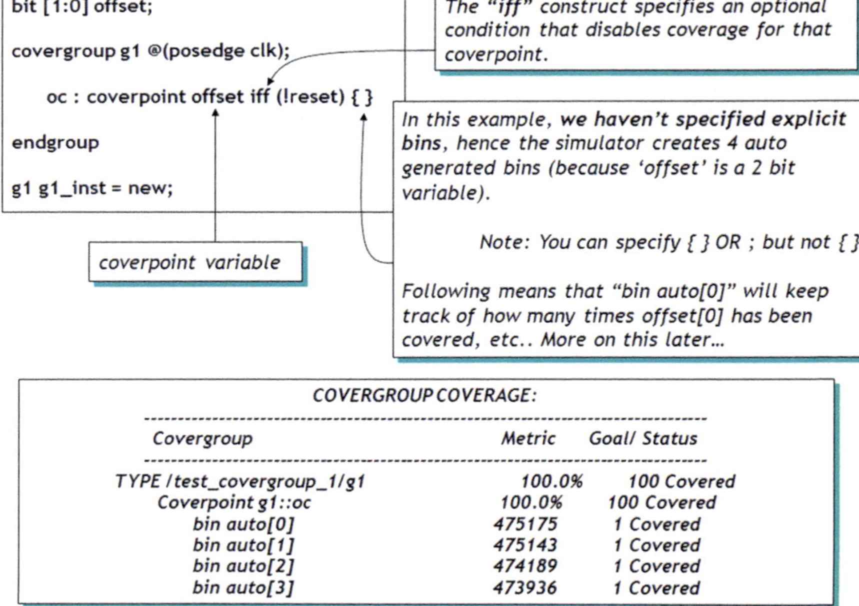

Fig. 26.2 "coverpoint"—Basics

If a data type is specified, then the coverpoint expression must be assignment compatible with the data type.

Values for the coverpoint will be of the specified data type and will be determined as though the coverpoint expression were assigned to a variable of the specified data type.

If no data type is specified, then the inferred data type for the coverpoint will be the self-determined type of the coverpoint expression.

The expression within the *iff* construct specifies an optional condition that disables coverage for that coverpoint.

If the guard expression evaluates to false at a sampling point, the coverage point is ignored.

```
covergroup AR;
      coverpoint s0 iff(!reset);
endgroup
```

In the preceding example, coverage point s0 is covered only if the value of "reset" is low.

26.4 Coverpoint Using a Function or an Expression

```
typedef enum {QUIET, BUSY} stateM;

function stateM cPoint_State_function(bit valid, ready);
      if (valid == 0) return QUIET;
      if ((valid & ~ready) == 0) return BUSY;
endfunction
...
covergroup cg @(posedge clk);
      cpStateM: coverpoint cPoint_State_function (valid, ready);
endgroup
cg cgInst = new(1'b0,1'b1);
```

In this example, we define a function "cPoint_State_function" of type stateM (which is a typedef enum). This function inputs "valid" and "ready." When "valid" == 0, we want to cover the state QUIET and when ((valid & ~ready) == 0), we want to cover the state BUSY.

In other words, this is an example of the number of "bins" to create *based on some condition*. There are two "bins" auto-generated by the simulator (since you did not specify explicit "bins" for the function and the function returns two explicit values (QUIET and BUSY) of the typedef enum "stateM"). And these two bins of the function will be covered based on the conditions provided in the function.

Here's an example of coverpoint of an expression:

```
bit cp_parity: coverpoint $countones(serial_word) ;
```

covers only those bits of "serial_word" that have been set. If the "serial_word" is 8-bits wide, the coverpoint will create 8 "bins"—one for each set bit.

One more example:

```
cExample: coverpoint (my_variable+your_variable) % 4 ;
```

covers the bits created by a *mod* 4 of (my_variable+your_variable).

26.5 Coverpoint: Other Nuances

You can also use part-select for a coverpoint.

```
cExample: coverpoint(address[31:16]);
```

You can cover a Ref variable as well.

```
covergroup (ref int ref_addr) cGroup;
     cPoint: coverpoint ref_addr;
endgroup
```

You can also filter coverpoint based on a specific condition, as in,

```
covergroup cGroup;
     cPoint: coverpoint cp_addr iff (!reset);
endgroup
```

26.6 Covergroup/Coverpoint Example…

PCI protocol consists of many different types of bus cycles. We want to make sure that we have covered each type of cycle. The enum type pciCommands (Fig. 26.3) describes the cycle types. The covergroup specifies the *correct* sampling edge (that being the "negedge FRAME_") for the PCI Commands. In other words, the sampling edge is quite important for performance reasons. If you sampled the same covergroup @ (posedge clk), there will be a lot of overhead because FRAME_ will fall only when a PCI cycle is to start. Sampling unnecessarily will indeed affect simulation performance.

In this example, we have not specified any bins. So, the simulator creates 12 auto bins for the 12 bus cycles types in the enum. Every time FRAME_ falls that the

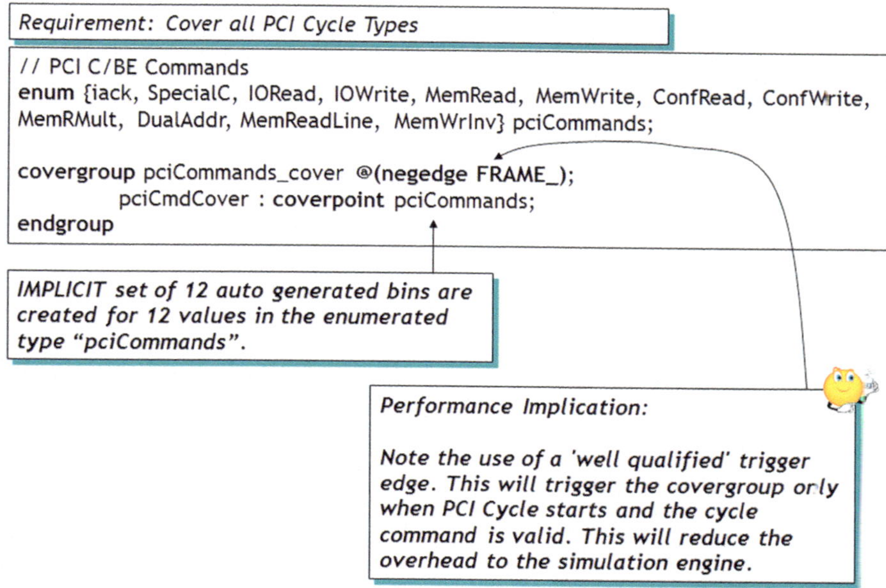

Fig. 26.3 "covergroup"/"coverpoint" Example

simulator will see if any of the cycle types in "enum pciCommands" is simulated. Each of the 12 auto bin corresponds to each of the enum type. When a cycle type is exercised, the auto bin corresponding to that cycle (i.e., that "enum") will be considered covered. Now, you may ask why wouldn't code coverage cover this. Code coverage will indeed do the job. But I am building a small story around this example. We'll see how this covergroup will be then reused for transition coverage which *cannot* be covered by code coverage.

The following example shows a very simple but complete test-bench with usage of covergroup/coverpoint. I am not showing the usage of "bins" in this example since we haven't covered that yet.

```
module testCover;

bit [3:0] CycleType;
bit[7:0] CPUMode;

bit clk;
real coverPercent;
event stimulusDone;

covergroup coverg @ (posedge clk);
     CycleTypeCP: coverpoint CycleType;
```

```
        CPUMode1CP: coverpoint CPUMode[7:4];
        CPUMode2CP: coverpoint CPUMode[3:0];
endgroup

coverg covergInst = new( );

always #10 clk=~clk;

initial
begin
        clk = 1'b0;
        for (int j=0; j<5; j++)
        begin
                @(negedge clk);
                CycleType = $random;
                CPUMode = $random;
                $display ($stime,,," CycleType = 0x%0h CPUMode =
                0x%0h", CycleType, CPUMode);
        end
       -> stimulusDone;
end

initial
begin
        @(stimulusDone); @(posedge clk);
        coverPercent = coverageInst.get_coverage( );     //get
        coverage on  covergroup 'coverg'
        $display("\nCoverage for covergroup 'coverg' = %f% ",
        coverPercent);
$finish(2);
end
```

Simulation Log:

```
20 CycleType = 0xa CPUMode = 0x5
40 CycleType = 0x3 CPUMode = 0xb
60 CycleType = 0x6 CPUMode = 0x2
80 CycleType = 0x1 CPUMode = 0x4
100 CycleType = 0xf CPUMode = 0x7

Coverage for covergroup 'coverg' = 30.0%
```

In module "testCover" we define a simple covergroup called "coverg." This covergroup contains 3 coverpoints. One for CycleType and two for CPUMode. We run a "for" loop and stimulate CycleType and CPUMode with $random numbers. We stimulate @(negedge) clk (for 5 clocks) and trigger a named event called "stimulusDone."

In another "initial" block we simply wait for trigger of event "stimulusDone" and query the percentage (real number) coverage of "coverg" covergroup using the built-in function get_coverage ( ). The simulation log explains the rest. We will cover instance specific options, covergroup specific options, and built-in functions later in Sect. 28.1.

26.7 System Verilog "bins": Basics…

What's a "bin"? A "bin" is something that collects coverage information (collect in a "bin"). bins are created for coverpoints. A coverpoint that is covering a variable (let's say the 8 bit "adr" as shown in Fig. 26.4) and would like to have different values of that variable be collected (covered) in different collecting entities, the "bins" will be those entities. "bins" allows you to organize the coverpoint sample (or transition) values.

A coverpoint includes a set of explicit or implicit bins that allow you to organize the coverpoints sample (or transition) values.

bins BinName = {Range in coverpoint variable};

```
bit[7:0] adr;
covergroup g1 @(posedge clk);

  ac: coverpoint adr iff (!reset)
{
   bins adrbin1 = {[0:3]};

   bins adrbin2 [] = {[4:5]};

   bins adrbin3 [2] = {[6:8]};

   bins adrbin4 [3] = {[9:10]};

   bins adrbin5 [] = {[9:12],[11:16]};

   bins heretoend = {[31:$]};

   bins others = default;
}
endgroup
g1 g1_inst = new;
```

SINGLE bin to cover adr values 0,1,2,3

INDIVIDUAL bins adrbin2[1] and adrbin2[2] for EACH adr value (4,5) ([] implies auto-generated by SV).

Fixed array of bins (here with # of bins less than range value :: 'adr' value 6 is in bin 'adrbin3[1]' while 'adrbin3[2]' contains 7,8

Fixed array of bins (here with # of bins greater than range value :: 'adr' value 9 will be covered in bin 'adrbin4[1]'; 'adr' value 10 in 'adrbin4[2]' while 'adrbin4[3]' will remain empty.

Overlapping values OK. Will create an array of 8 bins for 'adr' values from 9 to 16

From 31 to eternity ☺

and everything else that's not covered (for the coverpoint 'adr') by bins above, ends up in bins 'others'

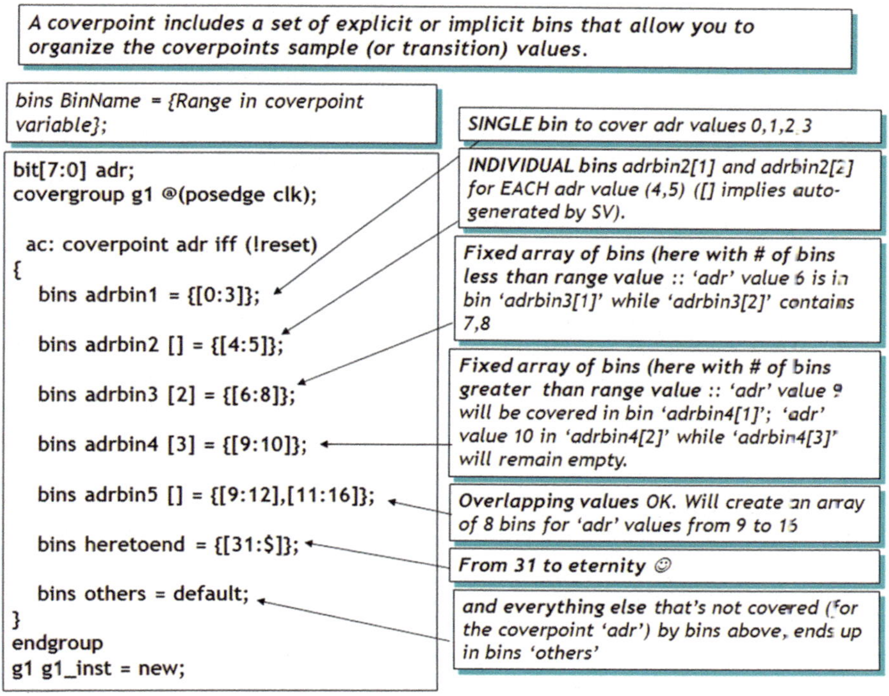

Fig. 26.4 "bins"–Basics

You can declare bins many different ways for a coverpoint. Recall that bins collect coverage. From that point of view, you have to carefully choose the declaration of your bins.

OK, here's the most important point *very* easy to misunderstand. In the following statement, how many bins will be created? 16 or 4 or 1 and what will it cover?

```
bins adrbin1 = {[0:3]};
```

Answer: 1 bin will be created to cover "adr" values equal to "0" or "1" or "2" or "3."

Note that "bins adrbin1" is without the [ ] brackets. In other words, "bins adrbin1" will *not* auto-create 4 bins for "adr" values {[0:3]}, it will rather create *only 1 bin* to cover "adr" values "0," " 1," "2," "3."

Very important point: Do not confuse {[0:3]} to mean that you are asking the bin to collect coverage for adr0 to adr15. {[0:3]} literally means "adr" value =0, =1, =2, =3. Not intuitive, but that's what it is!

Another important point. What "bins adrbin1 = {[0:3]};" also says is that if we hit *either* of the "adr" value ("0," "1," "2," or "3") that the single bin will be considered *completely* covered. Again, you don't have to cover all four values to have "bins adrbin1" considered covered. You hit any one of those 4 values and the "adrbin1" will be considered 100% covered.

But what if you want each value of the variable "adr" be collected in separate bins so that you can indeed see if each value of "adr" is covered explicitly. That's where "bins adrbin2[ ] ={[4:5]};" comes into picture. Here "[ ]" tells the simulator to create two explicit bins called adrbin2[1] and adrbin2[2] each covering the 2 "adr" values =4 and =5. adrbin2[1] will be considered covered if you exercised adr==4 and adrbin2[2] will be considered covered if adr==5 is exercised.

Other ways of creating bins are described in Fig. 26.4 with annotation to describe the nuances. Note that you can have "less" or "more" # of bins than the "adr" values on the RHS of a bins assignment. How will "bins" be allocated in such cases is explained in the figure. Note also the case {[31:$]} called "bins heretoend." What does "$" mean in this case? It means [32:255] since "adr" is an eight-bit variable.

Rest of the semantics is well described with annotation in the figure. Do study them carefully, since they will be very helpful when you start designing your strategy to create "bins."

Here's an example of how "bins" are created. Takes a bit of intuition to understand its semantics.

```
bins Dbins [4] = { [1:10], 11,12,14};
```

Let's see how this works. First, 4 bins are created since we are explicitly asking for 4 bins (Dbins[4]). Next, how many values are present on the right-hand side? 13. That is values 1 to 10, value 11, value 12, and value 14. Now, since there are only 4 bins and 13 values, how will these values be spread out among 4 bins?

```
Dbins[1] will be considered covered when we hit any of the
values 1,2,3
Dbins[2] will be considered covered when we hit any of the
values 4,5,6
Dbins[3] will be considered covered when we hit any of the
values 7,8,9
Dbins[4] will be considered covered when we hit any of the
values 10,11,12,14
```

Study this carefully so that you are sure how bins are created and how they get covered with different values.

Now, here's an example of how you may end up making mistakes and get Warnings from a simulator (courtesy—IEEE SystemVerilog LRM) (LRM, 2012)

```
bit [2:0] p1;   //values 0 to 7
bit signed [2:0] p2;   //values -4 to 3

covergroup g1 @(posedge clk);
coverpoint p1 {
bins b1 = { 1, [2:5], [6:10] };
bins b2 = { -1, [1:10], 15 };
}
coverpoint p2 {
bins b3 = {1, [2:5], [6:10] };
bins b4 = { -1, [1:10], 15 };
}
endgroup
```

Exercise: See if you can figure out why the following Warnings are issued? Note that p2 is "signed."

- For b1, a warning is issued for the range [6:10]. b1 is treated as though it had the specification {1, [2:5], [6:7]}.
- For b2, a warning is issued for the range [1:10] and for the values –1 and 15. b2 is treated as though it had the specification { [1:7] }.
- For b3, a warning is issued for the ranges [2:5] and [6:10]. b3 is treated as though it had the specification { 1, [2:3] }.
- For b4, a warning is issued for the range [1:10] and for the value 15. b2 is treated as though it had the specification { -1, [1:3] }.

26.8 "bins" with Expressions

Note also that the coverpoint and/or "bins" is not restricted to a simple variable or signal name. It can also cover any expression.

In the following example only the count of bits set in a vector is considered. Note that both the coverpoint and the bins are using an expression. The coverpoint covers those bits of "serial_word" that have been set. The bins "set_bins_count" will create explicit set of bins in the range from 0 to the whatever the range that comes out of "$countones(serial_word)."

Let us say that the serial_word is 2 bits wide and its value is 2'b10, then the "bins set_bits_count [ ] " will have the range [0:2'b10]. This means that 3 bins will be created to cover the values 00,01,10.

```
bit cp_parity: coverpoint $countones(serial_word)
{
bins set_bits_count[ ] = {[0:$bits(serial_word)]};
}
```

26.9 Bin Filtering Using the "with" Clause

Following shows how the "with" clause comes in handy to further restrict the creation of bins.

The "with" clause specifies that only those values that satisfy the given expression are included in the bins. In the expression, the name "*myValue*" is used to represent the candidate value. The candidate value is of the same type as the coverpoint.

Consider the following example:

```
int myValue, x;
a: coverpoint x
{
bins mod6[ ] = {[0:255]} with (myValue % 6 == 0);
}
```

This bin definition selects all values from 0 to 255 that are evenly divisible by 6. One more:

```
a: coverpoint x
{
bins mod3[ ] = {[0:255]} with ((myValue % 2) || (yourValue % 3)
|| no_value);
}
```

One more:

```
a: coverpoint x
{
bins func[ ] = x with (myValue %6 == 0) iff (!reset);
}
```

In this example, note the use of the coverpoint name "*x*" to denote that the *with (expression)* will be applied to all values of the coverpoint. Yes, you can use the name of the coverpoint itself in the bins definition with the "with" clause. And also note the use of "iff (!reset)" as a condition to the bins.

The "with" clause is actually a SystemVerilog construct (e.g., its usage in constraint random verification). Please refer to SystemVerilog LRM to understand its further nuances.

In short, filtering expressions using "with" clause provides a powerful and convenient new way to define the bins of your coverpoint. They are especially useful for cross points, where the language rules specify that bins are automatically created for any values that are not otherwise mentioned in the cross-point definition. This means that, in all realistic situations, it is essential for your cross-point definition to specify bins (some of which may be "ignore_bins" or "illegal_bins") that cover every possible combination of values. If you do not do this, you will find the cross contains unwanted auto-generated bins that will distort your coverage figures. In a regular coverpoint it is possible to use the "default" specification to create an ignored bin that collects all otherwise unspecified values of the coverpoint, but there is no such "default" bin specification for cross points.

Following is another example. Without the "with" clause you will get Warnings for overlapping bins. The "with" clause comes to rescue.

Problem Statement:

```
covergroup burstSize (int max, bSize);
cPburst_size : coverpoint bSize
{
bins small = {1};
bins mid = { [2 : max - 1] };
bins large = { max };
}
endgroup
burstSize  bSizeInst = new(2, 8);
      Warnings:
      If 'max' is 2; 'bins small' 'bins mid', 'bins large' overlap
      If 'max' is 1: all three will again overlap
      If 'max' is 0: that's illegal because 'bins mid' = [2 : -1]
```

In this example, we are covering a variable named "bSize" which is a parametrized argument. For this "bSize" coverpoint, we create 3 bins with certain values.

```
bins small = {1}; will create a bin for the value '1' of bSize.
bins mid = { [2:max-1]}; will create a bin for values 2 to
max-1
bins large = {max}; will create a bin for the 'max' value
```

So, what's wrong with that?
Take for example, max= 2. This will cause the bins to have following values.

```
bins small = {1};
bins mid = { [2:1] };
bins large = {2};
```

As you notice, all three small, mid, and large overlap with each other. This will cause compiler to give a Warning.

Now assume, max = 1. So, the bins will look like the following

```
bins small = {1};
bins mid = { [2:0] };
bins large = {1};
```

Again, all 3 bins overlap, and a Warning will be issued.

Finally, assume max=0;

```
bins small = {1};
bins mid = { [2:-1] };
bins large = {0};
```

Well, the "bins mid" now has a negative value. So, another Warning.

We need a mechanism to avoid such overlap. That's where the "with" clause comes to rescue.

Solution:

```
covergroup burstSize (int max, bSize);
cPburst_size ; coverpoint bSize)
{
bins small = {1};
bins mid = { [2 : max - 1] }  with (max >= 3);
bins large = { max } with (max > 3);
}
endgroup
burstSize bSizeInst [4, 8];
 // None of the bins will overlap
```

The same code but this time "bins large" defined using the "with" clause. With this code the bins won't overlap. Here's the analysis.

Take for example, max= 4. This will cause the bins to have following values (see, no overlap).

```
bins small = {1};
bins mid = { [2:3] };
bins large = {4};
```

As you notice, all three small, mid, and large do not overlap with each other.

The other values of "max" (3,2,1) won't create "bins mid" and "bins large" at all. So, no question of overlap.

I agree, this is a trivial example. But it gives you an idea of how to effectively use the "with" clause for bins filtering.

26.10 Covergroup/Coverpoint with bins: Example…

Recall the example on PCI that we started in the previous section. Its story continues here.

In Fig. 26.5 we are assigning different groups of PCI commands to different bins.

Recall that [ ] means auto-generated bins. Hence, for example, since "bins pcireads[ ]" in Fig. 26.5 has 5 variables (enum type in this example), 5 bins will be created—one for each enum type. This way of creating bins is very useful because

Requirement: Cover all PCI Cycle Types.

```
// PCI C/BE Commands
enum {iack, SpecialC, IORead, IOWrite, MemRead, MemWrite, ConfRead, ConfWrite,
MemRMult, DualAddr, MemReadLine, MemWrInv} pciCommands;

covergroup pciCommands_cover @(negedge FRAME_);

        pciCmdCover : coverpoint pciCommands

        {
        bins pcireads []={IORead, MemRead, ConfRead, MemRMult, MemReadLine};
        bins pciwrites [] = {IOWrite, MemWrite, ConfWrite, MemWrInv};
        bins pcimisc [] = {iack, SpecialC};
        }

endgroup
```

EXPLICIT bins to categorize PCI cycles in different bins. So, for example, when pcireads bins are 100% covered, we know that all PCI Read type cycles have been exercised.

Fig. 26.5 "covergroup"/"coverpoint" example with "bins"

in the complex maze of features to cover, it is sure nice to have a clear distinction among different functional groups. Here pcireads[ ] (5 bins) covers all PCI Read Cycles; pciwrites[ ](4 bins) covers all PCI Write Cycles and for the special cycles, there is the pcimisc[ ] (2 bins).

26.11 "covergroup": Formal and Actual Arguments

Figure 26.6 outlines the following points.

1. Covergroup can be parameterized for reuse
2. "Ref" type is required when you pass variables as actual to a formal. In other words, if you were passing a constant you would *not* need a "Ref" type as shown in Fig. 26.6. Note that a clocking event can be based on "ref" arguments of the covergroup. Finally, you cannot pass an automatic variable by "ref"rence.
3. Actual arguments are passed when you instantiate a covergroup. When the covergroup specifies a list of formal arguments, its instances shall provide to the new operator all the actual arguments that are not defaulted. Actual arguments are evaluated when the new operator is executed.
4. This is an example of *reusability*. Instead of creating two covergroups; 1 for "adr1" and another for "adr2," we have created only 1 covergroup called "gc" which has a formal called "address." We pass "adr1" to "address" in the instance gcadr1 and pass "adr2" to "address" in the instance gcadr2. We also pass the range of adr1 and adr2 to be covered with each instance. In short, it is a good idea to create parameterizable covergroups, as the situation permits. They can be useful not only within a project but also across projects.

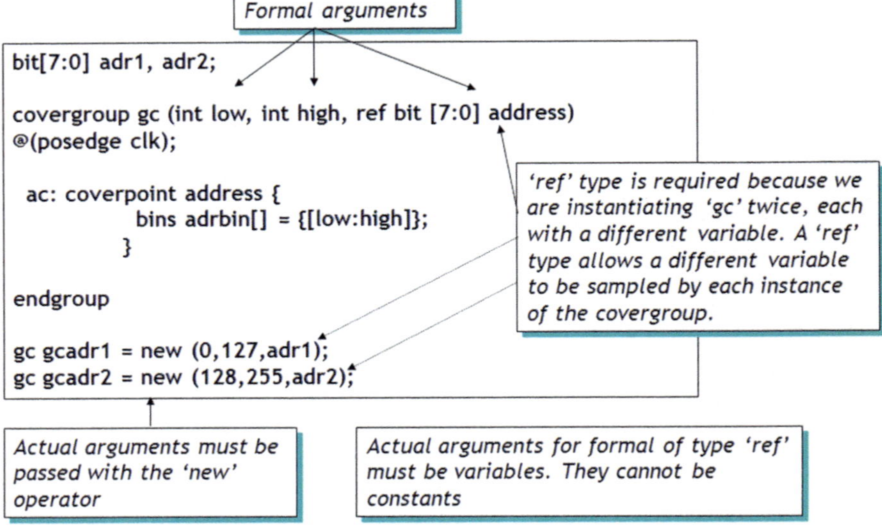

Fig. 26.6 "covergroup"—formal and actual arguments

5. An output or inout is illegal as a formal argument.
6. Since a covergroup cannot modify any argument to the new operator, a ref argument will be treated the same as a read-only const ref argument.
7. The formal arguments of a covergroup cannot be accessed using a hierarchical name (the formals cannot be accessed outside the covergroup declaration).
8. If an automatic variable is passed by reference, behavior is undefined.

Exercise: How many bins will be created for “bins adrbin[ ]” for each instance (“gcadr1” and “gcadr2”) of covergroup “gc”?

26.12 Coverpoint: Hierarchical References

Another useful feature is the ability of a coverpoint to reference variables hierarchically. See the following example.

```
covergroup gc @(posedge clk);
     coverpoint top.test.count;
     coverpoint top.adrGen.address;
endgroup
```

You can also pass references as arguments, as shown below.

```
covergroup gc (ref logic [7:0] address, ref int data);
     coverpoint a;
     coverpoint b;
endgroup

gc gc_inst = new(testbench.a, testbench.data); //Hierarchical
references
```

However, you cannot reference another coverpoint hierarchically, as shown in example below.

```
covergroup gc @(posedge clk);
     coverpoint top.test.covergroup_inst.cPa; //ILLEGAL
     //CANNOT reference 'coverpoint cPa' in the hierarchical
     covergroup instance
     // 'covergroup_inst'.
endgroup
```

IMPORTANT NOTE: Covergroups can contain coverpoints to hierarchical references.

In general, when we start using hardcoded hierarchical references, our covergroup (and consequently, our test-bench) is not as flexible or reusable as it could be. Instead, we could define arguments to our covergroup and then pass hierarchical references into the covergroup when it is instantiated. The instantiation could be done in a testcase or elsewhere so that now the covergroup is much more flexible. I highly recommend this approach.

26.13 Class: Embedded "covergroup" in a "class"

So where do you use or declare this "covergroup"? One of the best places to embed a coverage group is within a "class." Why a class? Here are some reasons. (Note—discussion of "class" is beyond the scope of this book. The author is assuming familiarity with SystemVerilog "class.")

- An embedded covergroup can define a coverage model for protected and local class properties without any changes to the class data encapsulation.
- Class members can be used in coverpoint expressions, coverage constructs, option initialization (we'll see option initialization in Chap. 28), etc.
- By embedding a coverage group within a class definition, the covergroup provides a simple way to cover a subset of the class properties.

This style of declaring covergroups allow for modular verification environment development.OK, let us see what Fig. 26.7 depicts.

```
class xyz;
  bit [3:0] m_x;
  int m_y;
  bit m_z;

  covergroup xyzCover @(m_z);
          coverpoint m_x;
          coverpoint m_y;
  endgroup

  function new();
    xyzCover xyzCovInst = new;
  endfunction
endclass
```

• *By embedding a 'covergroup' within a class definition, the 'covergroup' provides a simple way to cover as part of class definition the class properties (modular development)*

• *A 'class' can have more than one 'covergroup'*

Fig. 26.7 "covergroup" in a SystemVerilog class (courtesy LRM 1800-2005)

"covergroup xyzCover" is sampled on any change on variable "m_z." This covergroup contains two coverpoints, namely "m_x" and "m_y." Note that there are no explicit bins specified for the coverpoints. How many bins for each coverpoint will be created? As an exercise, please refer to previous sections to figure out.

Note that covergroup is instantiated within the "class." That makes sense since the covergroup is embedded within the class. Obviously, if you do not instantiate a covergroup in the "class," it will not be created and there will not be any sampling of data.

26.14 Class: Embedded covergroup: Hierarchical Accessibility

Here's a simple example of accessing a variable (or other class properties) in a class hierarchy.

```
class HLp;
    bit m_ev;
endclass

class Mclass;
    HLp m_obj;
    int m_a;

    covergroup Cov @(m_obj.m_ev);
        coverpoint m_a;
    endgroup

function new( );
    m_obj = new; // coverage group 'Cov' uses m_obj. So, m_obj
    must be constructed before 'Cov'
    Cov = new;
endfunction

endclass: HLp
```

In this example, "class HLp" declares an "bit m_ev" and "class Mclass" instantiates HLp and uses the class handle m_obj to access m_ev from "class HLp." "covergroup Cov" is embedded within "class Mclass." The clocking event for the embedded coverage group refers to data member m_ev of m_obj. The important point to note here that because the "covergroup Cov" uses m_obj, *m_obj must be instantiated before Cov in the class constructor.*

In previous sections we saw how to parameterize a covergroup. The following example takes that a bit further for a covergroup embedded in a class. In the following example, in the class constructor (function new) we declare two actuals, namely "int ha, int hb." We then pass these arguments to the construction of myCov (myCov = new (ha,hb)). This way when you instantiate the class (myC myCInst), you can pass arguments (formal) to the class instantiation which in turn passes these formal values to covergroup myCov.

```
class myC;
int address;

covergroup myCov (int high, int low) @(posedge clk);
 ac: coverpoint address {
                    bins adrbin[ ] = {[high:low]};
                }
endgroup

function new (int ha, int hb);
     myCov = new (ha,hb);
endfunction

endclass

initial begin
     myC myCInst = new(7,0);
end
```

26.15 Class: Multiple Covergroups in a Class

A "class" can indeed have more than one "covergroup" as shown below.

```
class multi;
bit [3:0] m_1;
int m_2;
bit m_z, m_a;

 covergroup m_xCover @(m_z) coverpoint m_1; endgroup
 covergroup m_yCover @(m_a) coverpoint m_2; endgroup

  function new ( );
     m_xCover m_x_CovInst = new;
      m_yCover m_y_CovInst = new;
  endfunction

endclass : multi
```

26.16 Class: Overriding Covergroups in a Class

A covergroup defined in a class can indeed be overridden in an extended (child) class. Here's an example of that.

First, let us define a "class parent" and declare a "covergroup pCov." Note that the "bins parentbins [ ]" creates 256 bins for the "byte pByte."

```
class parent;
rand byte pByte;

    covergroup pCov;
        coverpoint pByte
        { bins parentBins [ ] = {[0:255]}; } //256 bins for
        256 values of pByte.
    endgroup

    function new ( );
        pCov = new;
    endfunction

endclass
```

Next, define a "class child" which extends the "class parent." In this class, we redefine (i.e., override) "covergroup pCov" of "class parent." Note that the "bins childBinsAllinOne" creates only 1 bin for all values of the byte pByte. In other words, if any of the values of pByte is hit, the bin will be covered 100%.

```
class child extends parent;

    covergroup pCov;  //'class child' overrides 'covergroup
    pCov' of 'class parent'
        coverpoint pByte
        { bins childBinsAllinOne = {[0:255]}; } //ONE bin
        for all values
    endgroup

    function new ( );
        super.new( );
        pCov = new;
    endfunction

endclass
```

Let us verify this code with a simple test-bench.

```
module testOverride;
parent p1;
child c1;

      initial begin
      p1 = new( );
      c1 = new ( );

            for (int i=0;  i < 10; i++) begin
                  p1.randomize( );
                  c1.randomize ( );

                  p1.pCov.sample( );
                  c1.pCov.sample( );
                  #1;
            end

      $display("p1.pCov Instance Coverage = %0.2f %%", p1.pCov.
      get_inst_coverage( ) );
      $display("c1.pCov Instance Coverage = %0.2f %%", c1.pCov.
      get_inst_coverage( ) );

      end
```

Simulation Log:

```
p1.pCov Instance Coverage = 25 %
c1.pCov Instance Coverage = 100 %
```

Here's the interpretation of the simulation log. "class parent" has "**bins** parentBins [ ] = {[0:255]};" which means you need to hit all 256 values of pByte to cover all 256 bins. With our randomization of pByte from the "for loop," we hit only 25% of the 256 values and hence the instance coverage for "p1.pCov" is 25%.

In "class child," we overrode the "covergroup pCov" and used "**bins** childBinsAllinOne = {[0:255]};" to declare a single bin that covers all the 256 values of "pCov." In other words, our randomization hit at least one value of "pCov" and the "c1.pCov" instance was 100% covered.

This proves that we can indeed override a covergroup definition in an extended class.

What if you want to shape the coverage bins as the covergroup is constructed (instantiated)?

You will have greater flexibility by adding constructor arguments to the covergroup (parameterized covergroups). These arguments are then populated at the moment the covergroup is constructed and are used at that time to configure the covergroup's bins. In this way, values computed at runtime can be used to shape coverage bins. Here's an example,

```
class configurable_cGroup_class;
int x;
covergroup construct_bins_cg(input int min, max, num_bins);
coverpoint x {
      bins lowest = {min};
      bins highest = {max};
      bins middle[num_bins] = {[min+1:max-1]};
}
endgroup

function new(int min, int width);
      int max = min + width - 1;
      int n_bins = (width - 2)/4 ; // create 2 bins
      construct_bins_cg = new(min, max, n_bins);
endfunction
endclass
configurable_cGroup_class cfInst = new(4,10);
```

26.17 Class: Parameterizing Coverpoints

This section describes a simple way of parameterizing coverpoints through a parameterized class. Note that these parameters do not affect the way the component fits into its enclosing environment. They affect only the way its covergroup is built and used. This class can then be instantiated with different LBound and HBound parameters that will affect how coverpoint bins are created.

```
class useParams #(parameter int LBound=1, HBound=16);
bit [15:0]  addr;
bit [7:0] data;
covergroup cGroup;
      caddr: coverpoint addr {
      bins b1 [ ] = {LBound : HBound};
      illegal_bins i1 = {LBound-1:HBound-8};
      }
      cdata: coverpoint data;
endgroup
```

```
//constructor
function new (string name, uvm_component parent = null);
      super.new (name, parent);
      cGroup = new;
endfunction
......
endclass

useParams useP1 = new (32,64);
```

Here's another example in the same line of thought.

```
class configurable_cGroup_class;
int x;
covergroup construct_bins_cg (input int min, max, num_bins);
      coverpoint x {
        bins lowest = {min};
        bins highest = {max};
        bins middle[num_bins] = {[min+1:max-1]};
      }
endgroup
function new(int min, int width);
      int max = min + width - 1;
      int n_bins = (width - 2)/4 ; // aim for 4 values per bin
      construct_bins_cg = new(min, max, n_bins);
endfunction
endclass
configurable_cGroup_class  cgClass = new(16, 32);
```

26.18 Class: Creating Array of Instances of a "covergroup"

This is a powerful feature. You may need to create an array of covergroup instances in a class. For example, you want a different coverage report for each instance of the covergroup. Each instance requires a coverage log report of its own. But there is a caveat on *where* you create/define a covergroup so that you can create an array of instances of a covergroup from within a class.

First let us look at the incorrect way to create an array of covergroup instances. *Incorrect because this is a limitation of SystemVerilog*. I'll explain once I show you an example.

INCORRECT way of declaring a covergroup to create an array of its instances.

```
class ex_class extends uvm_component;
int myInt[10];
covergroup myCG (@posedge clk); //covergroup embedded in the
class 'ex_class'
      coverpoint myInt
      {
      bins myInt_range_1 = {124:0]};
      bins myInt_range_2 = {325:125]};
      }
endgroup: myCG
function new (string name, uvm_component parent=null); //class
constructor (UVM style)
super.new(name , parent);
      foreach (myInt[ i ])
            myCG myCG[ i ]=new(myInt[ i ]);     //ERROR
endfunction
endclass: ex_class
```

The covergroup "myCG" is declared *inside* the class "ex_class." This looks very benign, but this will produce a compile time ERROR. This is because the covergroup declared in a class is an anonymous type and the covergroup name becomes the instance variable. In other words, we haven't (cannot be) declared a variable that is an array of the covergroup "myCG" in the class. Without this when you try to create an array of instances of covergroup "myCG," the simulator has no idea where to store (i.e., in which variable) the array of instances. Confusing? A bit, mainly because this should be allowed and the current version (as of the publication of this book) of SystemVerilog does not; it's a limitation.

So, what's the solution? Well, simply declare the covergroup *outside* of the class, as shown below.

CORRECT way of declaring a covergroup to create an array of its instances.

```
covergroup myCG (@posedge clk); //covergroup declared outside of
the class 'ex_class'
      coverpoint myInt
      {
      bins myInt_range_1 = {124:0]};
      bins myInt_range_2 = {325:125]};
      }
endgroup: myCG
class ex_class extends uvm_component;
int myInt[10];
myCG myCG_val[10]; //An array of  class 'myCG'
```

```
function new (string name, uvm_component parent=null); //class
constructor(UVM style)
super.new(name , parent);
     foreach (myInt[ i ])
          myCG_val myCG[ i ]=new(myInt[ i ]); //NO ERROR
endfunction
endclass: ex_class
```

In this example, we declare the covergroup "myCG" *outside* of the class "ex_class." This allows us to declare a variable that is an array of covergroup "myCG" in the class "ex_class." Once this is done, you can instantiate the array of instances of covergroup "myCG" since the compiler has a handle to where the instances can be referenced.

26.19 Further Methodology Guidelines

Finally, here are further guidelines on functional coverage methodology.

- *Collect functional coverage in the UVM verification environment using the SystemVerilog covergroup construct.*
  We have discussed this point above to some extent. It is sometimes necessary or more convenient to process or transform the values coming from the DUT to create derived values that are actually sampled as coverpoints. Here is an example:

```
class my_coverage extends uvm_subscriber #(bus_tx);
  `uvm_component_utils(my_coverage)

  bus_tx m_item;  //m_item of type bus_tx
  int    m_address_delta;

  covergroup m_cov;
    cp_address_delta: coverpoint m_address_delta {
      bins zero = {0};
      bins one  = {1};
      bins Lower_Qword = { [1: 127] };
    }
  endgroup

  function new(string name, uvm_component parent);
    super.new(name, parent);
    m_cov = new;
  endfunction : new
```

```
    function void write (input bus_tx t);
      m_item = t;
      m_address_delta = m_item.current_address - m_item.
      previous_address;
      m_cov.sample( );
    endfunction : write

  endclass : my_coverage
```

The point is, we first derived m_address_delta coverpoint in function "write" and then sampled it. This technique can overcome the fundamental limitation that covergroups are defined at instantiation time and the definitions of the coverpoints cannot change dynamically.

- *Either place a covergroup in a class as an embedded covergroup or place a covergroup in a package and parameterize the covergroup so that it can be instantiated from classes in that package.*
  The embedded covergroup is the most straightforward way to use a covergroup in a class, but several classes can reuse the same covergroup by placing the covergroup declaration in a package outside of any class and having the classes instantiate the covergroup with appropriate parameters.
- *Covergroups should be instantiated within UVM component classes as opposed to within transaction or sequence classes.*
  Coverage should be collected from quasi-static objects that endure throughout the simulation, not from objects that come-and-go dynamically over time.
- *Covergroups should be instantiated within UVM subscribers or scoreboards that are themselves instantiated within a UVM environment class and are connected to the analysis ports of monitors/agents.*
  Use monitors to gather and assemble information in the form of transactions that are sent out through their analysis ports, but do not place covergroups inside the monitors themselves. This separation between data gathering in the monitor and data analysis in the subscriber/scoreboard is important for reuse.
- *Instantiate the covergroup in the constructor of the coverage collector class.*
  It is a SystemVerilog rule that embedded covergroups must be instantiated from the constructor. This goes against the general rule in UVM of keeping the constructor empty and creating sub-objects from the build_phase method.
- *In order to collect functional coverage information for internal signals within the DUT, encapsulate references to hierarchical paths to the DUT in a single SystemVerilog module (or interface), then access that module from the UVM environment using a virtual interface and SystemVerilog interface in the usual way.*
  This is a rehash of a point I've made earlier. You can indeed use hierarchical references to signals inside a DUT module. Internal signals within the DUT can be accessed using SystemVerilog hierarchical references or using the bind statement. Encapsulating all hierarchical references within a single module (or interface) allows the verification environment to be kept clean.

- *Where coverage collection spans multiple DUT interfaces and thus depends on analysis transactions received from more than one agent, use the `uvm_analysis_imp_decl macro to provide multiple analysis exports in the coverage collector class.*
  The uvm_subscriber class only has a single analysis export. The `uvm_analysis_imp_decl macro offers the most convenient way to write a subscriber class that accepts multiple incoming transaction streams, each with their own distinct write method.
- *Where appropriate, collect functional coverage information in SystemVerilog interfaces using the cover property statement.*
  This point is not related to class-based methodology but still worth mentioning from methodology point of view. Property-based coverage using the cover property statement can be a good way to collect functional coverage information for temporal sequences in interface protocols (as opposed to sample-based coverage using the covergroup statement) but note that the cover property statement cannot be used within a class-based environment.
- *Group coverpoints into multiple covergroups in order to separate coverage of specification features from coverage of implementation features.*
  Keeping specification coverage separate from implementation coverage will help at the point when the coverage model is re-used.
- *Use a variable coverage_enable within the configuration object of the coverage collector to enable or disable coverage collection.*
  Coverage collection incurs a performance and memory cost, and some use cases for the verification component may not require coverage collection. The UVM User Guide recommends the use of a variable named coverage_enable for this purpose.
- *Sample covergroups by calling their sample method as opposed to specifying a clocking event for the covergroup.*
  This can be the built-in sample method or an overridden sample method with a list of arguments, i.e., covergroup IDENTIFIER with function sample(…). Calling the sample method allows values to be sampled when and only when transactions arrive at the coverage collection component from the DUT.
- *Do not sample covergroups more frequently than necessary. Consider using a conditional expression iff (expression) with each coverpoint to reduce the sampling frequency.*
  Sampling too often will unnecessarily inflate the volume of coverage data that needs to be stored and analyzed. It might not be necessary or meaningful to sample each coverpoint every time the covergroup is sampled. Any conditional expression should be kept simple: complex iff conditions can be hard to debug.
- *Sample values within the DUT or at the outputs of the DUT. Do not sample the stimulus applied to the inputs of the DUT. Sample DUT registers when the register value is changed by the DUT, not when it is changed directly by the stimulus.*
  Sampling stimulus does not tell you anything about the behavior of the DUT itself, only about the behavior of the stimulus generator.

- *Consider setting the option.at_least of each covergroup and coverpoint to some value other than the default value of 1.*
  The default value of option.at_least only ensures that each state is hit once, which in general is insufficient to test whether or not the state has become deadlocked.
- *Be cautious in setting option.weight or option.goal of a covergroup or coverpoint.*
  There are two potential problems. First, the methodological problem that giving a greater or lesser weight to certain states might distort the coverage reporting, and second, the practical problem that these options are not implemented consistently across simulators.
- *Design coverpoint bins carefully to ensure that functionally significant cases are covered.*
  Since 100% coverage of the state space is unrealistic, careful design of coverage bins can be critical to verification quality. Part of the solution can be to create separate bins for typical values, special values, and boundary conditions. The choice of bins should relate back to the verification plan.
- *When designing coverpoints, specify any illegal values or values to be excluded for coverage as ignore_bins. Do not use illegal_bins.*
  Covergroups should be confined to collecting functional coverage information and not linked directly to error reporting. Illegal values should be trapped either using assertions or using the UVM report handler.

26.20 "cross" Coverage

"cross" is a very important feature of functional coverage. This is where code coverage completely fails. Figure 26.8 describes the syntax and semantics.

Syntax:

```
cover_cross ::=  [ cross_identifier : ] cross list_of_cross_items
[ iff ( expression ) ] cross_body
```

Two variables "offset" and "adr" are declared. Coverpoint for "offset" creates four bins called ofsbin[0] … ofsbin[3] for the four values of "offset," namely, 0,1,2,3. Coverpoint "adr" also follows the same logic and creates adrbin[0]… adrbin[3] for the four values of "adr," namely, 0,1,2,3.

adr_ofst is the label given to the "cross" of ar, ofst. First of all, the "cross" of "ar" (label for coverpoint adr) and "ofst" (label for coverpoint offset) will create another set of 16 bins (four bins of "adr" $*$ four bins of "offset"). These "cross" bins will keep track of the result of "cross." However, what does "cross" mean?

Four values of "adr" need to be covered (0,1,2,3). Let us assume adr==2 has been covered (i.e., adrbin[2] is covered). Similarly, there are four values of "offset" that need to be covered (0,1,2,3) and that offset==0 has also been covered (i.e., ofsbin[0] has been covered). However, have we covered "cross" of adr=2 (adrbin[2])and offset=0

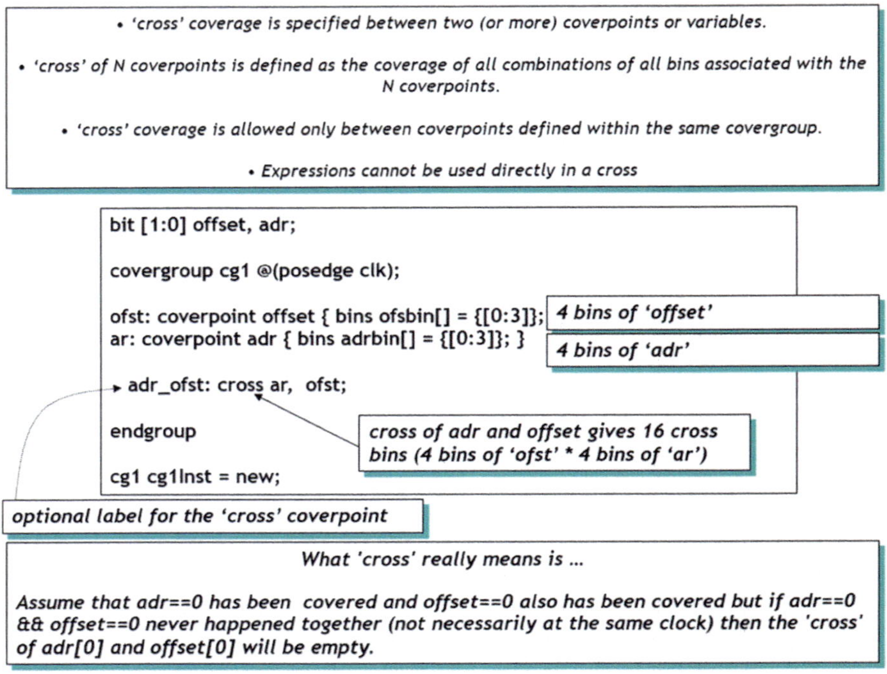

- 'cross' coverage is specified between two (or more) coverpoints or variables.
- 'cross' of N coverpoints is defined as the coverage of all combinations of all bins associated with the N coverpoints.
- 'cross' coverage is allowed only between coverpoints defined within the same covergroup.
- Expressions cannot be used directly in a cross

```
bit [1:0] offset, adr;

covergroup cg1 @(posedge clk);

ofst: coverpoint offset { bins ofsbin[] = {[0:3]}; }
ar: coverpoint adr { bins adrbin[] = {[0:3]}; }

adr_ofst: cross ar, ofst;

endgroup

cg1 cg1Inst = new;
```

4 bins of 'offset'

4 bins of 'adr'

cross of adr and offset gives 16 cross bins (4 bins of 'ofst' * 4 bins of 'ar')

optional label for the 'cross' coverpoint

What 'cross' really means is ...

Assume that adr==0 has been covered and offset==0 also has been covered but if adr==0 && offset==0 never happened together (not necessarily at the same clock) then the 'cross' of adr[0] and offset[0] will be empty.

Fig. 26.8 "cross" coverage—Basics

(ofsbin[0])? Not necessarily. "cross" means that adr=2 and offset=0 must be true "together" at some point in time. This does *not* mean that they need to be "covered" at the *same* time. It simply means that (e.g.) if adr==2 that it should remain at that value until offset==0 (or vice-versa). This will make both of them true "together." If that is the case, then the "cross" of adrbin[2] and ofsbin[0] will be considered "covered."

In the simulation log in Fig. 26.9, we see that both adrbin[2] and ofsbin[0] have been individually covered 100%. However, their "cross" has not been covered.

Let us look at the simulation log further.

First, you will see the 4 bins (ofsbin[0] to ofsbin[3]) of coverpoint cg1::ofst. All 4 bins are covered and hence coverpoint cg1::ofst is 100% covered. Next, you will see the 4 bins (adrbin[0] to adrbin[3]) of coverpoint cg1::ar. All bins are covered here as well and so is the coverpoint cg1::ar.

Now let us look at the "cross" of 4bins∗4bins=16 bins coverage. Both "ofst" and "ar" are 100% covered—but—the 3 cases that follow (among many others) are not covered because whatever values the test-bench drove, these bins never had the same value at any given point in time (e.g., adrbin[2] is "2" at time t, then ofsbin[0] should be "0" either at time t or any time after that, as long as adrbin[2] = "2").

Hence,

```
bins <adrbin[2],ofsbin[0] > 0 1ZERO
```

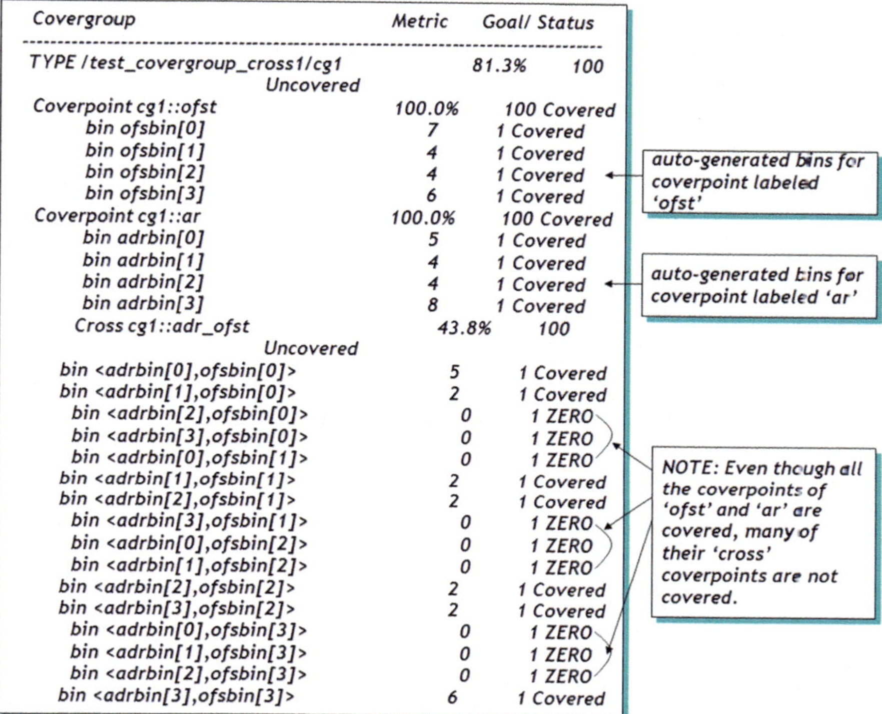

Fig. 26.9 "cross" coverage—simulation log

Similarly, there are other cases of "cross" that are not covered as shown in the simulation log. Such a log will clearly identify the need to enhance your test-bench. To reiterate, such "cross" cannot be derived from code coverage.

Figure 26.10 shows how cross is achieved between an enum type and a bit type. Idea is to show how cross coverpoint/bins are calculated. The figure describes different ways "cross" bins are calculated.

Recall that a "cross" can be defined only between N coverpoints, meaning you must have an explicit coverpoint for a variable in order to cross with another coverpoint. That is where there is an anomaly when it comes to enum type. The enum type "color" has no coverpoint defined for it. Yet, we are able to use it in "cross." That is because the language semantics implicitly creates a coverpoint for the enum type "color" and track its cross coverage.

Please study this and other figures carefully since they will act as guideline for your projects. A simulation log of this example is presented in Fig. 26.11. The annotations describe what's going on.

IMPORTANT NOTE: *Cross coverage is allowed only between coverage points defined within the same coverage group. Coverage points defined in a coverage group other than the one enclosing the "cross" cannot participate in a cross. Attempts to cross items from different coverage groups shall result in a compiler error.*

```
enum { r, g, b } color;
bit [3:0] adr;
bit [1:0] offset;

covergroup cg1 @(posedge clk);

  ofst: coverpoint offset;
  ar: coverpoint adr
     {
          bins ar1 = {[0:7]};
          bins ar2 = {[8:15]};
     }

   clr_adr: cross color, ar;
   adr_ofst: cross ar, ofst;
endgroup

cg1 cg1Inst = new;
```

auto-generated bins (4 bins because offset is 2bit wide)

Explicit set of bins. 'ar1' covers 'adr' values 0 to 7 while 'ar2' covers 'adr' values 8 to 15.

'cross' between enumerated variable color and coverpoint ar. Total cross bins = 6 (3 of 'color' crossed with 2 of 'adr')

'cross' between coverpoints 'ar' and 'ofset'. Total cross bins = 8 (4 of 'offset' crossed with 2 of 'adr')

SV implicitly creates a coverpoint for the variable 'color' in order to track it's cross coverage.

Fig. 26.10 "cross"—Example (further nuances)

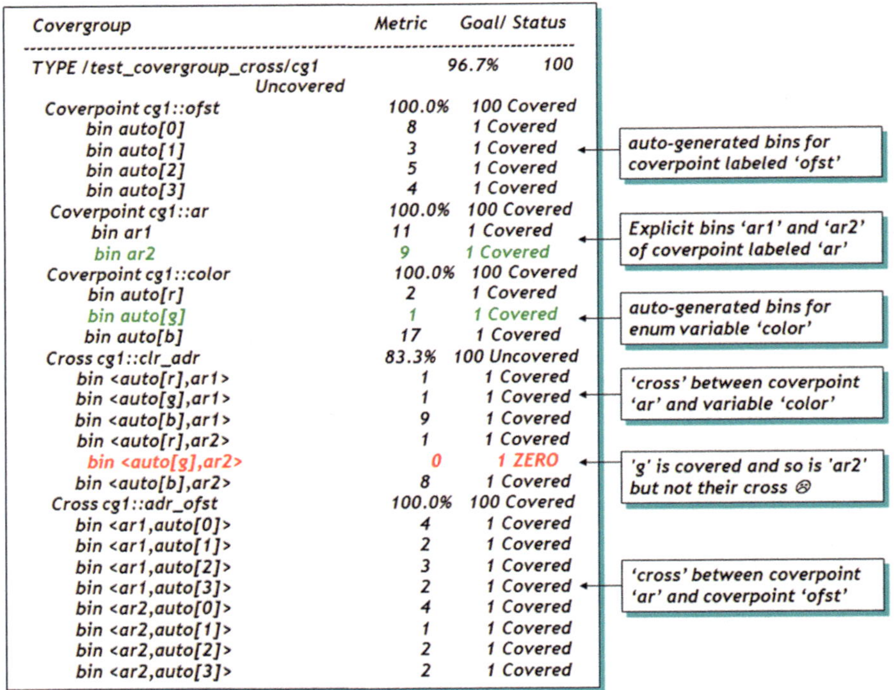

Fig. 26.11 "cross" example—simulation log

Note that "cross" is not limited to only 2 coverpoints/variables. You can have multiple coverpoints/variables crossed with each other.

For example, if you want to make sure that bits of a control_reg are all covered together at some point in time, you can create a "cross" as follows.

```
covergroup control_reg_cg( ) with function sample(bit[4:0]
control_reg);
option.name = "control_reg_cg";

CR1: coverpoint control_reg[0];
CR2: coverpoint control_reg [1];
CR3: coverpoint control_reg [2];
CR4: coverpoint control_reg [3];
CR5: coverpoint control_reg [4];

multiCross: cross CR1, CR2, CR3, CR4, CR5;
endgroup:
```

In this example, we are creating a "cross" of 5 coverpoints. This will ensure that each bit of control_reg is covered and that they are all covered together.

Here's one more example of cross among multiple coverpoints. Note that such crosses will explode in the number of cross bins that will be created. You will have to judiciously use ignore_bins to limit the number of cross bins that are created.

```
covergroup multiCrossCG( ) with function sample(bit[5:0] CR1,
bit[1:0] CR2);

  CR1_10: coverpoint CR1[1:0] {
    bins bits_5 = {0};
    bins bits_6 = {1};
    bins bits_7 = {2};
    bins bits_8 = {3};
  }

  CR1_2: coverpoint CR1[2] {
    bins stop_1 = {0};
    bins stop_2 = {1};
  }

  CR1_53: coverpoint CR1[5:3] {
    bins bits0246 = {3'b000, 3'b010, 3'b100, 3'b110}.
    bins bits3 = {3'b011};
    bins bits1 = {3'b001};
    bins bits5 = {3'b101};
    bins bits7 = {3'b111};
  }
```

```
    CR2_10: coverpoint CR2 {
      bins one = {0};
      bins four = {1};
      bins eight = {2};
      bins fourteen = {3};
    }

    MultiCross: cross CR1_10, CR1_2, CR1_53, CR2_10;

endgroup
```

Exercise: How many cross bins will be created in the above example?
Further examples of "cross" coverage:
Example 1:

```
bit [7:0] b1, b2, b3;
covergroup cov @(posedge clk);
     b2b3: coverpoint b2+b3;;
     b1Crossb2b3 : cross b1, b2b3;
endgroup
```

The thing to note in this example is that b2b3 coverpoint is an expression. It is indeed possible to "cross" between a coverpoint (which is an expression) and a variable.

Example 2:

```
bit [3:0] bNibble;
covergroup myCov @(posedge clk);
     A: coverpoint a_var { bins yy[ ] = { [0:9] }; }
     CC: cross bNibble, A;
endgroup
```

In this example, coverpoint of a_var (labeled "A") creates explicitly created 10 bins (yy[0], yy[1]…yy[9]), while bNibble creates auto-generated 16 bins (auto[0], auto[1], …. Auto[15]).

So, the cross of bNibble and A produces 160 cross bins (16 auto-generated bins of "bNibble" × 10 bins of "a_var").

```
<auto[0], yy[0]>
<auto[0], yy[1]>
...
<auto[0], yy[9]>

<auto[1], yy[0]>
...
<auto[15], yy[9]>
```

Example 3:
Some simple do's and don'ts of "cross." Consider the following code.

```
A: coverpoint a {
bins b1[ ] = {2, [4:5]};
bins b2[ ] = {6, 7};
bins b3[ ] = default;
}
B: coverpoint b {
bins b1[ ] = {3, [7:9]};
}
C: coverpoint c {
bins b1[ ] = {1,4};
}
cross A, B {
bins cb1 = binsof (A); // correct use
bins cb2 = binsof (A.b1); // correct use
bins cb3 = binsof (B.b2); // incorrect use as b2 is not a bin
of B (duh!)
bins cb4 = binsof (C); // incorrect use as cross does not
include C
}
//The expression within the binsof construct should not be a
default bin.
bins cb1 = binsof(A.b3); // incorrect use because b3 is a
default bin
//You cannot declare a vector bin inside a cross.
cross A, B {
bins cb1[ ] = binsof (A); // incorrect use as cb1 is declared
as vector
bins cb2 = binsof (A.b1); // correct use
}
//You cannot declare default bins inside a cross.
cross A, B {
bins cb1 = default; // incorrect use
}
//You cannot declare bins with the same name inside a cross.
cross A, B { bins cb1 = binsof (A);
bins cb1 = binsof (A.b2);
}
```

Example 4:

Consider the following code. Study it carefully since it uses the "with" clause and conditionals. See how cross bins are created.

```
module top;
logic [ 0 : 3 ] a, b;

covergroup cg @(posedge clk);
coverpoint a{
bins low[ ] = {[ 0 : 7 ]} ;
bins high = {[ 8 : 15 ]} ;
}
coverpoint b {
bins two[ ] = b with (item% 2 == 0 ) ;
bins three[ ] = b with (item% 3 == 0 ) ;
}
X: cross a, b
{
bins xab = binsof ((b.two) with (b > 5)) || binsof ((a.low)
with (a <7 ));
}
endgroup
............
endmodule
```

First, here are the tuples of "a" crossed with "binsof (b.two) with (b > 5)"

```
<low[0], two[6]>
<low[0], two[8]>
...
<low[0], two[14]>

<low[1], two[6]>
...
<low[7], two[14]>

<high, two[6]>
...
<high, two[14]>
```

Next, the tuples created by cross of "binsof (a.low) with a < 7)" and "b"

```
<low[0], two[0]>
<low[0], two[2]>
<low[0], two[14]>
...
```

```
<low[0], three[0]>
<low[0], three[3]>
...
<low[0], three[15]>

<low[1], two[0]>
<low[1], two[2]>
...
<low[1], three[15]>

<low[2], two[6]>
...
<low[6], three[15]>
```

Finally, the tuples created for "bins xab" of "cross a, b" are

```
<low[0], two[0]>
<low[0], two[2]>
<low[0], two[14]>
...
<low[0], three[0]>
<low[0], three[3]>
...
<low[0], three[15]>

<low[1], two[0]>
<low[1], two[2]>
...
<low[1], three[15]>

<low[2], two[6]>
...
<low[6], three[15]>
<low[7], two[6]>
...
<low[7], two[14]>
<high, two[6]>
...
<high, two[14]>
```

There are a lot more examples on "cross" in Sect. 26.25 where we use "intersect" to limit the number of bins created by a "cross." Also, more examples in Sect. 26.21 where we explore "transition" coverage.

26.21 "bins" for Transition Coverage

As noted in Fig. 26.12, this is a very useful feature of functional coverage. Transaction level transitions are very important to cover. For example, did CPU issue a read followed by write-invalid? Did you issue a D$miss followed by a D$hit cycle? Such transitions are where the bugs occur, and we want to make sure that we have indeed exercised such transitions.

Figure 26.12 explains how the semantics work. Note that we are addressing both the "transition" and the "cross" of "transition" coverage.

There are two transitions in the example.

bins ar1 = (8'h00 => 8'hff); which means that adr1 should transition from "0" to "ff" on two consecutive posedge of clk. In other words, the test-bench must exercise this condition for it to be covered.

Similarly, there is the "bins ar2" that specifies the transition for adr2 (1 => 0)..

The cross of transitions is shown at the bottom of the figure. Very interesting how this works. Take the first values of each transition (namely, adr1=0 && adr2=1). This will be the start points of cross transition at the posedge clk. If at the next (posedge clk) values are adr1="ff" && adr2=0, the cross transition is covered.

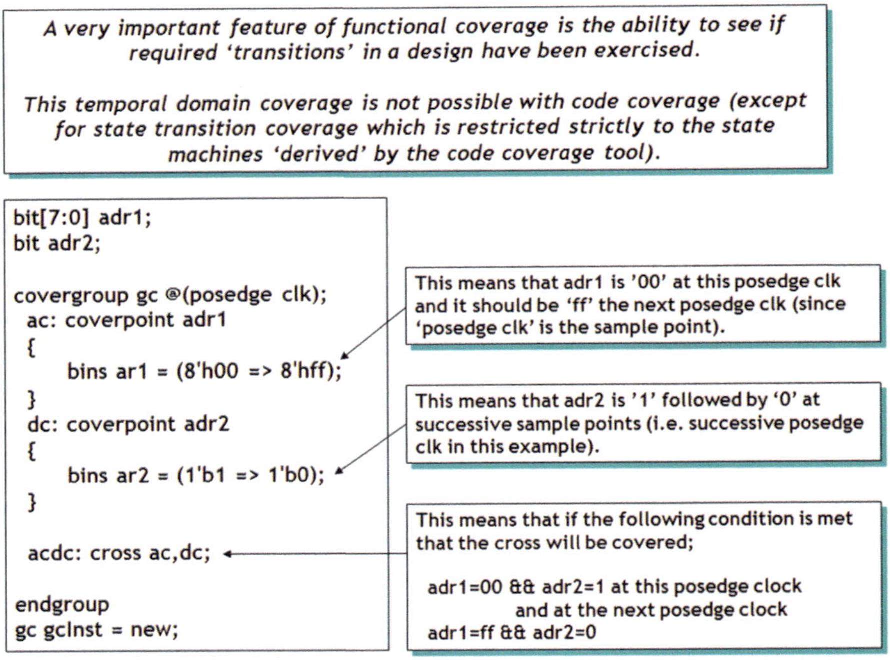

A very important feature of functional coverage is the ability to see if required 'transitions' in a design have been exercised.

This temporal domain coverage is not possible with code coverage (except for state transition coverage which is restricted strictly to the state machines 'derived' by the code coverage tool).

```
bit[7:0] adr1;
bit adr2;

covergroup gc @(posedge clk);
  ac: coverpoint adr1
  {
      bins ar1 = (8'h00 => 8'hff);
  }
  dc: coverpoint adr2
  {
      bins ar2 = (1'b1 => 1'b0);
  }

  acdc: cross ac,dc;

endgroup
gc gcInst = new;
```

This means that adr1 is '00' at this posedge clk and it should be 'ff' the next posedge clk (since 'posedge clk' is the sample point).

This means that adr2 is '1' followed by '0' at successive sample points (i.e. successive posedge clk in this example).

This means that if the following condition is met that the cross will be covered;

adr1=00 && adr2=1 at this posedge clock
and at the next posedge clock
adr1=ff && adr2=0

Fig. 26.12 "bins" for transition coverage

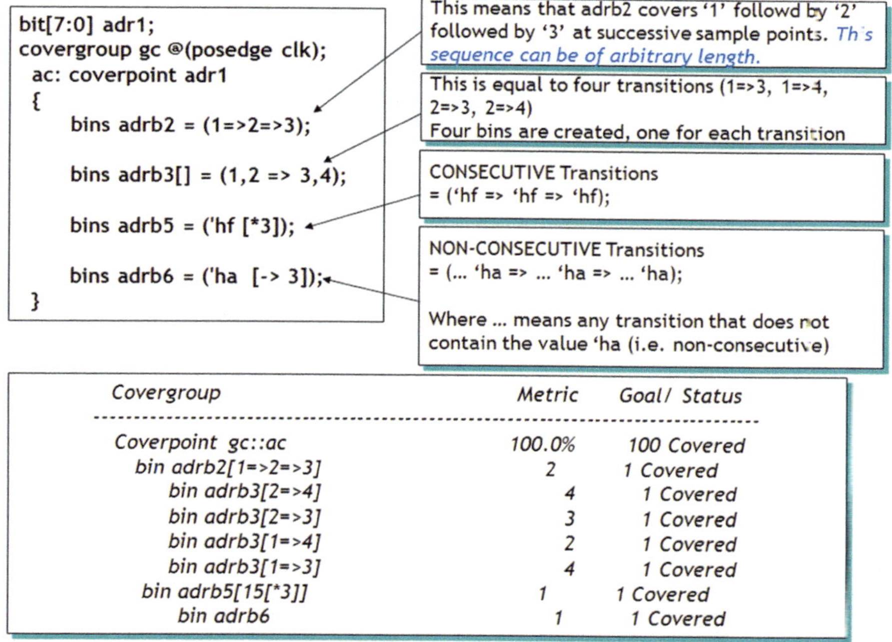

Fig. 26.13 "bins"—transition coverage further features

More on the "bins" of transition is shown in Fig. 26.13. In the figure, different styles of transitions have been shown. "bins adrb2" requires that "adr1" should transition from 1=>2=>3 on successive posedge clk. Of course, this transition sequence can be of arbitrary length. "bins adrb3[ ]" shows another way to specify multiple transitions. The annotation in the figure explains how we get 4 transitions.

"bins adrb5" is (in some sense) analogous to the consecutive operator of assertions. Here 'hf [∗3] means that adr1='hf should repeat 3 times at successive posedge clk.

Similarly, the non-consecutive transition ('ha [->3]) means that adr1 should be equal to 'ha, 3 times and not necessarily at consecutive posedge clk. Note that just as in non-consecutive operator, here also 'ha need to arrive 3 times with arbitrary number of clocks in-between their arrival and that "adr1" should *not* have any other value in-between these 3 transitions. The simulation log shows the result of a test-bench that exercises all the transition conditions.

One more example of transition coverage in Fig. 26.14.

This figure shows a very interesting semantic feature of transition. Note that "bins adrb3[ ]=(1,2 => 3,4)" is *not* the same as "bins adrb4[ ] = (1=>3, 1=>4, 2=>3, 2=>4);".

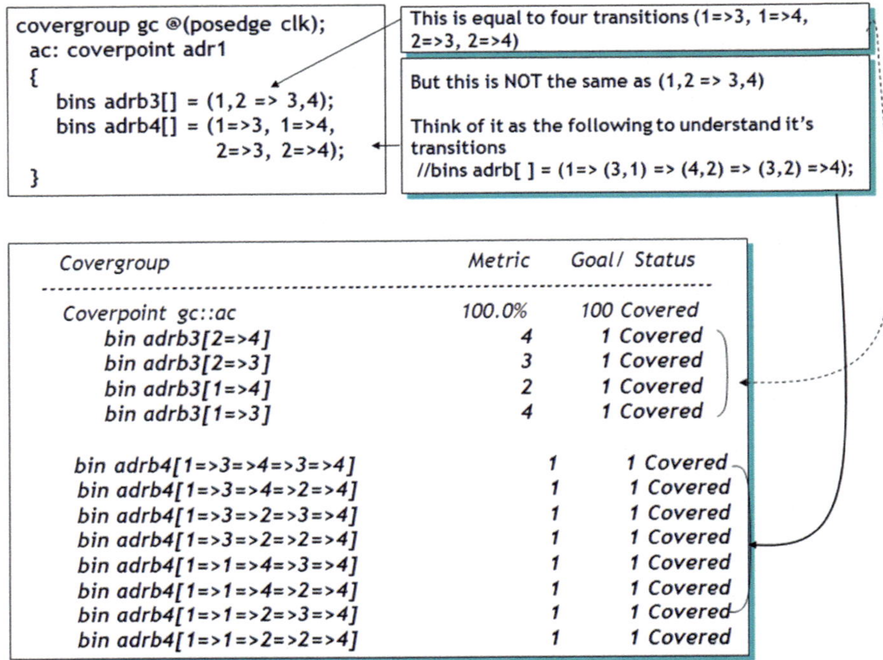

Fig. 26.14 "bins" for transition—example with simulation log

"bins adrb3[ ]=(1,2 => 3,4)" means transitions ***(1=>3, 1=>4, 2=>3, 2=>4)***. BUT "bins adrb4[ ]=(1=>3, 1=>4, 2=>3, 2=>4)" means

```
[1=>3=>4=>3=>4]
[1=>3=>4=>2=>4]
[1=>3=>2=>3=>4]
[1=>3=>2=>2=>4]
[1=>1=>4=>3=>4]
[1=>1=>4=>3=>4]
[1=>1=>2=>3=>4]
[1=>1=>2=>2=>4]
```

Is that intuitive? I don't think so. However, the following will explain.

```
'bins adrb4[ ]=(1=>3, 1=>4, 2=>3, 2=>4)' is equivalent to
'bins adrb4[ ]=(1=>(3, 1)=>(4, 2)=>(3, 2)=>4)'
```

If you see the equivalent definition, you will be able to understand the transition. Study them carefully, you will figure out why the transitions look the way they do.

In short, you need to be careful how you specify transition bins.

Here's an example of the use of "goto" repetition operator (covered in the SVA part of the book) in specifying transition bins.

```
'ha [-> 4]
```

Means, the following transitions will be covered.

```
… => 'ha … => 'ha … => 'ha … => 'ha
```

'ha will occur non-consecutively 4 times and strictly no 'ha in between those 4 transitions. This definition is identical to what we have discussed on "goto" operator in the SVA chapter on "goto" operator (Sect. 8.13).

Some more examples of transition bins.

Example 1:

```
bins Tbin1 = (7 => 8 => 9), ([10:11], 12=> 13,14);
```

This will specify the following transitions.

```
7=> 8 => 9
10 => 13
10 => 14
11 => 13
11 => 14
12 => 13
12 => 14
```

Example 2:

```
bins Tbin2 = (1 => [->4:6] => 1;
1 => …=>4 … =>5 … => 6 => 1
```

Now let us turn back to our favorite PCI example that we have been following. We started with simple coverage, moved to "bins" coverage, and now we will see the "transition" coverage. This is the reason we were building on the same example showing how such a coverage model can be derived starting with a simple model.

Figure 26.15 shows the same enum type and same bins. But in addition, it now defines two transition bins named R2W (for Read to Write) and W2R (for Write to Read).

bins R2W means that all possible Read cycles types are followed by all possible Write cycles. Each of the transition covers a separate transition to cover all possible cycle transitions on a PCI bus.

Requirement: Cover all PCI Cycle Types and transitions among Read and Write cycles.

```
// PCI C/BE Commands
enum {iack, SpecialC, IORead, IOWrite, MemRead, MemWrite, ConfRead, ConfWrite,
MemRMult, DualAddr, MemReadLine, MemWrInv} pciCommands;

covergroup pciCommands_cover @(posedge clk);
        pciCmdCover : coverpoint pciCommands
        {
        bins pcireads [] = {IORead, MemRead, ConfRead, MemRMult, MemReadLine};
        bins pciwrites [] = {IOWrite, MemWrite, ConfWrite, MemWrInv};
        bins pcimisc [] = {iack, SpecialC};

        bins R2W [] = (IORead, MemRead, ConfRead, MemRMult, MemReadLine =>
IOWrite, MemWrite, ConfWrite, MemWrInv);
        bins W2R [] = (IOWrite, MemWrite, ConfWrite, MemWrInv => IORead,
MemRead, ConfRead, MemRMult, MemReadLine);
        }
endgroup
```

Fig. 26.15 Example of PCI Cycles transition coverage

In this example that means we must cover the following transactions.

```
IORead => IOWrite; IORead => MemWrite; IORead => ConfWrite;
IORead => MemWrInv;
MemRead => IOWrite; MemRead =>MemWrite; MemRead => ConfWrite;
MemRead => MemWrInv;
ConfRead => IOWrite; ConfRead => MemWrite; ConfRead =>
ConfWrite; ConfRead => MemWrInv;
MemRMult => IOWrite; MemRMult => MemWrite; MemRMult =>
ConfWrite; MemRMult => MemWrInv;
MemReadLine => IOWrite; MemReadLine => MemWrite; MemReadLine =>
ConfWrite; MemReadLine => MemWrInv;
```

Same type of transitions are covered by bins W2R[ ].

As you notice, this is quite powerful. Many bugs occur when there is a transition from one transaction type to the next. You have to make sure that your test-bench indeed covers such transitions.

26.22 “wildcard bins”

Since no one likes to type a sequence repeatedly, we create don’t care (or as the functional coverage lingo calls it “wildcard” bins). Self-explanatory. As shown in Fig. 26.16, you can use either an “x” or a “z” or “?” (doesn’t this look familiar to

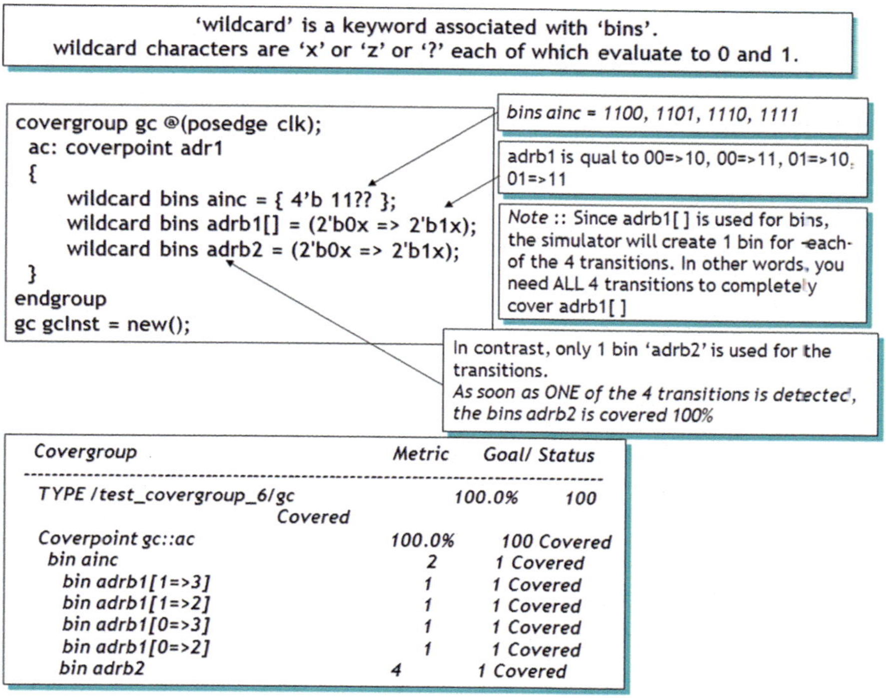

'wildcard' is a keyword associated with 'bins'.
wildcard characters are 'x' or 'z' or '?' each of which evaluate to 0 and 1.

```
covergroup gc @(posedge clk);
  ac: coverpoint adr1
  {
       wildcard bins ainc = { 4'b 11?? };
       wildcard bins adrb1[] = (2'b0x => 2'b1x);
       wildcard bins adrb2 = (2'b0x => 2'b1x);
  }
endgroup
gc gcInst = new();
```

bins ainc = 1100, 1101, 1110, 1111

adrb1 is qual to 00=>10, 00=>11, 01=>10, 01=>11

Note :: Since adrb1[ ] is used for bins, the simulator will create 1 bin for each of the 4 transitions. In other words, you need ALL 4 transitions to completely cover adrb1[ ]

In contrast, only 1 bin 'adrb2' is used for the transitions.
As soon as ONE of the 4 transitions is detected, the bins adrb2 is covered 100%

```
 Covergroup                                     Metric     Goal/ Status
 ------------------------------------------------------------------------
 TYPE /test_covergroup_6/gc                          100.0%      100
                                     Covered
 Coverpoint gc::ac                              100.0%      100 Covered
  bin ainc                                          2        1 Covered
    bin adrb1[1=>3]                                 1        1 Covered
    bin adrb1[1=>2]                                 1        1 Covered
    bin adrb1[0=>3]                                 1        1 Covered
    bin adrb1[0=>2]                                 1        1 Covered
   bin adrb2                                     4        1 Covered
```

Fig. 26.16 wildcard "bins"

Verilog?) to declare "wildcard" bins. Note that such bins must precede with the keyword "wildcard."

"wildcard bins ainc" specifies that adr1 values 1100, 1101, 1110, 1111 need to be covered.

Note also "wildcard bins adrb1[ ]" and "wildcard bins adrb2." One creates implicit ([ ]) 4 bins while the other creates only 1 bin. The one that creates 4 explicit bins will check to see that—*each of the*—4 transitions take place. While adrb2 that creates only 1 bin will be considered covered if—*any*—of the 4 transitions take place.

26.23 "ignore_bins"

"ignore_bins" is very useful when you have a large set of bins to be defined. Instead of defining every one of those, if you can identify the ones that are not of interest, then you can simply define "ignore_bins." The ones that are specified to be ignored will indeed be ignored and rest will be covered.

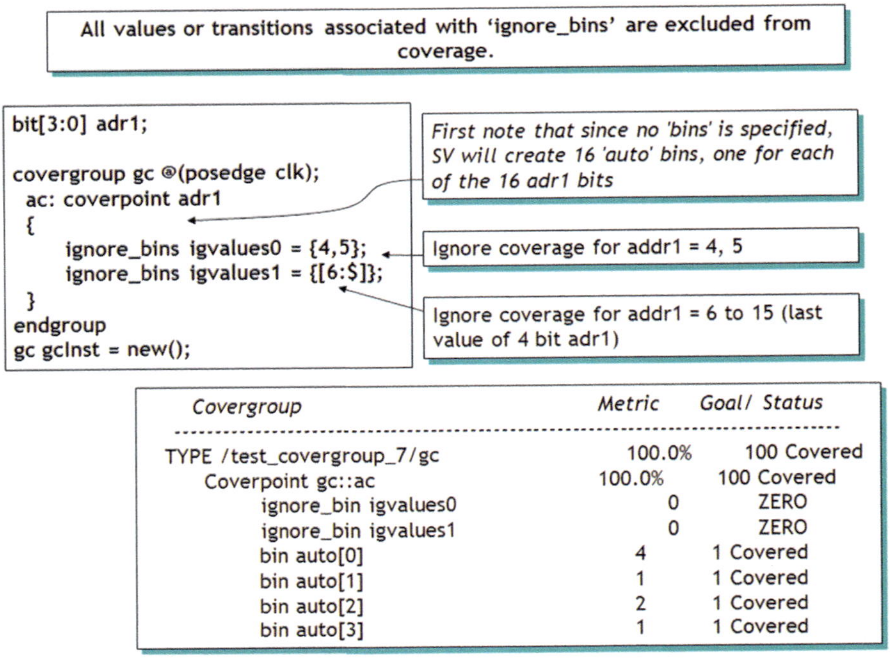

All values or transitions associated with 'ignore_bins' are excluded from coverage.

```
bit[3:0] adr1;

covergroup gc @(posedge clk);
  ac: coverpoint adr1
  {
      ignore_bins igvalues0 = {4,5};
      ignore_bins igvalues1 = {[6:$]};
  }
endgroup
gc gcInst = new();
```

First note that since no 'bins' is specified, SV will create 16 'auto' bins, one for each of the 16 adr1 bits

Ignore coverage for addr1 = 4, 5

Ignore coverage for addr1 = 6 to 15 (last value of 4 bit adr1)

Covergroup	Metric	Goal/ Status
TYPE /test_covergroup_7/gc	100.0%	100 Covered
Coverpoint gc::ac	100.0%	100 Covered
ignore_bin igvalues0	0	ZERO
ignore_bin igvalues1	0	ZERO
bin auto[0]	4	1 Covered
bin auto[1]	1	1 Covered
bin auto[2]	2	1 Covered
bin auto[3]	1	1 Covered

Fig. 26.17 "ignore_bins"—Basics

As we noted at the onset of this chapter, when there is no explicit "bins" defined for a coverpoint, the simulator creates all possible bins that the "covered" variable requires. In Fig. 26.17, only "ignore_bins" are specified in the coverpoint adr1 but no "bins." This means that the simulator will first create all 16 bins (adr1 is 4 bit wide) for adr1. Then it will ignore value 4,5 and 6,7,8,9,10,11,12,13,14,15 and cover only adr1=0,1,2,3. This is reflected in the simulation log at the bottom of Fig. 26.17.

Some examples of ignore_bins.

Example 1:

If "ignore_bins" is specified with a guard expression (iff), then it is effective only if the guard expression is true at the time of sampling. Consider the following code:

```
covergroup cg1 @(posedge clk);
coverpoint a {
bins b1[ ] = {0, 1, 2, 3};
ignore_bins ignore_vals = {0, 3} iff c+d ;
}
endgroup
```

If during a simulation run, coverpoint a takes values 0, 1, 2, and 3, values 0 and 3 are ignored if and only if c+d is true at the point of sampling. If c+d is true throughout the simulation, the coverage percentage is reported as 2/4, which is 50%. This is what the simulators that I have tried have reported. My personal take is that if certain bins are ignored then they should not be part of the coverage report and the report in this case should be 100%. But then, I don't design simulators!!!

If the value of expression c+d is false at the time of sampling and the sampled values are 0 or 3, then these values are not ignored and the count corresponding to bins b1[0] and b1[3] is incremented. In this situation, coverage percentage is reported as 4/4, which is 100%.

Example 2:

The ignore_bins specification in the bin declaration of a coverpoint may result in "empty" bins for a coverpoint. An empty bin is essentially a user-defined/automatic bin of a coverpoint for which all values are ignored using ignore_bins. If ignore_bins specification results in empty bins for a coverpoint, then empty bins will not be dumped in the coverage database. For example:

```
covergroup cg1 @(posedge clk);
coverpoint a {
bins b1[ ] = {0, 1, 2, 3};
bins b2 = {[5:8], 9};
ignore_bins ignore_vals = {0, 3, 1, 2};}
}
endgroup
```

In the above example, bin "b1" is an empty bin and will not be dumped because all of the values associated with the bin "b1" are specified as ignore_bins.

However, this is not true for following cases when you use scalar wildcard bins (this is based on Cadence—Incisive simulator results)

In the case of scalar wildcard bins, even if the bin becomes empty due to ignore_bins or illegal_bins specification, it is still dumped to the coverage database For example, the following wildcard bin definition results in "empty" bins.

```
covergroup CG @(posedge clk);
coverpoint var1 {
wildcard bins b1 = { 1'bx };
ignore_bins ign = {0,1};
}
endgroup // CG
```

In the above description, even though the scalar wildcard bin `b1` becomes empty due to `ignore_bins` specification, it is *still* dumped to the coverage database (again, the observation is based on Cadence—Incisive simulator).

Example 3:

```
covergroup cg1 @(posedge clk);
coverpoint a {
bins b1[ ] = {0, 1, 2, 3};
ignore_bins ignore_vals = {0, 3};
}
endgroup
```

In the above example, if the sampled value of coverpoint "a" is 3, it is ignored and the count for bin b1[3] is not incremented. If during a simulation run, coverpoint a takes values 0, 1, 2, and 3, the coverage percentage is reported as 2/2, which is 100%. The values specified with ignore_bins have been removed from the total bin count.

Example 4:

In the case of a fixed size vector bin, all the values are first distributed across fixed size bins, and then the values specified with the `ignore_bins` are removed from the range list. For example:

```
A: coverpoint a
{
bins a1[3] = {0,2,3,[4:6],8,9};
ignore_bins ig = {2,3};
}
```

To start with the bin values will be spread across bins a1[3] as follows (taking into account ignore_bins)

a1[1] will have value 0
a1[2] will have value 4 (since 2,3 are ignored)
a1[3] will have values 5,6,8,9

Example 5:

For transition bins, even if a transition occurs, it cannot be considered hit if it is specified with ignore_bins. Consider the following code:

```
A: coverpoint a {
bins b1[ ] = (2=>5=>1), (1=>4=>3);
ignore_bins ignore_trans = (2=>5);
}
```

In this case, even if the transition 2=>5=>1 takes place it won't be considered a hit, since transition 2=>5 is ignored.

Example 6:

A simple example showing the use of "with" clause and ignore_bins and how they interact.

```
module top;
reg[15:0] CC;
covergroup cg@(posedge clk);
withC : coverpoint CC {
      bins bin_b1[ ] = {[0:20]} with (item % 2 == 0)
      ignore_bins gCC = CC with (item % 3 == 0)
}
endgroup
…………
endmodule
```

The "bins_b1" of coverpoint "CC" will create the following bins.

```
bins bin_b1[ ] = {0,2,4,6,8,10,12,14,16,18,20};
```

and ignore_bins "gCC" will ignore following bins.

```
ignore_bins gCC = {0,3,6,9,12,15,18,21,24,27 ….};
```

So, after ignore_bins is applied to coverpoint "CC," the following bins will remain and will be used in coverage report.

```
cg (covergroup)
    CC (coverpoint)
        bins of 'CC' (after ignore_bins) = bins_b1[2], bins_
        b1[4], bins_b1[8], bins_b1[10], bins_b1[14], bins_
        b1[16], bins_b1[20]
```

Example 7:

This is an example simply to illustrate that "ignore_bins" is very useful for weeding out bins from a "cross." As you know "cross" can create a large number of bins (depending on what you are crossing, of course) and ignore_bins will help you keep only the bins of interest. For example, you declare two coverpoints for variables x and y. Their cross will create a large number of bins. But you are interested in only those bins where x > y. Here's how you do it.

```
int x, y;
covergroup xyCG;
  x_c : coverpoint x;
  y_c : coverpoint y;
```

```
    x_y_cross: cross x_c, y_c {
      ignore_bins ignore_x_GT_y = x_y_cross with (x_c > y_c);
    }
  endgroup
```

26.24 "illegal_bins"

"illegal_bins" is interesting in that it will complain, if you *do* cover a given scenario. In Fig. 26.18, we refer to the same old "adr1." 16 auto bins are created for coverpoint "adr1" since we don't explicitly declare any bins for it. And we say that if the test-bench ever hits adr1==0, that it should be considered illegal. Coverage of adr1==0 should not occur. As seen from the simulation log, coverage of adr1=0 results in an Error.

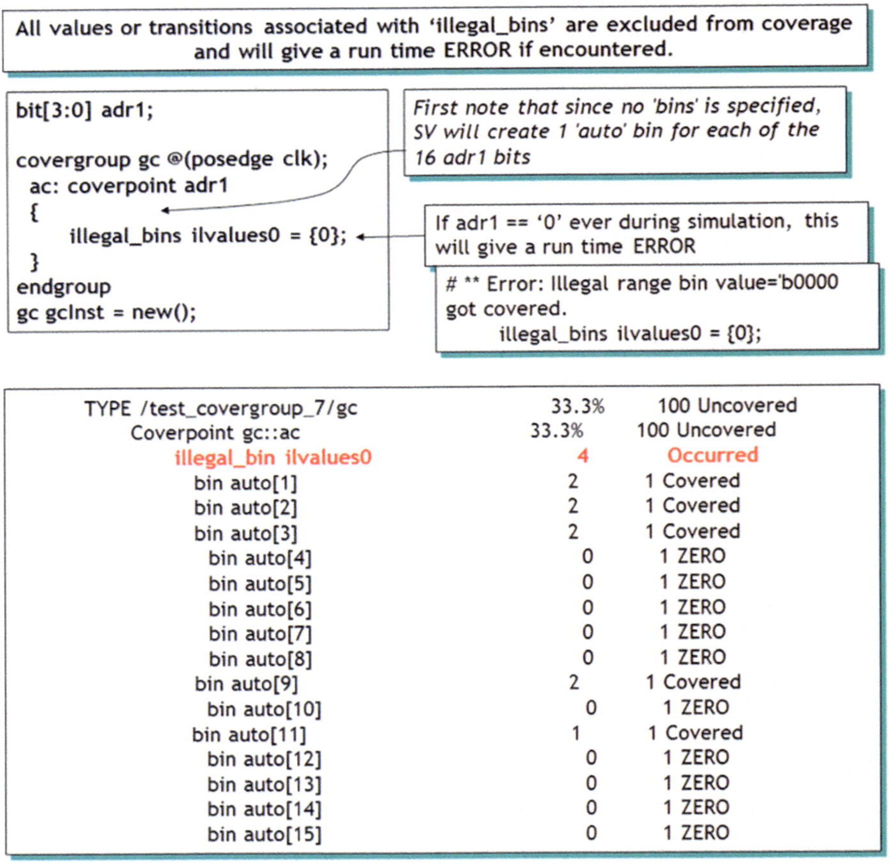

All values or transitions associated with 'illegal_bins' are excluded from coverage and will give a run time ERROR if encountered.

```
bit[3:0] adr1;

covergroup gc @(posedge clk);
  ac: coverpoint adr1
  {
      illegal_bins ilvalues0 = {0};
  }
endgroup
gc gcInst = new();
```

First note that since no 'bins' is specified, SV will create 1 'auto' bin for each of the 16 adr1 bits

If adr1 == '0' ever during simulation, this will give a run time ERROR

```
# ** Error: Illegal range bin value='b0000
got covered.
        illegal_bins ilvalues0 = {0};
```

TYPE /test_covergroup_7/gc	33.3%	100 Uncovered
Coverpoint gc::ac	33.3%	100 Uncovered
illegal_bin ilvalues0	4	Occurred
bin auto[1]	2	1 Covered
bin auto[2]	2	1 Covered
bin auto[3]	2	1 Covered
bin auto[4]	0	1 ZERO
bin auto[5]	0	1 ZERO
bin auto[6]	0	1 ZERO
bin auto[7]	0	1 ZERO
bin auto[8]	0	1 ZERO
bin auto[9]	2	1 Covered
bin auto[10]	0	1 ZERO
bin auto[11]	1	1 Covered
bin auto[12]	0	1 ZERO
bin auto[13]	0	1 ZERO
bin auto[14]	0	1 ZERO
bin auto[15]	0	1 ZERO

Fig. 26.18 "illegal_bins"

Illegal bins take precedence over any other bins, that is, they will result in a run-time error even if they are also included in another bin. Specifying an illegal value has no effect on a transition that includes the value. Illegal transition bins cannot specify a sequence of unbounded or undetermined varying length.

Note that "illegal_bins" takes precedence of "ignore_bins." Here's an example.

```
covergroup cg1 @(posedge clk);
coverpoint a {
     bins b1[ ] = {0, 1, 2, 3};
     illegal_bins ill = {1, 2};
     ignore_bins ign = {2,3}
}
endgroup
```

In this example, value 2 is specified both as ignore_bins and as illegal_bins. When simulation hits the value 2, an Error will be reported in the Coverage report, since illegal_bins takes precedence over "ignore_bins."

26.25 "binsof" and "intersect"

Ok, now we are in some seriously esoteric territory! Figure 26.19 shows that "coverpoint b" has two bins; one is called "bb" and has value from 0 to 12 of "b." bins cc on the other hand has values 13,14,15,16 of "b" that need to be covered. Then we declare a "cross" of a, bc. So, let us first see what this cross looks like.

"a" has 4 implicit bins a[0], a[1], a[2], a[3] for four values 0, 1, 2, 3 and "bc" has two bins "bb" and "cc." So, the "cross a, bc" produces

```
<a[0],bb>  <a[0],cc>
 <a[1],bb>  <a[1],cc>
 <a[2],bb>  <a[2],cc>
 <a[3],bb>  <a[3],cc>
```

We are moving along just fine—right? But now it gets interesting. We don't want to cover all "cross" bins into our "cross" of "a" and "bc." We want to "intersect" them and derive a new subset of the total "cross" set of bins.

```
bins isect_ab  = binsof (bc) intersect {[0:1]};
```

says that take all the bins of "bc" (which are "bb" and "cc") and intersect them with the "values" 0 and 1. Now note carefully that "bb" carries values {[0:12]} which includes the values 0 and 1, while "cc" does not cover 0 or 1. Hence, only those cross products of "bc" that cover "bb" are included in this intersect. Note that the below-mentioned subset is selected from the "cross" product shown above.

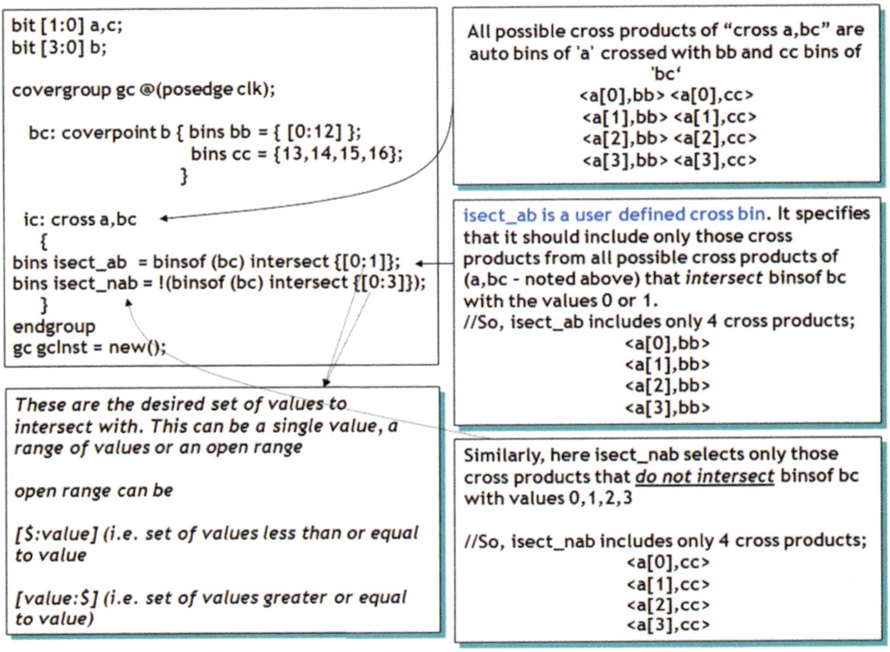

```
bit [1:0] a,c;
bit [3:0] b;

covergroup gc @(posedge clk);

  bc: coverpoint b { bins bb = { [0:12] };
                     bins cc = {13,14,15,16};
                   }

  ic: cross a,bc
  {
bins isect_ab  = binsof (bc) intersect {[0:1]};
bins isect_nab = !(binsof (bc) intersect {[0:3]});
  }
endgroup
gc gcInst = new();
```

All possible cross products of "cross a,bc" are auto bins of 'a' crossed with bb and cc bins of 'bc'
<a[0],bb> <a[0],cc>
<a[1],bb> <a[1],cc>
<a[2],bb> <a[2],cc>
<a[3],bb> <a[3],cc>

isect_ab is a user defined cross bin. It specifies that it should include only those cross products from all possible cross products of (a,bc - noted above) that *intersect* binsof bc with the values 0 or 1.
//So, isect_ab includes only 4 cross products;
<a[0],bb>
<a[1],bb>
<a[2],bb>
<a[3],bb>

These are the desired set of values to intersect with. This can be a single value, a range of values or an open range

open range can be

[$:value] (i.e. set of values less than or equal to value

[value:$] (i.e. set of values greater or equal to value)

Similarly, here isect_nab selects only those cross products that ***do not intersect*** binsof bc with values 0,1,2,3

//So, isect_nab includes only 4 cross products;
<a[0],cc>
<a[1],cc>
<a[2],cc>
<a[3],cc>

Fig. 26.19 "binsof" and "intersect"

```
<a[0],bb>,
<a[1],bb>,
<a[2],bb>,
<a[3],bb>
```

Now onto the next intersect.

```
bins isect_nab = ! (binsof (bc) intersect {[0:3]})
```

Similarly, first, note that here we are using negation of binsof. So, this "bins" statement says, take the intersect of binsof (bc) and {[0:3]} and discard them from the "cross" of a, bc. Keep only those that do not intersect. Note that {[0:3]} again fall into the bins "bb" and would have resulted in exactly the same set that we saw for the non-negated intersect. Since this one is negated it will ignore cross with "bb" and only pick the "bins cc" from the original "cross" of a, bc.

```
<a[0],cc>,
<a[1],cc>,
<a[2],cc>,
<a[3],cc>
```

Here's another example (courtesy SystemVerilog LRM).

```
bit [7:0] v_a, v_b;
covergroup cg @(posedge clk);
a: coverpoint v_a
{
bins a1 = { [0:63] };
bins a2 = { [64:127] };
bins a3 = { [128:191] };
bins a4 = { [192:255] };
}
b: coverpoint v_b
{
bins b1 = {0};
bins b2 = { [1:84] };
bins b3 = { [85:169] };
bins b4 = { [170:255] };
}
c : cross a, b
{
bins c1 = ! binsof(a) intersect {[100:200]};// 4 cross
products
bins c2 = binsof(a.a2) || binsof(b.b2);// 7 cross products
bins c3 = binsof(a.a1) && binsof(b.b4);// 1 cross product
}
endgroup
```

Let us see how this example works.

First, coverpoint "v_a" contains 4 bins covering all possible values of variable "v_a." These are bins a1, a2, a3, a4. Similarly, the coverpoint "v_b" contains 4 bins covering all possible values of variable "v_b." These are bins b1, b2, b3, b4.

The cross of a,b includes the following bins (before taking into account further conditions specified in the cross labeled "c").

```
<a1,b1>,
<a1,b2>,
<a1,b3>,
<a1,b4>
...
<a4,b1>,
<a4,b2>,
<a4,b3>,
<a4,b4>
```

Ok, so far so good. Now let's look at "bins c1" (of "cross a,b"). It says that create bins for only those values of coverpoint "a" that do *not* intersect with values ranging from 100 to 200. But first let us see what would "c1" create before the negation of the cross intersection.

Bins a2 (of "a") intersects with value 100. Bins "a3" intersects with the values from 128 to 191 which is in the range of 100:200. bins "a4" intersects with the values 192 to 200 which is again in the range of 100:200.

So, bins a2, a3, a4 intersect with the values 100 to 200.

Now, we take the inversion of this intersect, meaning which bins do *not* intersect with the values from 100 to 200. That leaves only "a1" as the non-intersecting bin. Hence, bins c1 will create following cross products.

```
<a1,b1>,
<a1,b2>,
<a1,b3>,
<a1,b4>
```

Now let's look at

```
bins c2 = binsof(a.a2) || binsof(b.b2);
```

In this bin, binsof(a.a2) means the bin "a2" of coverpoint labeled "a" (coverpoint v_a). The binsof (b.b2) means bin "b2" of coverpoint labeled "b" (coverpoint "v_b"). So, "binsof(a.a2) || binsof(b.b2)" will create 7 cross products as follows (note that this is a logical OR):

```
<a2,b1>,
<a2,b2>,
<a2,b3>,
<a2,b4>,
<b2,a1>,
<b2,a3>,
<b2,a4>
```

Note that <b2,a2> is already included with "a2" combinations and hence not repeated with "b2" combinations since this is a logical OR.

Finally, let's look at

```
bins c3 = binsof(a.a1) && binsof(b.b4);
```

cross bin, c3, specifies that c3 should include only cross products whose values are covered by bin a1 of coverpoint "v_a" and cross products whose values are covered by bin b4 of coverpoint "v_b." So, this generates only one cross product (note that this is a logical AND):

```
<a1,b4>
```

Study this example carefully. It has a lot of good information on how a "cross" is formed.

Chapter 27
Performance Implications of Coverage Methodology

Introduction: This chapter describes the methodology components of Functional Verification. What you should cover, when you should cover, performance implications and applications on how to write properties that combine the power of assertions with power of Functional Coverage.

A. B. Mehta, *SystemVerilog Assertions and Functional Coverage*,
https://doi.org/10.1007/978-3-030-24737-9_27

27.1 Know *What* You Should Cover

- Don't cover the entire 32-bit address bus.
  - Cover only the addresses of interest (e.g., Byte/word/dword aligned; start/end address; bank crossing address)
- Don't cover the entire counter range but cover only those that are of importance. For example,
  - Cover only the rollover counter values (transition from all 1's to all 0's)
- No need to cover the entire 2K Fifo
  - Cover only fifo full, fifo empty, fifo full crossed with fifo_push, fifo empty crossed with fifo read, etc.
- Auto-generated bins are both a convenience and a nuisance. They may create a lot of clutter with auto-generated bins that may not be relevant. Be judicious in usage of auto generated "bins."
- Use "cross" and "intersect" to weed out unwanted "bins." Also, "illegal_bins" and "ignore_bins."

27.2 Know *When* You Should Cover for Better Performance

- Enable your cover points only when they are meaningful
  - Disable coverage during "reset"
  - Cover "test mode" signals only when in test mode (for example, JTAG TAP Controller TMS asserted)
  - Make effective use of coverage methods such as "start," "stop," "sample" (more on this later…)
  - Do not repeat with covergroups what you have covered with SVA "cover"
  - Make effective use of covergroup "trigger" condition
  - Make effective use of the PASS "action" block associated with SVA "cover" to activate a covergroup

If some of these points (e.g., "trigger") don't quite make sense, please hold on. We will be covering such features in upcoming sections.

27.3 sample( ) Method

Functional coverage should be carefully collected as discussed above. The language does allow tasks that allow you to control when to start collecting coverage and when to stop. These tasks can be associated with an instance of a covergroup and invoked from procedural block.

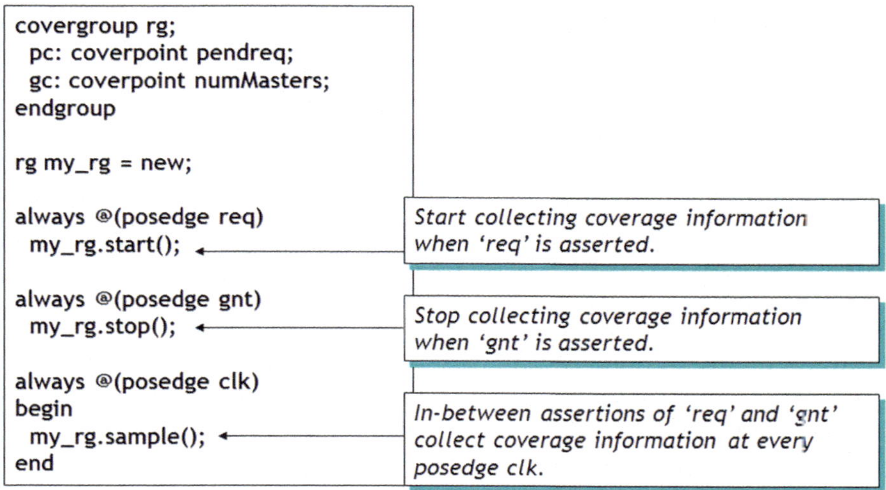

Fig. 27.1 Functional coverage—performance implication

Figure 27.1 shows the covergroup "rg" with two coverpoints "pc" and "gc." "pc" covers all the pending requests and "gc" covers the number of masters on the bus when those requests are made. "my_rg" is the instance of this covergroup.

Since we want to start collecting pending requests at the assertion of req, when the requests are granted, we don't want to cover pending requests and number of masters anymore. "gnt" related cover can be another covergroup.

Simple control but very good performance improvement. Use it wisely to speed up your simulation and a more meaningful coverage log.

Lastly, there is the sampling edge task sample( ) which derives its sampling edge from "**always** @ (posedge clk)" and applies it to "my_rg" as its sampling edge. This also tells us that we can have covergroup specific sampling edges. Very good feature. Note that my_rg.sample( ) will "start" when my_rg.start( ) is executed and will stop when my_rg.stop( ) is executed. This is about as easy as it gets when it comes to controlling collection of coverage information.

Note that the pre-defined "sample( )" method cannot accept any arguments. But what if you *do* want to pass arguments with the sample( ) method? Read on…

27.4 User Defined sample( ) Method

As we saw in previous sections, you can explicitly call a sample( ) method in lieu of providing a clocking event directly when a covergroup is declared. But what if you want to parameterize the built-in sample( ) method and pass it exactly the data that you want sampled? In other words, you want a way to sample coverage data from contexts other than the scope enclosing the covergroup declaration.

For example, an overridden sample method can be called with different arguments to pass directly to covergroup the data to be sampled from within a task or function or from within a sequence or property of a concurrent assertion.

As we have seen, concurrent assertions have special sampling semantics, i.e., data values are sampled in the preponed region. This helps passing the sampled data from a concurrent assertion as arguments to an overridden sample method which in turn facilitates managing various aspects of assertion coverage, such as sampling of multiple covergroups by one property, sampling of multiple properties by the same covergroup, or sampling different branches of a sequence or property (including local variables) by any arbitrary covergroup.

The syntax for using your own sample ( ) method is:

```
coverage_event ::=
clocking_event
| with function sample ( [ tf_port_list ] )
```

Let us see a simple example of this. Note the use of "*with function*" semantics in the example.

```
covergroup coverSample with function sample (int X);
    coverpoint X;
endgroup : coverSample

coverSample cS = new ( );

property propertySample;
    int pX;    //local variable

    //iSig1, iSig2, iSig3 are some internal signals of your design
    @(posedge clk)  (iSig1, pX = iSig2) |=> (iSig3, cS.sample(pX));

endproperty : propertySample
```

In this example, we first declare a covergroup called coverSample and declare a function called sample( ) using the key words "with function sample." This means that the function sample is user defined and has formal parameter "int X." With such a declaration you can now call the sample( ) function from within a task or function or from within a sequence or property of a concurrent assertion.

The example further defines a property named *propertySample* where we declare a local variable called "int pX." This is the variable we want to sample in the covergroup *coverSample.* So, we call the function *sample*( ) from *propertySample* on some condition and pass the variable pX (actual) to the formal "X" of the covergroup *coverSample*. This way, we can cover the variable data that we want to cover *at the time* we want to cover it. The user-defined sample( ) method can have any kind of procedural code to manipulate the data we pass to sample( ) both in combinatorial and temporal domain.

Here's an example of calling sample ( ) from a SystemVerilog function.

```
function fSample;
int fS;
....
cS.sample (fS);
endfunction
```

Note that you cannot use a formal argument both as the formal of a *covergroup* as well as the formal argument of the *sample* ( ) function. Here's an example:

```
covergroup X1 (int cV) with function sample (int cV);
    coverpoint cV;  //Compile ERROR
endgroup
```

Compile Error: As you notice, "int cV" is declared both as the formal argument of "*covergroup* X1" as well as the *sample* method. *This will give a compiler Error.*

Here's another example (Prakash) that shows how to pass a class object to a covergroup for sample. The issue is if the covergroup argument is a class object then that object cannot be pointing to NULL when the covergroup is instantiated. data_obj is a class in your code.

Here's the code that will give a NULL pointer Error.

```
class coverage;
  covergroup cg_abc (data_obj obj);
    coverpoint obj.mode {
      bins range = {[3:'hB]};
    }
  endgroup

  function new ( );
      data_obj obj;
    cg_abc = new (obj);
  endfunction
endclass

module tb;
  initial begin
    coverage m_cov = new( );
    for (int i = 0; i < 10; i++) begin
      data_obj obj = new( );
      obj.randomize( );    // How do I assign obj to covergroup
      inst ?
```

```
      m_cov.cg_abc.sample( );
    end
    $display ("Coverage : %0.2f %%", m_cov.cg_abc.get_coverage());
  end
endmodule
```

Simulation Log:

```
ncsim> run
ncsim: *E,TRNULLID: NULL pointer dereference.
```

So, the issue is when you call sample( ) without the "with function" and your covergroup requires the data_obj handle (obj), how do you pass it? As shown in the code when you call sample( ), you cannot pass an argument required by the covergroup "cg_abc." This will give a NULL pointer error at run time.

Solution: Use user defined sample( ) method ("with function"). Here's the solution code.

```
class coverage;
    covergroup cg_abc with function sample (data_obj obj);
    coverpoint obj.mode {
      bins range = {[3:'hB]};
    }
  endgroup

  function new ( );
      data_obj obj;
      cg_abc = new( );
  endfunction
endclass

module tb;
  initial begin
    coverage m_cov = new( );
    for (int i = 0; i < 10; i++) begin
      data_obj obj = new( );
      obj.randomize( );
      m_cov.cg_abc.sample(obj);
    end
    $display ("Coverage : %0.2f %%", m_cov.cg_abc.get_coverage());
  end
endmodule
```

Simulation Log:

```
ncsim> run
Coverage : 100.00 %
ncsim: *W,RNQUIE: Simulation is complete.
```

27.5 Querying for Coverage

There are a few ways you can query for coverage. In other words, you can "sample" for coverage; get coverage information; and then stop coverage. This helps in not taking up valuable simulation time once you have reached your coverage goals. Here's an example of how you can do that for a covergroup.

```
covergroup rg;
  pc: coverpoint pendreq;
  gc: coverpoint numMasters;
endgroup

rg my_rg = new;

real cov;

always @(posedge req)
  my_rg.start( );

always @(posedge grant)
begin
  my_rg.sample( );
  cov = my_rg.get_inst_coverage ( );    //Coverage information
  on my_rg instance of 'covergroup rg'
  if (cov > 90) my_rg.stop;
end
```

Built-in to all covergroups, coverpoints and 'cross'es is a function called get_coverage( ). This function returns a real number of the percentage coverage. In the above example, we "start" coverage for the covergroup instance "my_rg." After that, we "sample" the coverage of that instance at every posedge of grant. If the coverage is greater than 90% we stop collecting coverage. This helps tremendously with unnecessary coverage collection and thereby reducing simulation overhead.

However (isn't there always a *however* in life!!),

You cannot do the following, assuming Cbins1 is defined as a 'bins' in a covergroup.

```
cov = rg_inst.Cpoint.Cbins1.get_coverage ( );    //ILLEGAL:
get_coverage( ) on 'bins' not allowed.
```

You *cannot* query coverage on individual bins. But not to fret. There's a solution. Just make each "bins" a coverpoint. Yep, not an elegant solution, but at least there is one. First let's see the example with "bins"

```
bit[7:0] adr1;
covergroup gc @(posedge clk);
  ac: coverpoint adr1
  {
      bins adrb2 = (1=>2=>3);
      bins adrb3[ ] = (1,2 => 3,4);
      bins adrb5 = ('hf [*3]);
  }
endgroup
```

Since, you cannot collect coverage on a bin, we convert each bin into a coverpoint.

```
bit[7:0] adr1;

covergroup gc @(posedge clk);
    Cpointadrb2: coverpoint adr1 { bins adrb2 = (1=>2=>3);}
    Cpointadrb3: coverpoint adr1 { bins adrb3[ ] = (1,2 => 3,4);}
    Cpointadrb5: coverpoint adr1 {bins adrb5 = ('hf [*3]);
endgroup
```

As you notice, we took each of the "bins" of the covergroup "gc" and got rid of "coverpoint adr1" since each bin will now be individually covered. Then, we converted each "bins" into a coverpoint. This will allow you to collect coverage information (using get_coverage ( ) function) on each coverpoint which in turn essentially gives you coverage information on underlying "bins."

27.6 strobe( ) Method

Note that optionally there is also a "strobe" option (see Chap. 28) that can be used to modify the sampling behavior. When the strobe option is not set (the default), a coverage point is sampled the *instant* the clocking event takes place. If the clocking event occurs multiple times in a time step, the coverage point will also be sampled multiple times. The "strobe" option can be used to specify that coverage points are sampled in the Postponed region, thereby filtering multiple clocking events so that only one sample per time slot is taken. The strobe option only applies to the scheduling of samples triggered by a clocking event.

27.7 Application: Have You Transmitted All Different Lengths of a Frame?

This application combines local variables, subroutine calls, covergroups and interaction with procedural code outside of the assertion. Here's how it works (Fig. 27.2).

Read this example bottom up.

Property frameLength says that on the rising edge of TX_EN we should check the length of the transmitted frame using sequence frmLength.

This application exemplifies the use of
- *local variables*
- *subroutine call associated with an expression to update a variable*
- *covergroup triggered from an explicit event.*

```
logic [7:0] FrameLngth = 0;
event measureFrameLength;

covergroup length_cg @(measureFrameLength );
        coverpoint FrameLngth;
endgroup

task store_Frame_Lngth;
input [7:0] x;
        FrameLngth = x;
        -> measureFrameLength;
endtask

sequence frmLength;
int cnt;
        (TX_EN, cnt=1)  ##1  ((TX_EN, cnt++)[*0:$])
        ##1 (!TX_EN, store_Frame_Lngth(cnt))
endsequence

property frameLength;
        @(posedge TX_CLK) $rose(TX_EN) |-> frmLength;
endproperty

fLength: assert property (frameLength);
```

Fig. 27.2 Application—Have you transmitted all different lengths of a frame?

Sequence frmLength declares a local variable "cnt" and at TX_EN==1, initializes cnt=1. One clock later (##1) it increments cnt forever ((TX_EN, cnt++)[∗0:$]) until TX_EN de-asserts (falls). At that time, we call a task (i.e., a subroutine) called store_Frame_Lngth(cnt) and provide it the final count as a parameter. This final count is the length of the Frame that started with TX_EN assertion.

The task store_Frame_Lngth takes the "cnt" as input and assigns it to "logic" type FrameLngth and triggers a named event called measureFrameLength.

Now the covergroup length_cg triggers at "measureFrameLength" edge, which we just triggered explicitly from task store_Frame_Lngth. The coverpoint covers FrameLngth.

In short, we measure the Frame Length starting assertion of TX_EN until de-assertion of it. We measure the frame length between assertion and de-assertion of TX_EN and cover it. With every new assertion of TX_EN, we measure the length of a new frame.

Note that "coverpoint FrameLngth" does not specify any explicit bins. That will create 256 explicit bins each containing a frame length. This way we make sure that we have covered all (i.e., 256) different frame lengths.

Chapter 28
Coverage Options

Introduction: This chapter describes the Coverage Options offered by the language. Options for 'covergroup' type (both instance specific and instance specific per-syntactic level) are described. Practical project methodology-based examples are provided that you can directly deploy in your project.

A. B. Mehta, *SystemVerilog Assertions and Functional Coverage*,
https://doi.org/10.1007/978-3-030-24737-9_28

The syntax for specifying these options in the covergroup definition is
option.option_name = expression;

Option Name	Default	Description
weight = *number*	1	If set at the **covergroup** syntactic level, it specifies the weight of this **covergroup** instance for computing the overall instance coverage of the simulation. If set at the **coverpoint** (or **cross**) syntactic level, it specifies the weight of a **coverpoint** (or **cross**) for computing the instance coverage of the enclosing **covergroup**.
goal = *number*	90	Specifies the target goal for a **covergroup** instance, or a **coverpoint** or a cross of an instance.
name = *string*	unique name	Specify a name of the **covergroup** instance. If unspecified, a unique name for each instance is automatically generated by the tool.
comment = *string*	" "	A comment that appears with the instance of a **covergroup**, or a **coverpoint** or **cross** of the **covergroup** instance. The comment is saved in the coverage database and included in the coverage report.
at_least = *number*	1	Minimum number of hits for each bin. A bin with a hit count that is less than *number* is not considered covered.
detect_overlap = *boolean*	0	When true, a warning is issued if there is an overlap between the range list (or transition list) of two bins of a **coverpoint**
auto_bin_max = *number*	64	Maximum number of automatically created bins when no bins are explicitly defined for a **coverpoint**
cross_auto_bin_max = *number*	unbounded	Maximum number of automatically created cross product bins for a **cross**.
cross_num_print_missing = *number*	0	Number of missing (not covered) cross product bins that must be saved to the coverage database and printed in the coverage report.
per_instance = *boolean*	0	Each instance contributes to the overall coverage information for the **covergroup** type. When true, coverage information for this **covergroup** instance is tracked as well.

LRM: SystemVerilog 3.1a, Table 20-1

Fig. 28.1 Coverage options—reference material

28.1 Coverage Options: Instance Specific: Example

Here's another simple example on how you can *exclude* coverage of a coverpoint from total coverage.

```
logic [15:0] addr, data;
covergroup cov1;
cov1_covpoint: coverpoint addr;
      {
      bins zero = {0}; //1 bin for value 0
      bins low = {1:3}; //1 bin for values 1,2,3
      bins high [ ] = {4:$}; //explicit number of bins for all
      remaining values
      option.weight = 5; //coverpoint 'cov1_covpoint' has a
      weight of 5
      }
cov2_covpoint: coverpoint data;
      {
      bins all = {0:$}; //1 bin for all values
      optin.weight = 0; //No weight; Exclude coverage of this
      coverpoint towards total
      }
cross cov1_copoint, cov2_covpoint;
      {
      option.weight = 10;  //Higher weight for this 'cross'
      towards total
      }
endgroup
```

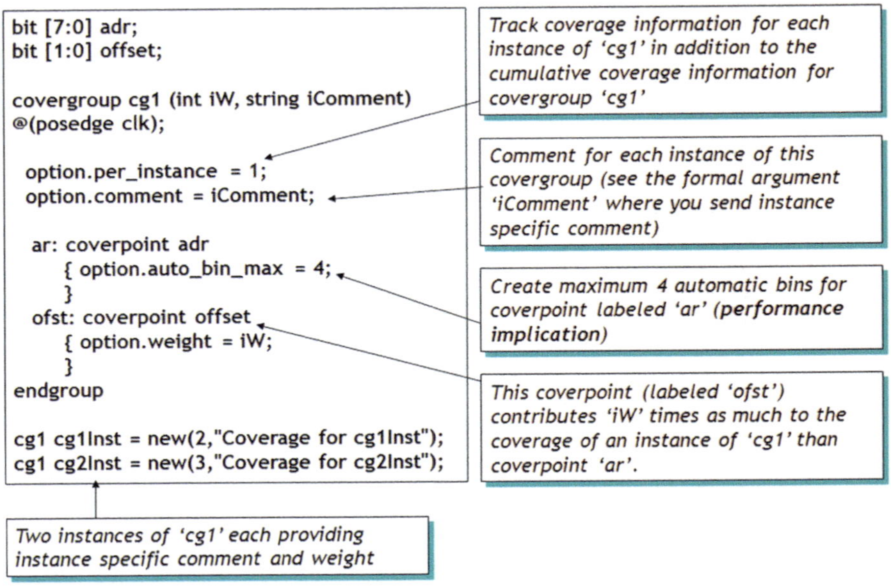

```
bit [7:0] adr;
bit [1:0] offset;

covergroup cg1 (int iW, string iComment)
@(posedge clk);

  option.per_instance = 1;
  option.comment = iComment;

   ar: coverpoint adr
       { option.auto_bin_max = 4;
       }
   ofst: coverpoint offset
       { option.weight = iW;
       }
endgroup

cg1 cg1Inst = new(2,"Coverage for cg1Inst");
cg1 cg2Inst = new(3,"Coverage for cg2Inst");
```

Track coverage information for each instance of 'cg1' in addition to the cumulative coverage information for covergroup 'cg1'

Comment for each instance of this covergroup (see the formal argument 'iComment' where you send instance specific comment)

*Create maximum 4 automatic bins for coverpoint labeled 'ar' (**performance implication**)*

This coverpoint (labeled 'ofst') contributes 'iW' times as much to the coverage of an instance of 'cg1' than coverpoint 'ar'.

Two instances of 'cg1' each providing instance specific comment and weight

Fig. 28.2 Coverage options—instance specific—example

28.2 Coverage Options: Instance Specific Per-syntactic Level

*The following table summarizes the syntactical level (**covergroup**, **coverpoint**, or **cross**) at which instance options can be specified. All instance options can be specified at the covergroup level. Except for the **weight**, **goal**, **comment**, and **per_instance** options, all other options set at the covergroup syntactic level act as a default value for the corresponding option of all coverpoints and crosses in the covergroup. Individual coverpoint or crosses can overwrite these default values. When set at the covergroup level, the **weight**, **goal**, **comment**, and **per_instance** options do not act as default values to the lower syntactic levels.*

Option Name	Allowed in Syntactic Level		
	covergroup	coverpoint	cross
name	Yes	No	No
weight	Yes	Yes	Yes
goal	Yes	Yes	Yes
comment	Yes	Yes	Yes
at_least	Yes (default for coverpoints & crosses)	Yes	Yes
detect_overlap	Yes (default for coverpoints)	Yes	No
auto_bin_max	Yes (default for coverpoints)	Yes	No
cross_auto_bin_max	Yes (default for crosses)	No	Yes
cross_num_print_missing	Yes (default for crosses)	No	Yes
per_instance	Yes	No	No

LRM: SystemVerilog 3.1a, Table 20-2

Fig. 28.3 Coverage options—instance specific per-syntactic level

The following table lists options that **describe a feature of the 'covergroup' type as a whole.**

type_option.option_name = expression;

Option Name	Default	Description
weight = *constant_number*	1	If set at the **covergroup** syntactic level, it specifies the weight of this **covergroup** for computing the overall cumulative (or type) coverage of the saved database. If set at the **coverpoint** (or **cross**) syntactic level, it specifies the weight of a **coverpoint** (or **cross**) for computing the cumulative (or type) coverage of the enclosing covergroup.
goal = *constant_number*	90	Specifies the target goal for a **covergroup** type, or a **coverpoint** or cross of a **covergroup** type.
comment = *string_literal*	" "	A comment that appears with the **covergroup** type, or a **coverpoint** or cross of the **covergroup** type. The comment is saved in the coverage database and included in the coverage report
strobe = *constant_number*	0	If set to 1, all samples happen at the end of the time slot, like the $strobe system task.

LRM: SystemVerilog 3.1a, Table 20-3

Coverage type-options per Syntactic Level

LRM: SystemVerilog 3.1a, Table 20-4

Option Name	Allowed Syntactic Level		
	covergroup	coverpoint	cross
weight	Yes	Yes	Yes
goal	Yes	Yes	Yes
comment	Yes	Yes	Yes
strobe	Yes	No	No

Fig. 28.4 Coverage options type specific per syntactic level

28.3 Coverage Options for "covergroup" Type: Example

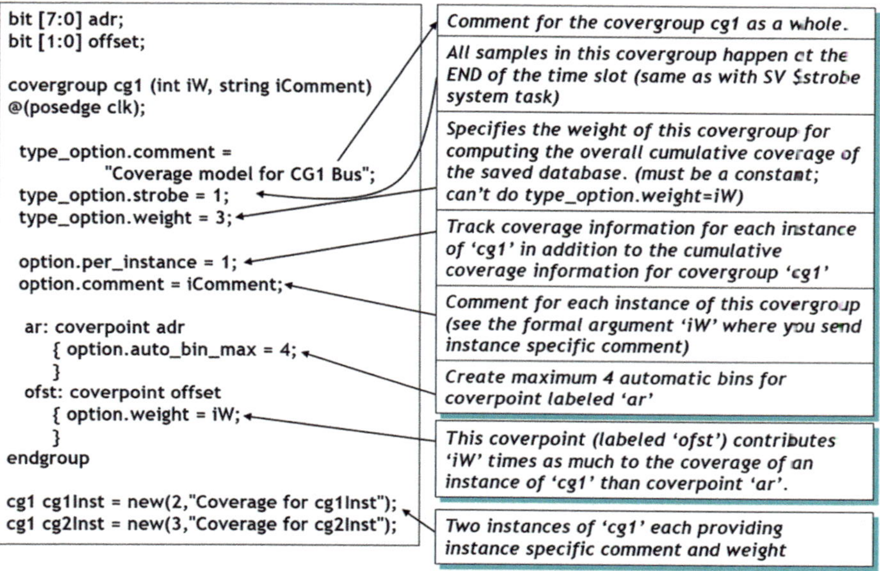

Fig. 28.5 Coverage options for "covergroup" type specific—comprehensive example

28.4 Coverage System Tasks, Functions, and Methods

Predefined coverage system tasks and functions	
$set_coverage_db_name (<name>);	Sets the filename of the coverage database into which coverage info. is saved at the *end* of a simulation run
$load_coverage_db (<name>);	Load from a given filename the cumulative coverage information for all the coverage group types.
$get_coverage ();	Returns a real number in the range of 0 to 100 the overall coverage of *all* covergroups.

Pre-defined coverage methods used in procedural code

LRM: SystemVerilog 3.1a, Table 20-5

Method	Can be called on			Description
	covergroup	coverpoint	cross	
void sample()	Yes	No	No	Triggers sampling of the covergroup
real get_coverage()	Yes	Yes	Yes	Calculates type coverage number (0...100)
real get_inst_coverage()	Yes	Yes	Yes	Calculates coverage number (0...100)
void set_inst_name (string)	Yes	No	No	Sets the instance name to the given string
void start()	Yes	Yes	Yes	Starts collecting coverage information
void stop()	Yes	Yes	Yes	Stops collecting coverage information
real query()	Yes	Yes	Yes	Returns the cumulative coverage information (for the coverage group type as a whole)
real inst_query()	Yes	Yes	Yes	Returns the per-instance coverage information for this instance.

Fig. 28.6 Predefined coverage system tasks and functions

28.4.1 sample ( ) Method

Explicit "sample" of a covergroup. You call this method exactly when you want to sample a covergroup. In other words, you can design conditional logic to trigger the sampling of a covergroup. Does not apply to a coverpoint or a "cross."

28.4.2 real get_coverage (ref int, ref int)

The get_coverage( ) method returns the cumulative coverage, which considers the contribution of all instances of a particular coverage item and it is a static method that is available on both types (via the :: operator) and instances (using the "." operator). The get_coverage( ) method both accept an optional set of arguments and a pair of int values passed by reference. When the optional arguments are specified, the get_coverage( ) method assign to the first argument the value of the covered bins, and to the second argument the number of bins for the given coverage item.

28.4.3 real get_inst_coverage (ref int, ref int)

get_inst_coverage ( ) method returns the coverage of the specific instance on which it is invoked; thus, it can only be invoked via the "." operator. The get_inst_coverage ( ) method both accept an optional set of arguments, a pair of int values passed by reference. When the optional arguments are specified, the get_inst_coverage ( ) method assign to the first argument the value of the covered bins, and to the second argument the number of bins for the given coverage item.

Here's an example:

```
covergroup cg (int xb, yb) ;
     coverpoint x {bins xbins[ ] = { [0:xb] }; }
     coverpoint y {bins ybins[ ] = { [0:yb] }; }
endgroup
cg cv1 = new (1,2); // cv1.x has 2 bins, cv1.y has 3 bins
cg cv2 = new (3,6); // cv2.x has 4 bins, cv2.y has 7 bins
initial begin
cv1.x.get_inst_coverage(covered,total);  // total = 2 ::
Coverage for coverpoint instance 'x'
cv1.get_inst_coverage(covered,total); // total = 5 :: Coverage
for covergroup instance 'cv1'
cg::x::get_coverage(covered,total); // total = 6 :: Coverage
for all instances of Coverpoint 'x'
cg::get_coverage(covered,total); // total = 16 :: Coverage for
all instances of covergroup 'cg'
end
```

28.4.4 void set_inst_name (string)

Sets the instance name to the given "string." Note that you cannot set_inst_name of a coverpoint (obviously). Applies only to a covergroup instance name, since covergroup is the only thing you instantiate.

Index

A. B. Mehta, *SystemVerilog Assertions and Functional Coverage*,
https://doi.org/10.1007/978-3-030-24737-9

Printed by Printforce, the Netherlands